BOUNDARY VALUE PROBLEMS AND FOURIER EXPANSIONS

BOUNDARY VALUE PROBLEMS AND FOURIER EXPANSIONS

Revised Edition

Charles R. MacCluer
Michigan State University

DOVER PUBLICATIONS, INC.
Mineola, New York

Copyright

Bibliographical Note

This Dover edition, first published in 2004, is a newly revised and expanded edition of *Boundary Value Problems and Orthogonal Expansions: Physical Problems from a Sobolev Viewpoint,* originally published by the IEEE Press, Piscataway, NJ, in 1994.

Library of Congress Cataloging-in-Publication Data

MacCluer, C. R.
Boundary value problems and Fourier expansions / Charles R. MacCluer.—Rev. ed.
p. cm.
Rev. ed. of: Boundary value problems and orthogonal expansions. c1994.
Includes bibliographical references and index.
ISBN 0-486-43901-1 (pbk.)
1. Boundary value problems. 2. Fourier analysis. I. MacCluer, C. R. Boundary value problems and orthogonal expansions. II. Title.

QA379.M28 2004
515'.35—dc22

2004056023

Manufactured in the United States of America
Dover Publications, Inc., 31 East 2nd Street, Mineola, N.Y. 11501

This book is dedicated to three great teachers:

- *H. B. Mann*
- *P. R. Halmos*
- *D. J. Lewis*

As a mathematical discipline travels far from its empirical source, or still more, if it is a second and third generation only indirectly inspired by ideas coming from "reality," it is beset with very grave dangers.

J. von Neumann

Contents

To the Student

This book takes several radical departures that will alter your method of study. First is its *brevity* — major concepts are presented with minimal possible detail. Details are pushed into the exercises, omitted, or postponed until later sections. This will require a more active role from you. I encourage you to consult other texts, fellow students, scientists, engineers, and mathematicians. Since this subject touches all the physical sciences and engineering, you will find a warm reception to your questions.

The second departure is the use of *Big Tools.* Rather than using ad hoc arguments, I have instead designated certain sophisticated mathematical tools as unproven *Principles*, thereafter to be used at will. This will enable us to proceed more cleanly and at an advanced level. You may need to spend some time thinking about the big picture.

Third is the highly physical orientation of this book. This Mathematics arose from problems in heat transfer and vibration, and is best understood in terms of these notions and language. Specific real-world applications are discussed. You may need to consult others for help with some physical details and intuition.

The fourth is the integration of *numerical methods* and *symbolic manipulation.* You may need to review MatLab or your favorite compiler or spreadsheet.

Fifth, this book is based on the modern *Sobolev methods.* This viewpoint is elegant and consonant with implementation by digital computer. This will require you to rethink the notions of limit and convergence.

The first 5 sections form an informal introduction that develop your physical and mathematical intuition. The next section introduces you to Hilbert space where this subject naturally belongs. In the following 6 sections, we pose and solve the standard problems of the subject. The last 7 sections are short introductions to selected topics.

I am interested in your comments and suggestions. Email me at the address *maccluer@msu.edu*

To the Instructor

Two sections follow: *How this Book Differs* and *Using this Book.*

How this Book Differs

This book reflects three fundamental 'recent' shifts in paradigm — to the digital computer, to norm-based methods, and to Sobolev methods.

This book de-emphasizes pointwise convergence, the least useful and most pathological of all notions of convergence. Instead, deviations in shapes, signals, or temperature regimes are most naturally measured by their L^2 or L^1 norms, since these norms are physically natural measurements of power or energy. With a norm viewpoint, L^2, L^1, and uniform convergence can all be discussed in a simple and unified way. The Galerkin/Sobolev weak solutions are, after all, the practical computer implementable solutions.

In this book, the classical boundary and initial value problems are restated in a form more consistent with modern Dynamical Systems. PDEs are rewritten as linear 'ordinary' differential equations with constant (time-invariant) spatial differential operator coefficients; solutions are trajectories of undulating shapes, evolving from an initial shape over time, located with time varying coordinates with respect to a fixed and physically natural basis — the spatial eigenmodes. Solutions are moving points in Hilbert space.

A second departure from the standard texts is the use of *Big Tools.* The older texts attempt internal completeness at the expense of many ad hoc arguments. In this book, certain sophisticated tools are designated as *Principles*, thereafter to be used at will. These tools are referenced but not proven, stated in a way comprehensible to the intended audience. For example, completeness of the classical orthogonal expansions is deduced in all cases from Rellich's Theorem on the (compactness of the resolvent of the) Laplacian with zero boundary conditions. The approach is mature, elegant, and promotes interest in advanced course work.

Attempting internal completeness in a text at this level is a mistake. The Comap/Exxon survey shows that the bulk of advanced Mathematics is being taught outside mathematics departments. I suspect this is because we mathematicians insist on teaching what our clients do not need nor want to know. Our clients need powerful mathematical tools explained clearly. They are not interested in the inner details or pathologies at the fringe of the subject. Such a passion for internal completeness would make grade-school arithmetic inaccessible. So it is with BVPs. Older BVP texts assume the student exists in a vacuum, without background, experience, or resources. This text encourages students to consult with mathematicians, engineers, fellow students, and to visit the library.

A third departure is the highly physical orientation of this book. Such an approach appeals to the clientele of such a course and fixes ideas well. Dimensionless variables are employed yielding clean, detail-free problems. One may think of this use of the physical notions of heat flux, temperature, voltage, etc., as an alternative language, just as in algebraic geometry the geometric language is a means for remembering the essentially algebraic results.

I have taken a further step. Not only are problems often discussed in physical language, but specific real-world applications are described. For example, the Kelvin Line Source problem is motivated by a discussion of the seasonal temperature variations about the vertical heat exchanger of a ground-coupled heat pump. There will be descriptions of problems stemming from large flexible structures and structural damping, terminated ideal and lossy transmission lines, waveguides, acoustics, Schroedinger's hydrogen atom, topics in building technology, control of distributed systems, heat exchangers, ELF submarine communication, RF currents on a conductor, cavity resonators, variations in subsurface ground temperatures, diffusion of pollutants into an environment, cooling of curing composite materials, petroleum exploration, etc. It is important with the modern student to appear directly relevant. It is no longer enough to motivate a separation of variables problem with an off-handed, "this comes from a problem in heat transfer."

A fifth departure is to tacitly base the book on the modern Sobolev viewpoint. Ultimately all the theory rests on compact embeddings and weak solutions. This is not explicit until the last two chapters.

The intended audience is senior or first-year graduate students of applied mathematics, chemistry, physics, electrical, mechanical, or civil engineering. The prerequisite background is a solid grounding in linear algebra and a first course in ordinary differential equations.

Because simulation has largely replaced experimentation, the modern graduate engineering education is becoming far more analytical. Many beginning graduate engineering students, wishing to fill gaps in their background, will find this modern treatment ideal.

The thrust of this book is to present linear BV Problems in a unified, simple, yet modern fashion. This is not an encyclopedia of BVP solutions. The student, once armed with the solid foundation provided by this book, can intelligently search the famous collections for solutions to a particular application. The book is short, concise, and is limited to a short list of standard representative problems that are reprised again and again throughout the book.

Using this Book

This book can be used in standard lecture format or in various degrees of the R. L. Moore system. Many if not most details of the development are given as exercises. These exercises can be worked and presented in lecture for a continuous flow, or assigned for homework in some mix consonant with your own style. Other exercises are routine variations of the worked examples.

The first 5 sections form an informal and intuition-building introduction to boundary value problems. Nothing from Chapters 1–4 is truly essential. However all of Chapter 5 is essential material that cannot be omitted.

Chapter 6 introduces the mathematical tools: norms and inner products, orthogonal series. The essential ideas are orthogonal bases, eigenfunctions, and Rellich's theorem. Much of this theoretical material could be passed over, although this would be a disservice to the young scientist or engineer. The crux of the book (the following 6 sections) does not depend heavily on this theoretical material.

The next 6 chapters (7–12) consist of three pairs of chapters: Fourier series applied to rectangular problems, Bessel functions applied to cylindrical problems, and orthogonal polynomials applied to spherical problems. Conceivably a course could be taught using Chapter 5 and these 6 Chapters alone.

The last 7 chapters are short introductions to selected topics. The underpinning of the book in the Sobolev viewpoint is finally revealed in the last two sections. Chapters 13, 14, and 19 are best taught to stronger students.

Acknowledgements

Many students helped shape this book. Among the many I especially thank

> R. Randel, S. Gogate, K. Noren, J. G. Hershberger, S-D Chen, L. A. Scott, R. H. Mohtar, M. W. Landgraf, Y. R. Hicks, C-P Chao, M. T. Guenther, M. J. Twork, J. B. Schroeder, T. L. Scofield, S-L Chen, D. Bohl, Y-B Cho, J. Chou, E. Faulkner, J. Hargrove, R. Kappagantu, J. Rupp, Y-T Wang, K. Wu, D-G Zhu, H. Asthana, S. B. Mishra, and Jon D. Courtney.
>
> Best of luck in your careers.

I am indebted to Tammy Hatfield and Cathy Friess who placed much of this book into cyberspace. I also thank my wife Ann for her support and legal advice.

One way of learning new material is to make a list of naive questions then seek out an expert. If your ego can stand the abuse the method is quick and effective. I thank the experts

> D. P. Nyquist, Gregory M. Weirzba, R. Lal Tummala, C. J. Radcliffe, P. M. FitzSimons, R. C. Rosenberg, D. H. Y. Yen, Joel Shapiro, Sheldon Axler, J. D. Schuur, C. Chiu, N. L. Hills, C. C. Ganzer, C-Y Wang, C. E. Weil, H. Paul Shuch, Y. Jasuik, Q. Du, Timothy A. Grotjohn, Wade Ramey, P. A. Lappan, W. T. Sledd, B. L. MacCluer, T. H. Parker, V. P. Sreedharan, J. W. Kerr, and W. C. Brown.
>
> Many thanks to my engineering mentor Gerald L. Park, P.E.
>
> I especially wish to thank my gifted colleague Milan Miklavčič who suffered the brunt of the questions.

This book is a revision of an IEEE Press edition published in 1994. The reviewers of this new version offered many helpful suggestions. For their insights I thank Rafe Mazzeo of Stanford University, William Heller of University of Texas-Pan American, Frank Baginski of George Washington University, and two anonymous but helpful reviewers from North Carolina State and Duke Universities.

BOUNDARY VALUE PROBLEMS AND FOURIER EXPANSIONS

Chapter 1

Preliminaries

We begin with a review of partial derivatives and their two chain rules. Several examples of partial differential equations (PDEs) are solved. The relaxing temperatures within a block are obtained by a numerical model which reveals the PDE governing heat conduction. Finally, the important ∇ operator is introduced and the 'big six' PDEs are displayed.

§1.1 Partial Derivatives

The most interesting problems involve multiple degrees of freedom: vibrating structures, changing temperatures within solids, voltage potentials within regions, orbiting electrons, growing economies or populations, etc. Astonishingly accurate models of these complicated phenomena have been given using partial differential equations — as relations among the various partial derivatives of measured quantities.

Recall that the *partial derivative* at (x^0, y^0) of a function f of the two real variables x, y with respect to (say) the first variable x is the limit

$$\frac{\partial f(x^0, y^0)}{\partial x} = \lim_{h \to 0} \frac{f(x^0 + h, y^0) - f(x^0, y^0)}{h} \tag{1.1}$$

when it exists. The symbol f_x is a common alternate notation for the partial derivative $\partial f / \partial x$. So for instance, if $f(x, y) = x^3 + y^2 + x^7 y^5 + 1$, then $f_x = 3x^2 + 7x^6 y^5$, while $f_y = 2y + 5x^7 y^4$. The partial f_x represents the sensitivity of f to changes in x while holding y fixed.

More generally, the *directional derivative* of f at (x^0, y^0) in the direction of the unit vector $\mathbf{v} = (a, b)$ is the limit

$$D_{\mathbf{v}} f(x^0, y^0) = \lim_{h \to 0} \frac{f(x^0 + ha, y^0 + bh) - f(x^0, y^0)}{h} \tag{1.2}$$

when it exists. The existence of the directional derivative in all directions $\mathbf{v}$ from (x^0, y^0) is guaranteed when f is *differentiable* at (x^0, y^0), i.e., when f is well approximated by a plane nearby (x^0, y^0) — see [Apostol]. It is easy to derive the following formula for the directional derivative:

$$D_{\mathbf{v}} f(x^0, y^0) = a\frac{\partial f(x^0, y^0)}{\partial x} + b\frac{\partial f(x^0, y^0)}{\partial y}, \tag{1.3}$$

(Exercise 1.1). This formula for the directional derivative generalizes to n variables $x = (x_1, x_2, \ldots, x_n)$:

$$D_{\mathbf{v}} f(x^0) = \nabla f(x^0) \cdot \mathbf{v}, \tag{1.4}$$

where ∇f is the *gradient* of f, i.e, the vector field

$$\nabla f = (\frac{\partial f}{\partial x_1}, \frac{\partial f}{\partial x_2}, \ldots, \frac{\partial f}{\partial x_n}),$$

and where $\cdot$ is dot product. See §1.5. Because

$$D_{\mathbf{v}} f(x^0) = \nabla f(x^0) \cdot \mathbf{v} = |\nabla f(x^0)|\ |\mathbf{v}|\ \cos\theta = |\nabla f(x^0)|\ \cos\theta, \tag{1.5}$$

the gradient points in the direction of the maximal increase of f.

The formula for the directional derivative is, in turn, a special case of the first of two chain rules for partial derivatives.

First Chain Rule. Suppose $x = x(t)$ is a curve in n-space that is differentiable at $t = t^0$, and that $f(x)$ is a function of n variables differentiable at $x^0 = x(t^0)$. Then

$$\left.\frac{df(x(t))}{dt}\right|_{t^0} = \nabla f(x^0) \cdot \frac{dx(t^0)}{dt} = \sum_{i=1}^{n} \frac{\partial f(x^0)}{\partial x_i}\frac{dx(t^0)}{dt}. \tag{1.6}$$

In more familiar notation,

$$\frac{df}{dt} = \frac{\partial f}{\partial x}\frac{dx}{dt} + \frac{\partial f}{\partial y}\frac{dy}{dt} + \frac{\partial f}{\partial z}\frac{dz}{dt}.$$

Consequently,

the gradient is normal to all contour surfaces $f = c$.

The first chain rule is itself a special case.

Second Chain Rule. Suppose $f(u)$ is a differentiable function of the n variables $u = (u_1, u_2, \ldots, u_n)$, while each u_i is itself a differentiable function of the r variables $x = (x_1, x_2, \ldots, x_r)$. Then

$$\frac{\partial f}{\partial x_i} = \sum_{j=1}^{n} \frac{\partial f}{\partial u_j} \frac{\partial u_j}{\partial x_i}, \qquad i = 1, 2, \ldots, r. \tag{1.7}$$

Review these chain rules by working Exercises 1.2–1.6 and 1.19–1.22.

§1.2 Several Example PDEs

Let us work through several simple examples of partial differential equations, each of some physical importance.

Example 1. Consider the simple second order PDE

$$u_{xy} = 0, \qquad -\infty < x, y < \infty. \tag{1.8}$$

Since $(u_x)_y = 0$, u_x cannot depend on y, and so $u_x = h(x)$. But then integrating with respect to x yields that u must be of the form

$$u(x, y) = f(x) + g(y). \tag{1.9}$$

Conversely note that any function of the form (1.9) with f and g differentiable is indeed a solution of (1.8).

If we happen to know the values of $u(x, y_0)$ and $u(x_0, y)$ on the two lines $x = x_0$ and $y = y_0$, then we can recover the functions f and g uniquely within constants that add to 0 (Exercise 1.7). We will see this simple PDE again when we solve the all-important *wave equation* in Chapter 4.

Example 2. The first order PDE

$$au_x + bu_y = 0, \qquad -\infty < x, y < \infty, \tag{1.10}$$

where a and b are constants, is called the *transport* equation for reasons explained below. Observe that this equation is a geometric statement: The directional derivative

$$\nabla u \cdot (a, b) = 0$$

of u is zero in the direction (a, b), i.e., $u = u(x, y)$ is constant along curves with tangent vector (a, b). This means u is constant along each line $ay - bx = c$. But then $u = u(x, y)$ is determined solely by the value c, i.e.,

$$u(x, y) = f(ay - bx). \tag{1.11}$$

Conversely, note that any u of the form (1.11) clearly satisfies the PDE (1.10) as long as f is differentiable. These lines $ay - bx = c$ are called the *characteristic lines* of the PDE (1.10).

Physical realization of the transport equation. Think of fluid flowing through a pipe with velocity $v = v(x, t)$ at location x at time t. This fluid is carrying immiscible particles of density $\rho = \rho(x, t)$ per unit length at location x at time t. The total particle mass m within the tube between $x = a$ and $x = b$ is therefore

$$m = \int_a^b \rho \, dx. \tag{1.12}$$

By Leibniz's principle (§3.2) we may differentiate past the integral to find the instantaneous rate of change of the mass within this length of pipe:

$$m_t = \int_a^b \rho_t \, dx. \tag{1.13}$$

On the other hand, this gain of mass is the net flow in from the left less the flow out from the right, i.e.,

$$m_t = \rho(a, t)v(a, t) - \rho(b, t)v(b, t). \tag{1.14}$$

Equating and dividing by $b - a$ gives that

$$\frac{1}{b-a} \int_a^b \rho_t \, dx = -\frac{\rho(b, t)v(b, t) - \rho(a, t)v(a, t)}{b - a}. \tag{1.15}$$

In the limit, as $b \to a$, we have *conservation of mass*:

$$(\rho v)_x = -\rho_t, \tag{1.16}$$

or in the case velocity v is constant, we have *transport*:

$$v\rho_x + \rho_t = 0. \tag{1.17}$$

Example 3. Consider a generalization of transport:

$$a(x, y)u_x + b(x, y)u_y = 0. \tag{1.18}$$

This can be read as the directional derivative in the direction (a, b) is 0, i.e., u is constant along a curve with tangent vector (a, b). Such curves have normal $(-b, a)$, so locally satisfy

$$-b\,dx + a\,dy = 0. \tag{1.19}$$

Solutions $\phi(x, y) = c$ of the differential form (1.19) are called the *characteristic* curves of the PDE (1.18), one per integration constant c. Then

$$u(x, y) = f(\phi(x, y)) \tag{1.20}$$

will at least locally solve (1.18) since

$$au_x + bu_y = af'(\phi)\phi_x + bf'(\phi)\phi_y = f'(\phi)(-ba + ba) = 0.$$

The solution $u = u(x, t)$ is constant along the characteristic curves. Practice this method of characteristics by working Exercises 1.8 – 1.15.

§1.3 Transient Heat Flow in a Block

Let us investigate the flow of heat within a rectangular block of homogeneous isotropic material as shown in Figure 1.1. The top and bottom faces are insulated, thus preventing flow through these faces. Suppose initially the block is uniformly at temperature 1 when suddenly the remaining four side faces are placed and held in contact with heat sinks at temperature 0. Again refer to Figure 1.1.

Intuitively it is clear that the temperature u within the block will eventually relax to 0 as heat is forced through the four faces $x = 0$, $x = L$, $y = 0$, and $y = L$. What is not clear is the temperature regime during this relaxation. For instance, can overshoot occur — will portions of the block ever be at a negative temperature? Does this relaxation require an infinite amount of time? Must temperatures be symmetric in x and y? Is there but one trajectory that the regime of temperatures must follow during this relaxation to 0?

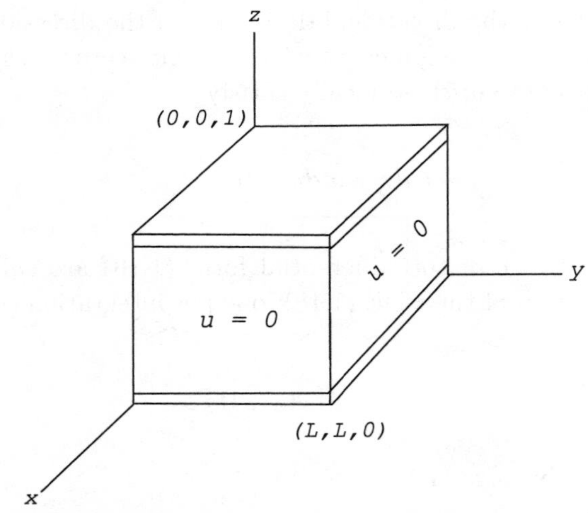

Figure 1.1 A rectangular solid homogeneous block, initially at temperature 1, with top and bottom faces $z = 0, 1$ insulated. Suddenly at $t = 0$, the remaining four faces are put in contact with sinks at temperature 0.

A numerical solution

A practical method for resolving these questions is *relaxation*, nowadays called *finite differences* [Ames; Golub and Ortega; Richmeyer and Morton]. Because the top and bottom faces are insulated, it is reasonable to assume no flow occurs in the vertical direction. Thus temperature within the block is of the form

$$u = u(x, y, t), \quad 0 \leq x, y \leq L \tag{1.21}$$

where

i) $u(x, y, 0) = 1$ for $0 < x < L,\ 0 < y < L,$

ii) $u(0, y, t) = 0 = u(L, y, t)$ for $0 \leq y \leq L,\ t \geq 0,$

iii) $u(x, 0, t) = 0 = u(x, L, t)$ for $0 \leq x \leq L,\ t \geq 0.$

The temporal condition i) is called an *initial condition* while the spatial conditions ii) and iii) are called *boundary conditions.*

Partition the face $0 \leq x, y \leq L$ in the usual way into n^2 congruent subsquares of side length $\Delta x = \Delta y = L/n$ indexed by their upper right-hand vertex (x_i, y_j) for $i, j = 1, 2, \ldots, n$.

Make the simplifying assumption that characterizes the method of finite differences: during the k-th small interval Δt of time the temperature of the (i,j)-th block is constant and uniformly $u_{i,j}^k$. (Traditionally time steps are indexed as superscripts while spatial indices are subscripts.)

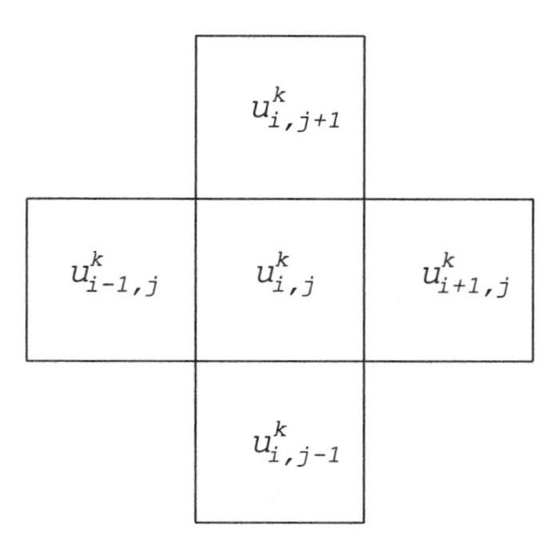

Figure 1.2 The *five point rule* for finite differences. During the *k*th interval of time, the center block loses heat to its four neighboring blocks; each block is imagined to be at a constant temperature throughout.

Let us perform a thermal power balance among the subblocks. On the one hand, the rate $\Delta \dot{Q}^k$ that heat is lost from the (i,j)-th block to its four neighbors is proportional to the temperature differences across the four faces (see Figure 1.2); the constant of this proportionality is the *conductivity* κ scaled inversely by the distance traveled by the heat and directly by the surface area of the membrane through which it passes:

$$\begin{aligned}\Delta \dot{Q}^k &= \frac{\kappa \Delta y}{\Delta x}[u_{i,j}^k - u_{i+1,j}^k + u_{i,j}^k - u_{i-1,j}^k] \\ &\quad + \frac{\kappa \Delta x}{\Delta y}[u_{i,j}^k - u_{i,j+1}^k + u_{i,j}^k - u_{i,j-1}^k].\end{aligned}$$

Conductivity κ is in units of thermal power by thickness per degree temperature gradient per area.

On the other hand, this total heat lost by the (i,j)-th block must be made up by a decrease in temperature during the k-th interval of time:

$$\Delta t \cdot \Delta \dot{Q}^k = c\rho\, \Delta x\, \Delta y(u_{i,j}^k - u_{i,j}^{k+1}),$$

where c is the *specific heat* of the material, the heat released by the material per unit mass per degree of temperature drop, and where ρ is the *density* of the material (mass per unit volume).

Solving for the subsequent temperature of the subblock yields the explicit *five point rule*:

$$u_{i,j}^{k+1} = u_{i,j}^k + \frac{\alpha \Delta t}{(\Delta x)^2}[u_{i+1,j}^k - 2u_{i,j}^k + u_{i-1,j}^k]$$

$$+\frac{\alpha \Delta t}{(\Delta y)^2}[u_{i,j+1}^k - 2u_{i,j}^k + u_{i,j-1}^k], \qquad (1.22)$$

where $\alpha = \kappa/c\rho$ is the *diffusivity* of the material (in units of area/time).

The initial condition i) is imposed by requiring that

i)′ $u_{i,j}^0 = 1$ for all $1 \leq i, j \leq n$.

The boundary conditions ii) and iii) are imposed by introducing fictitious boundary cells $u_{0,j}^k, u_{n+1,j}^k$ and $u_{i,0}^k, u_{i,n+1}^k$ and requiring

ii)′ $u_{0,j}^k = 0 = u_{n+1,j}^k$ for all j and k

and

iii)′ $u_{i,0}^k = 0 = u_{i,n+1}^k$ for all i and k.

The relation (1.22) together with conditions i)′, ii)′, and iii)′ are now implemented on a digital computer.

During World War II such "relaxation" problems were solved by placing a large number of people with good computing skills at desks, each desk representing a cell. Upon a command, each desk would perform the right hand side of (1.22), then pass copies of the result to the four neighboring desks. Nowadays of course a digital computer is employed. See Figure 1.3.

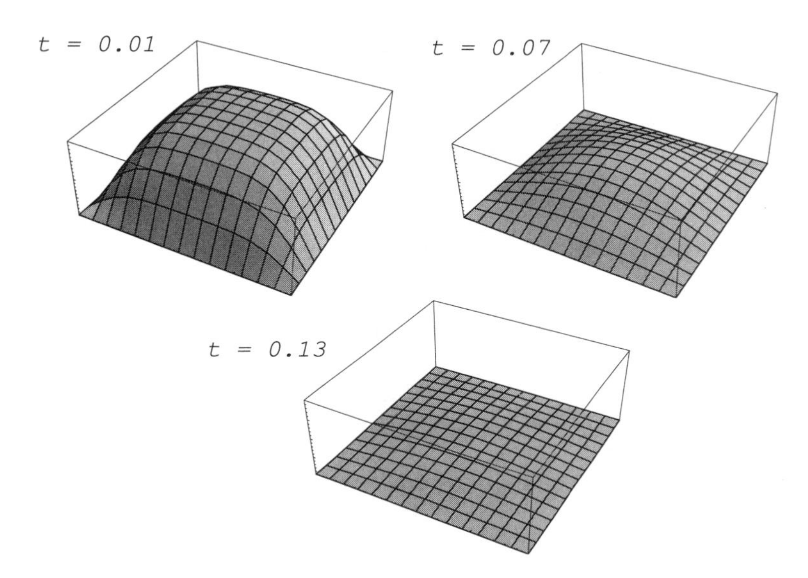

Figure 1.3 Snapshots of the relaxing temperatures of the block of Figure 1.1 with $L = \alpha = 1$ (via the symbolic manipulator *Mathematica*). See Chapter 15.

Warning. The explicit formula (1.22) suffers from instabilities unless the quantities $\alpha\Delta t/\Delta x^2$ and $\alpha\Delta t/\Delta y^2$ are small. Later we will discuss a stable implicit method known as the *Crank-Nicolson Method* that is now universally employed in such problems.

The analytic model

The equation (1.22) of relaxation can be rearranged into the form

$$\frac{u_{i,j}^{k+1} - u_{i,j}^{k}}{\Delta t}$$

$$= \alpha \left[\frac{\frac{u_{i+1,j}^{k} - u_{i,j}^{k}}{\Delta x} - \frac{u_{i,j}^{k} - u_{i-1,j}^{k}}{\Delta x}}{\Delta x} + \frac{\frac{u_{i,j+1}^{k} - u_{i,j}^{k}}{\Delta y} - \frac{u_{i,j}^{k} - u_{i,j-1}^{k}}{\Delta y}}{\Delta y} \right] \tag{1.23}$$

and so, in the limit, as $\Delta x, \Delta y \to 0$, it is reasonable that temperature u must satisfy in the interior of the square the partial differential equation (PDE)

$$\frac{\partial u}{\partial t} = \alpha \left(\frac{\partial^2 u}{\partial x^2} + \frac{\partial^2 u}{\partial y^2} \right), \tag{1.24}$$

an equation known as the *heat equation* or *equation* of *diffusion.* We will derive this famous equation by analytic means in §3.2.

§1.4 The ∇ Operator

Recall that the *del* operator

$$\nabla = \frac{\partial}{\partial x}\mathbf{i} + \frac{\partial}{\partial y}\mathbf{j} + \frac{\partial}{\partial z}\mathbf{k} \tag{1.25}$$

when applied to a scalar function f yields the *gradient*

$$\nabla f = \frac{\partial f}{\partial x}\mathbf{i} + \frac{\partial f}{\partial y}\mathbf{j} + \frac{\partial f}{\partial z}\mathbf{k}, \tag{1.26}$$

a vector field, while in contrast the *divergence* transforms a vector field

$$F = P\mathbf{i} + Q\mathbf{j} + R\mathbf{k}$$

into the scalar field

$$\nabla \cdot F = \frac{\partial P}{\partial x} + \frac{\partial Q}{\partial y} + \frac{\partial R}{\partial z}. \tag{1.27}$$

The *curl*

$$\nabla \times F = \begin{vmatrix} \mathbf{i} & \mathbf{j} & \mathbf{k} \\ \dfrac{\partial}{\partial x} & \dfrac{\partial}{\partial y} & \dfrac{\partial}{\partial z} \\ P & Q & R \end{vmatrix} \tag{1.28}$$

transforms vector fields to vector fields.

As will be seen from the list of the 'big six' PDEs in §1.5, the most physically important operator is the **Laplacian**, the divergence of the gradient

$$\nabla^2 f = \nabla \cdot \nabla f = \frac{\partial^2 f}{\partial x^2} + \frac{\partial^2 f}{\partial y^2} + \frac{\partial^2 f}{\partial z^2}, \tag{1.29}$$

transforming scalar fields to scalar fields. In particular, when $f = f(x, y)$ and $g = g(x)$,

$$\nabla^2 f = \frac{\partial^2 f}{\partial x^2} + \frac{\partial^2 f}{\partial y^2} \quad \text{and} \quad \nabla^2 g = g''. \tag{1.30}$$

The **Laplacian in cylindrical coordinates** (Exercise 1.19) is

$$\nabla^2 u = \frac{\partial^2 u}{\partial r^2} + \frac{1}{r}\frac{\partial u}{\partial r} + \frac{1}{r^2}\frac{\partial^2 u}{\partial \theta^2} + \frac{\partial^2 u}{\partial z^2}. \tag{1.31}$$

The **Laplacian in spherical coordinates** (Exercise 1.20) is

$$\nabla^2 u = \frac{\partial^2 u}{\partial \rho^2} + \frac{2}{\rho}\frac{\partial u}{\partial \rho} + \frac{1}{\rho^2 \sin^2\phi}\frac{\partial^2 u}{\partial \theta^2} + \frac{1}{\rho^2}\frac{\partial^2 u}{\partial \phi^2} + \frac{\cot\phi}{\rho^2}\frac{\partial u}{\partial \phi}. \tag{1.32}$$

§1.5 The Big Six PDEs

The most frequently encountered distributed problems are modeled by variations on one of the "Big Six" PDEs:

1. *The heat equation (or equation of diffusion)*

$$\frac{\partial u}{\partial t} = \nabla^2 u. \tag{1.33}$$

2. *The wave equation*

$$\frac{\partial^2 u}{\partial t^2} = \nabla^2 u. \tag{1.34}$$

3. *Laplace's equation*

$$\nabla^2 u = 0. \tag{1.35}$$

4. *Poisson's equation*

$$\nabla^2 u = f. \tag{1.36}$$

5. *Schrödinger's equation*

$$i\frac{\partial \psi}{\partial t} = -\nabla^2 \psi + v(x)\psi. \tag{1.37}$$

6. *The plate (beam) equation*

$$\frac{\partial^2 u}{\partial t^2} = -\nabla^2\nabla^2 u. \tag{1.38}$$

Some of the above have been brought to their *nondimensional* form by an appropriate rescaling of variables.

Exercises

A function is *harmonic* on a domain Ω if it satisfies Laplace's equation $\nabla^2 u = 0$ on Ω.

1.1 Prove the formula (1.3) for the directional derivative.

1.2 Show that $u = \log r = \log \sqrt{x^2 + y^2}$ is harmonic on the punctured plane $r > 0$.

1.3 Show that $u = 1/\rho = 1/\sqrt{x^2 + y^2 + z^2}$ is harmonic on $\rho > 0$. Is $v = 1/r = 1/\sqrt{x^2 + y^2}$ harmonic on $r > 0$?

1.4 Prove that $\nabla^2(\frac{1-x^2-y^2}{(1-x)^2+y^2}) = 0$ within the unit disk $x^2 + y^2 < 1$.

Hint: showing that $\nabla^2 u = 0$ directly is difficult. I recommend either

a) the use of a symbolic manipulator, or

b) an algebraic reduction to a simpler problem, or

c) noting that u is the real part of the analytic function $f(z) = (1 + z)/(1 - z)$.

1.5 Let $v(x, y) = u(\bar{x}, \bar{y}) = u(ax + by, ay - bx)$ for $a^2 + b^2 \neq 0$. This change of variables

$$\begin{pmatrix} a & b \\ -b & a \end{pmatrix} \begin{vmatrix} x \\ y \end{vmatrix} = \begin{vmatrix} \bar{x} \\ \bar{y} \end{vmatrix}$$

is a rotation and a dilation. Show using the chain rule that wherever sensible,

$$\nabla^2 v(x, y) = 0 \Leftrightarrow \nabla^2 u(\bar{x}, \bar{y}) = 0.$$

(More generally, harmonicity is conformally invariant for functions in the plane.)

1.6 Using the second chain rule, show that if $\alpha = x - ct$ and $\beta = x + ct$, then $u_{tt} = c^2 u_{xx}$ if and only if $u_{\alpha\beta} = 0$.

1.7 Show that the functions f and g of (1.9) are uniquely determined within constants that sum to zero.

1.8 Solve $u_t + x u_x = 0$ via the method of characteristics.

1.9 What is the physical interpretation in (1.17) of its characteristic lines?

1.10 Solve $u_x + u_{xy} = 0$.

1.11 Solve $u_x + 2u_t = 0$ given that $u(x, 0) = \sin x$.

1.12 Using the chain rule, show that by the change of variables $\bar{x} = ax + by$ and $\bar{y} = bx - ay$, transport $au_x + bu_y = 0$ becomes simply $u_{\bar{x}} = 0$.

1.13 Solve $yu_x + xu_y = 0$.

1.14 Solve $u_{xx} = 0$ when $u = u(x, y)$.

1.15 Solve $u_{xx} = y^2 u$ when $u = u(x, y)$.

1.16 Argue that if $\rho = \rho(x, t)$ is the density of freeway traffic flow at time t past point x and $v = v(x, t)$ its velocity, then $\rho_t + (v\rho)_x = 0$.

1.17* Let $u = u(x, t)$ be the population density of a country, i.e.,

$$P(b, t) - P(a, t) = \int_a^b u(x, t)\, dx$$

is the population at time t between the ages of a and b. Let $q(x)$ be the probability of death less net immigration at age x per year. Argue for *von Foerster's model*

$$u_t + u_x + qu = 0.$$

Solve analytically by multiplying through by the integrating factor $\exp[\int_0^x q(y)\, dy]$. Argue that the left boundary condition is of the form

$$u(0, t) = \int_0^\infty \beta(x) u(x, t)\, dx.$$

Simulate this model via the numerical model $u_{i+1}^{j+1} = u_i^j - hq_i u_i^j$ and experiment with various initial populations, birth, and death rates. Is this model reasonable?

1.18 Generically let f be a scalar field, F a vector field. Let ∇, $\nabla\cdot$, and $\nabla\times$ denote the gradient, divergence, and curl respectively [Schey]. Of the 18 possible triple combinations of symbols $\nabla * \nabla * *$ formed from the table

$$\nabla \quad \begin{matrix} \\ \cdot \\ \times \end{matrix} \quad \nabla \quad \begin{matrix} \\ \cdot \\ \times \end{matrix} \quad \begin{matrix} f \\ \\ F \end{matrix},$$

only five make sense. Which five? Of the five, two are identically 0. Which two?

1.19 Establish in two ways the **Laplacian in polar coordinates**

$$\nabla^2 u = \frac{\partial^2 u}{\partial r^2} + \frac{1}{r}\frac{\partial u}{\partial r} + \frac{1}{r^2}\frac{\partial^2 u}{\partial \theta^2},$$

a) first by applying the chain rule to the change of variables $x = r\cos\theta$ and $y = r\sin\theta$, and then

b) with physical intuition — rederive the heat equation by accounting for the flux leaving the four edges of $r_0 < r < r_0 + \Delta r$ and $\theta_0 < \theta < \theta_0 + \Delta\theta$. Take the limit.

Corollary. (Laplacian in Cylindrical Coordinates)

$$\nabla^2 u = \frac{\partial^2 u}{\partial r^2} + \frac{1}{r}\frac{\partial u}{\partial r} + \frac{1}{r^2}\frac{\partial^2 u}{\partial \theta^2} + \frac{\partial^2 u}{\partial z^2}.$$

1.20 Establish that the **Laplacian in Spherical Coordinates** is

$$\nabla^2 u = \frac{\partial^2 u}{\partial \rho^2} + \frac{2}{\rho}\frac{\partial u}{\partial \rho} + \frac{1}{\rho^2\sin^2\varphi}\frac{\partial^2 u}{\partial \theta^2} + \frac{1}{\rho^2}\frac{\partial^2 u}{\partial \varphi^2} + \frac{\cot\varphi}{\rho^2}\frac{\partial u}{\partial \varphi}$$

in two ways:

a) first via the chain rule on the change of variables $x = \rho\sin\varphi\,\cos\theta$, $y = \rho\sin\varphi\,\sin\theta$, $z = \rho\cos\varphi$, then again

b) via physical intuition — rederive the heat equation by accounting for the flux leaving the six faces of the box determined by spherical infinitesimals.

1.21 Show using the form of the Laplacian in polar coordinates that $u = \theta$ is harmonic, i.e., $\nabla^2 u = 0$ within any sector of opening less than 2π radians.

1.22 Redo Exercise 1.3 using Exercises 1.19–1.20.

1.23 Show that the units of diffusivity α must be *area/time* by two methods: first from examining $\alpha = \kappa/\rho c$ and second by balancing units in the heat equation (1.23) itself.

1.24 Find a vector field F with no potential, i.e., for which there is no scalar function φ such that $\nabla\varphi = -F$.

Hint: Give an infamous example like 'dQ' $= c_V dT + nRTdV/V$ or '$\nabla\theta$.'

1.25 Clausius's version of the Second Law of Thermodynamics is that *heat cannot of itself pass from a cold to a hot body.* [Planck, §112]. Attempt to deduce that no portion of the block of §1.4 can be at a negative temperature during the above relaxation. More strongly, attempt to show that temperature at each location is a decreasing function of time.

1.26 Argue that the dispersion of a spilled pollutant may in some circumstances be modeled by the three dimensional equation of diffusion

$$\frac{\partial u}{\partial t} = \alpha \left[\frac{\partial^2 u}{\partial x^2} + \frac{\partial^2 u}{\partial y^2} + \frac{\partial^2 u}{\partial z^2} \right].$$

1.27 Using a spreadsheet or MatLab, find the approximate temperatures of the block of Figure 1.1 at selected times $0 < t < 1$ for the case $\alpha = L = 1$ [Orvis].

1.28 Argue that the thickness of the block of Figure 1.1 is irrelevant, that heat transfer within the block is modeled by (1.3) for any thickness.

1.29 Consider a cylindrical rod of length $L = 1$ of homogeneous material of diffusivity α that is insulated everywhere but at the end faces. Initially the rod is uniformly at temperature 0, when suddenly the right end is put in contact with a source at temperature 1 while the left end is held at temperature 0. Derive Finite Difference and PDE models of the transient temperatures within the rod. Using a spreadsheet or MatLab, experimentally verify the intuitively clear eventual steady state. Experimentally discover for what value of $h = \alpha\Delta t/\Delta x^2$ your numerical model becomes unstable.

1.30 Argue that the results of the previous Exercise could equally well hold for an infinite slab of thickness L.

1.31 In the early days of the telegraph, undersea cables experienced what is today called *intersymbol distortion*: transmitted pulses would smear spatially as they traveled along the cable, combining with previous and subsequent pulses. William Thomson in 1854 postulated that the cable can be modeled as cascaded infinitesimal sections of distributed series resistance R (Ohms per meter) together with distributed shunt capacitance C (farads per meter), as shown in Figure 1.4. Recalling that voltage v and current i are related by the rules $v = Ri$ and $C dv/dt = i$ for a resistor and capacitor respectively, reconstruct Kelvin's deduction that the voltage V and current I on the cable must satisfy the system of coupled PDEs

$$\frac{\partial V}{\partial x} = -RI \text{ and } C\frac{\partial V}{\partial t} = -\frac{\partial I}{\partial x}$$

giving that the voltage V on the undersea cable is modeled by diffusion

$$\frac{\partial V}{\partial t} = \alpha\frac{\partial^2 V}{\partial x^2}$$

with diffusivity

$$\alpha = \frac{1}{RC}.$$

(The above problem is seminal for the field of Electrical Engineering. See the superb historical account by Paul J. Nahin entitled *Oliver Heaviside: Sage in Solitude.*)

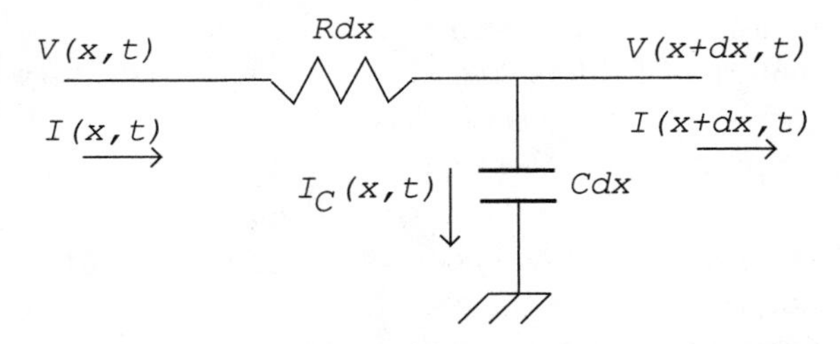

Figure 1.4 Kelvin's model of the undersea cable: cascaded infinitesimal sections of distributed series resistance and shunt capacitance.

1.32 A telegraph key is closed, applying (say) 1 volt to one end of an early very long undersea telegraph cable. By direct calculation substantiate Stokes's belief that the voltage $V(x,t)$ at x meters down the cable at time t has value (step response)

$$V(x,t) = \operatorname{erfc}\left(x\sqrt{\frac{RC}{4t}}\right),$$

where the *error function complement* is defined by

$$\operatorname{erfc}(x) = \frac{2}{\sqrt{\pi}} \int_x^\infty e^{-\beta^2}\, d\beta.$$

1.33 Verify analytically the observed *Kelvin's Rule* of *Squares*, that the *peak current* in the above undersea cable *experiences a delay time proportional to the square of the distance down the line.* Explicitly

$$t_{delay} = \frac{RC}{2}x^2.$$

1.34 What is the physical interpretation of Kelvin's Rule of Squares for heat conduction through a thick slab?

1.35 Although a remarkably accurate model of everyday heat conduction phenomena, the diffusion view must be flawed: show that Stokes's step response above implies instantaneous communication over arbitrarily long distances.

The next 6 exercises form a primer on point-set topology.

1.36 Let Ω be a subset of $\mathbf{R}^n$. A point x is a *boundary point* of Ω if every ball centered at x of positive radius meets both Ω and its complement $\Omega' = \mathbf{R}^n \setminus \Omega$. The *interior* Ω^0 of Ω is all the non-boundary points of Ω. Show that the entire space $\mathbf{R}^n$ is therefore the union of three disjoint sets: $\mathbf{R}^n = \Omega^0 \cup \partial\Omega \cup \Omega'^0$, the *interior* of Ω, the *boundary* of Ω and the *exterior* of Ω.

1.37 Give an example of an infinite set that consists entirely of boundary points.

1.38 A set is *open* if it is consists entirely of interior points. A set Ω is *connected* if it is impossible to find open non-empty disjoint sets A and B that together cover Ω where neither alone will cover Ω. Give examples of connected sets, and of disconnected sets. Show that any two points of a *domain* (*region*) Ω, an open connected set, can be connected by a curve made up of line segments parallel to the coordinate axes that lies completely within Ω.

1.39 A set C is *compact* if every sequence $\{x_n\}_{n=1}^{\infty}$ drawn from C possesses a convergent subsequence $\{x_{n_k}\}_{k=1}^{\infty}$ converging to a limit in C. A sequence $\{x_n\}_{n=1}^{\infty}$ *converges* to the limit x_0 if every open ball about x_0 contains all but a finite number of the terms x_n of the sequence. Show that compact subsets of $\mathbf{R}^n$ are closed and bounded. (A *closed* set contains all its boundary points). The converse is the basic

Heine-Borel Principle. Every closed bounded subset of $\mathbf{R}^n$ is compact.

And its companions:

Definition. A function f is *continuous on the set* E if for every convergent sequence $x_k \to x_0$ with terms and limit in E, $f(x_k) \to f(x_0)$.

Principle. The continuous image of a compact set is compact.

Principle. The continuous image of a connected set is connected.

1.40 Deduce that a real valued continuous function on a closed and bounded subset of $\mathbf{R}^n$ is bounded and in fact achieves its maximum and minimum values.

1.41 The **Completeness Axiom** posits that *the real numbers* $\mathbf{R}$ *are complete*, i.e., every Cauchy convergent sequence converges, or in an equivalent form, that *every set bounded from above possesses a supremum.* Prove however that the subset $\mathbf{Q}$ of rational numbers is not complete.

1.42 Suppose $u = u(x,t)$ is the density of cars (in cars per unit length) on a freeway at position x at instant t. Argue that

$$u_t + (vu)_x = 0,$$

where $v = v(x,t)$ is the speed of cars past the point x at time t. Does v in fact depend on u, i.e., $v = f(u)$? See [May] or [Klamkin].

Chapter 2
Steady Problems

A system is in *steady state* if no change occurs over time. Consider for instance a diffusion process (1.24) within a homogeneous isotropic material that has reached steady state u, i.e., $\partial u/\partial t \equiv 0$. Then u must satisfy *Laplace's Equation*

$$\nabla^2 u = 0, \tag{2.1}$$

where the operator ∇^2 is known as the *Laplacian* and is written with any one of the several notations

$$\nabla^2 u = \nabla \cdot \nabla u = \text{div grad } u = \Delta u = \frac{\partial^2 u}{\partial x^2} + \frac{\partial^2 u}{\partial y^2} + \frac{\partial^2 u}{\partial z^2}. \tag{2.2}$$

A solution u to Laplace's equation (2.1) is said to be a *harmonic function.*

The Laplacian is a central player in analytic models of many natural processes. For example, as established below in the Exercises, electrostatic potential v must satisfy Laplace's equation in charge-free regions. The same considerations demonstrate that gravitational potential must satisfy Laplace's equation in mass-free regions.

Laplace's equation $\nabla^2 u = 0$ is a statement of conservation — the 'stuff' carried by the flow ∇u entering a closed surface equals the 'stuff' leaving the surface. More analytically, $\nabla^2 u$ controls the error between the value of u at a point and the average value of u nearby (Exercise 2.14). We will examine these notions with precision in Chapter 3.

But in this chapter let us obtain intuitive insight into steady state problems by attempting to find steady temperatures on several selected regions with simple boundary condditions. Our goal is to solve

$$\nabla^2 u = 0$$

on the given regions — first with given boundary values (*Dirichlet problems*), then when the flux leaving through the boundary is specified (*Neumann problems*).

§2.1 Square Regions (SSP 1)

Consider the unit square of Figure 2.1. Each edge is at temperature (voltage) $u = 0$ except for the edge at $x = 1$ which is held at temperature (voltage) $u = 1$. What is the steady state temperature (voltage) within? (The vertices where an abrupt change of temperature occurs are to be thought of as points of insulation.)

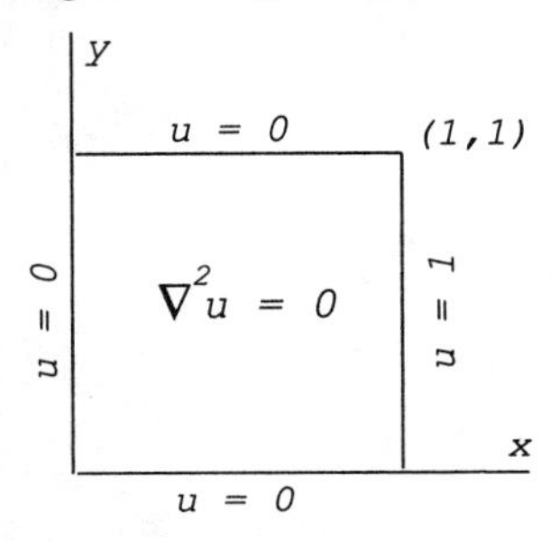

Figure 2.1 Steady state within a square with given boundary values.

For that matter, does a steady state solution exist? Is there only one such solution? We shall see later there is indeed exactly one bounded solution that satisfies these four Dirichlet boundary conditions. The solution takes the form

$$u(x, y) = \sum_{n=1}^{\infty} c_n \sinh n\pi x \, \sin n\pi y. \tag{2.3}$$

But now given such a solution, we may bootstrap to solve many related problems. For example, consider the boundary values of Figure 2.2.

A solution v of this new problem is (exercise)

$$v(x, y) = u(x, y) + u(y, x).$$

The solution v is the *superposition* of two steady state solutions, a use of the *additive* property of the Laplacian, a linear operator:

$$\nabla^2(\alpha u_1 + \beta u_2) = \alpha \nabla^2 u_1 + \beta \nabla^2 u_2.$$

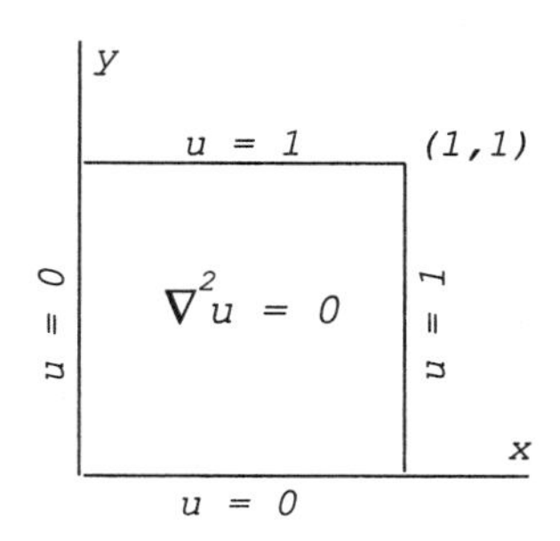

Figure 2.2 Steady state within a square with prescribed boundary values.

§2.2 Sectors (SSP 2)

Consider the two dimensional steady state boundary problem shown in Figure 2.3, a sector of opening angle θ_0.

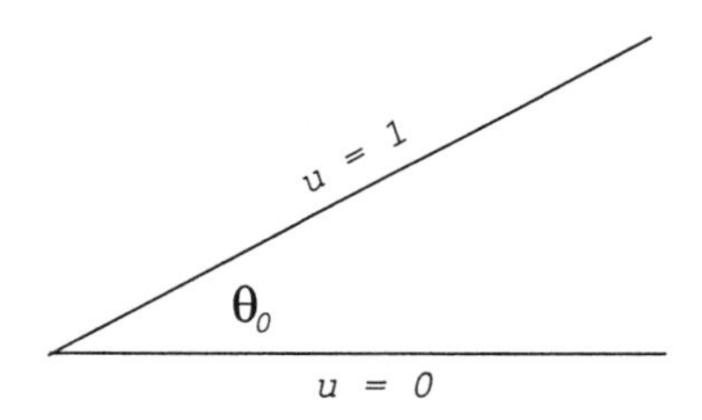

Figure 2.3 Steady state within a sector of opening θ_0 with boundary values specified.

As established below in the exercises,

$$u(x, y) = \frac{\theta}{\theta_0} \tag{2.4}$$

is harmonic and satisfies the given boundary conditions provided $0 < \theta_0 < 2\pi$ where θ is the argument of polar coordinates.

Alert. This solution is not the only solution! The problem is *ill posed* since it does not specify conditions at infinity. For example, in the halfplane case $\theta_0 = \pi$,

$$u(x, y) = \frac{\theta}{\pi} + \mu \sin x \sinh y$$

is also a solution for any μ (Exercise 2.22). However solution (2.4) is the only bounded solution as we shall see. Ill-posed models have led to catastrophic engineering failures.

§2.3 Infinite Strips (SSP 3)

Consider the cross section of an infinite slab of thickness L shown in Figure 2.4. Intuitively the correct physical solution should be a linear increase in temperature from left to right, i.e.,

$$u(x,y) = T_0 + (T_1 - T_0)\frac{x}{L}, \tag{2.5}$$

which indeed solves the problem. However this problem is ill posed since there are other (unbounded) solutions (Exercise 2.23). As we shall see subsequently, mathematics substantiates physical intuition — the above linear solution (2.5) is the only bounded solution. See Exercise 2.7.

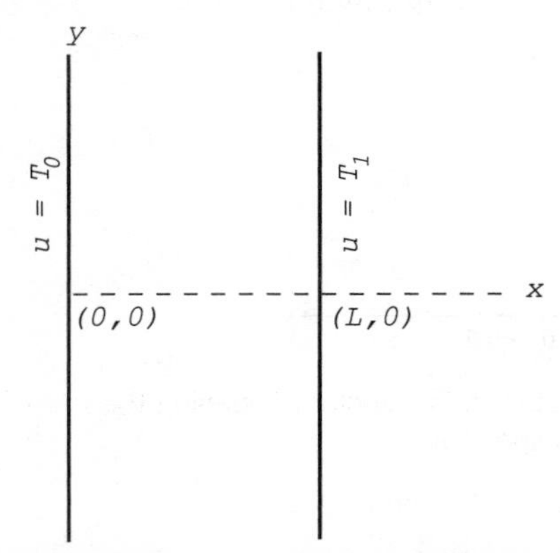

Figure 2.4 Steady state within an infinite strip of width L with specified boundary values.

§2.4 Annular Regions (SSP 4)

Waste heat from an urban industrial process is often used for heating nearby residential or commercial buildings [ASHRAE, S11]. The buildings extract their heat from a buried water pipe carrying hot water from the plant. A first step in the design of such closed-loop district heating systems is to analyze the rate of heat loss from the buried pipe and its effect on surrounding ground temperatures. The following calculations of Fourier are used to select pipe insulation levels.

Consider a cross-section of the pipe of radius a and the surrounding earth out to radius b with boundary temperatures as shown in Figure 2.5 — the inner circle is at temperature A while the outer circle is at temperature B.

Because of radial symmetry, it is intuitively clear that the steady state temperature u is a function of the radial distance r from the center, i.e., $u = u(r)$.

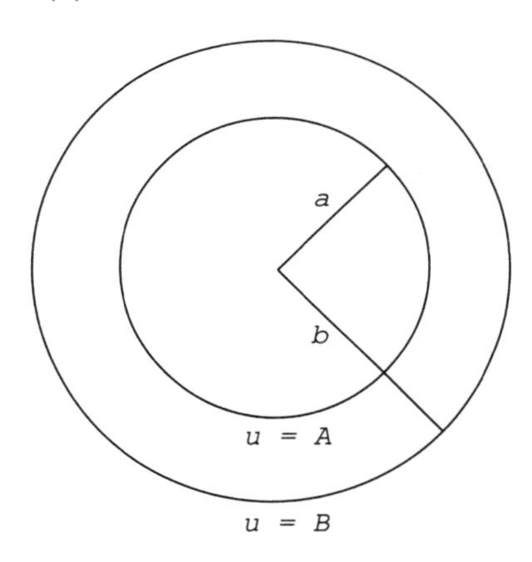

Figure 2.5 Cross section of a buried pipe of radius a carrying fluid at temperature $u = A$. Earth at a distance $r = b$ is unaffected and remains at temperature $u = B$.

If you jump to the conclusion that u is merely a linear function of r, you will be in error. The function $u = r$ is not harmonic (Exercise 2.8).

Because we are in steady state, the rate that heat leaves each disk of radius r is constant for all r. Thus for small Δr

$$2\pi r \frac{\kappa}{\Delta r}[u(r) - u(r + \Delta r)]$$

is essentially a constant $\dot{Q}$. In the above, conductivity κ is scaled by the area $2\pi r$ of the curved surface of a unit length pipe and inversely by the thickness Δr of the material traversed.

In the limit, as $\Delta r \to 0$,

$$r\frac{\partial u}{\partial r} = -\frac{\dot{Q}}{2\pi\kappa}. \tag{2.6}$$

Hence our steady solution u is of the form

$$u = C \log r + D, \tag{2.7}$$

where for design purposes it is important to retain that $C = -\dot{Q}/2\pi\kappa$. Direct calculation (Exercise 2.8) shows that indeed the u given by (2.7) is harmonic within the annular region.

Imposition of the boundary conditions (Exercise 2.6) yields Fourier's famous solution

$$u = \frac{u(b) - u(a)}{\log b - \log a} \log r + \frac{u(a) \log b - u(b) \log a}{\log b - \log a}, \tag{2.8}$$

where

$$C = \frac{u(b) - u(a)}{\log b - \log a} = -\frac{\dot{Q}}{2\pi\kappa} \tag{2.9}$$

is the *logarithmic mean temperature*, a standard parameter of heat exchanger design.

§2.5* Solutions via Brownian Motion

Whenever a steady Dirichlet problem can be transformed into a problem with boundary values 0 or 1 there is this stunning

Rule of Thumb. (Lévy) Suppose the boundary values specified for the bounded region Ω are either 0s or 1s. Laplace's equation can be solved on Ω for a bounded solution u as follows: The value of u at the point x equals the probability that a particle released at x, undergoing Brownian motion, will first exit the boundary at points with value 1 specified [Petersen].

Example. Consider the steady Dirichlet problem $\nabla^2 u = 0$ within the unit disk $x^2 + y^2 < 1$ with boundary conditions $u = 1$ for $0 < \theta < \pi$ and $u = 0$ for $-\pi \le \theta \le 0$. The solution by Lévy's rule of thumb must be the view angle α from (x, y) of the 1-boundary less its view from the 0-boundary, all normalized by the net view angle from the 1-boundary, i.e.,

$$u = \frac{\alpha - \pi/2}{3\pi/2 - \pi/2} = \frac{\alpha}{\pi} - \frac{1}{2}. \tag{2.10}$$

See Figure 2.6.

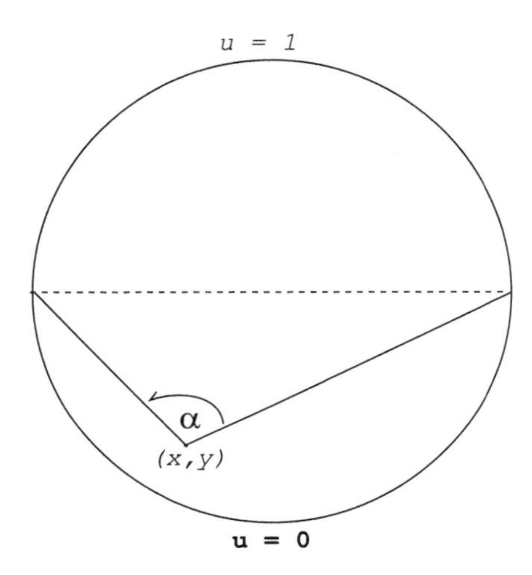

Figure 2.6 The probability *u* that a particle released at *(x,y)* escapes by Brownian motion successfully through the upper semicircle is the view angle α of success less the view from failure, all normalized by the net view from success: $u = (\alpha - \pi/2)/\pi$.

Remark. All boundary problems considered thus far have been of the *Dirichlet type* where the values of the solution u are specified on the boundary. However the majority of natural processes are modeled with more complicated boundary conditions. We now turn to problems of the *Neumann type* where the *rate* that heat enters or leaves the boundary is specified.

§2.6 A Rod (SSP 5)

The left end $x = 0$ of an isotropic homogeneous rod is in contact with a sink at temperature 0 while the right end $x = L$ is in contact with a heating element discharging heat at the rate $\dot{q}$ Watts into the rod. What is the steady state temperature profile within the rod?

The flux condition at the right end $x = L$ translates into the normal derivative condition

$$\dot{q} = a\kappa \frac{du}{d\mathbf{n}} = a\kappa\, u'(L)$$

where κ is conductivity and a the area of the end face of the rod.

This problem can be brought to the dimensionless form (Exercise 2.15)

$$\nabla^2 u = u''(x) = 0, \quad 0 < x < 1, \tag{2.11}$$

subject to the boundary conditions

i) $u(0) = 0$

ii) $u'(1) = 1$.

But harmonic functions of a single linear variable x are linear (Exercise 2.7), i.e.,

$$u(x) = Ax + B.$$

Imposing boundary condition i) gives $B = 0$. Imposing condition ii) yields $A = 1$ and the solution $u(x) = x$.

§2.7 The Rod Revisited (SSP 6)

Let us modify the previous problem by insulating the left end:

$$\nabla^2 u = u''(x) = 0, \quad 0 < x < 1, \tag{2.12}$$

subject to the boundary conditions

i) $u'(0) = 0$

ii) $u'(1) = 1$.

Intuitively this problem cannot have a steady solution since heat is entering but not leaving. This is substantiated mathematically since $u'(x) \equiv A$ above, and thus cannot at the same time equal 0 and 1. The problem has no steady state.

§2.8 Closed-Loop Heat Pumps (SSP 7)

An efficient means of heating buildings is by extracting low quality heat from the earth, then raising its quality to a usable level with an electrically driven heat pump. See Figure 2.7. Good design can achieve a coefficient of performance (COP) of 4, that is, 4 units of usable heat can be delivered while paying for only one unit of electrical energy.

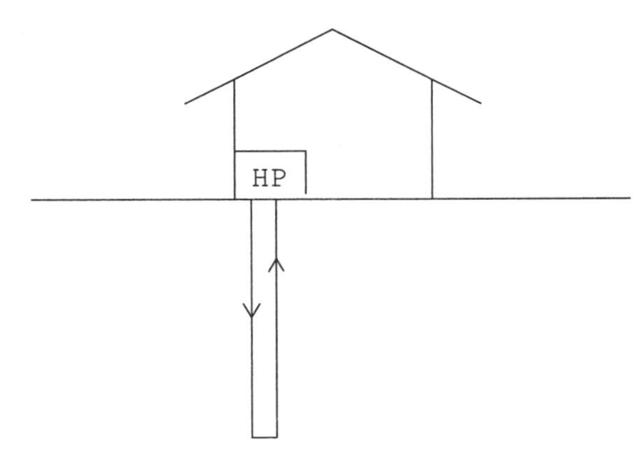

Figure 2.7 The closed-loop ground-coupled heat pump. Heat is extracted from the earth with a continuous loop of buried water pipe. An electrically driven heat pump raises the quality of the heat to a useful level.

Heat is extracted from the earth by circulating cooled water around a continuous closed loop of plastic water pipe buried either horizontally or vertically. The surrounding earth cools as its heat is extracted and delivered to the building. What should be the total length of the buried pipe? If too short, nearby ground temperatures will fall to a level where the heat pump is no longer economical [Bose et al.].

Let us approximate the ground temperatures as seen by the exchanger towards the end of the heating season, when conditions have more or less settled to steady state. We assume that earth temperatures at the far-field distance of $r = b$ from the exchanger have been unaffected and remain at the yearly average temperature B. The exchanger is of radius $r = a$, while the heat pump is extracting heat at the design heat load rate of $\dot{Q}$ per unit length of pipe. See Figure 2.8. The ground temperatures therefore satisfy

$$\nabla^2 u = 0, \qquad a < r < b \tag{2.13}$$

subject to

i) $u(b) = B$,

ii) $2\pi a \kappa \, u'(a) = \dot{Q}$.

As we have seen in Exercise 1.19, for radially symmetric problems, Laplace's equation becomes

$$ru'' + u' = 0,$$

where the prime denotes differentiation with respect to the radius r. As we saw in SSP 4, solutions are

$$u(r) = C \log r + D, \tag{2.14}$$

where imposing ii) gives

$$C = ru'(r) = au'(a) = \dot{Q}/2\pi\kappa.$$

Imposing i) yields the near earth temperatures

$$u(r) = B - \frac{\dot{Q}}{2\pi\kappa} \log b/r. \tag{2.15}$$

An actual design problem is posed in Exercise 2.16. More accurate transient versions of this problem will be solved in later sections.

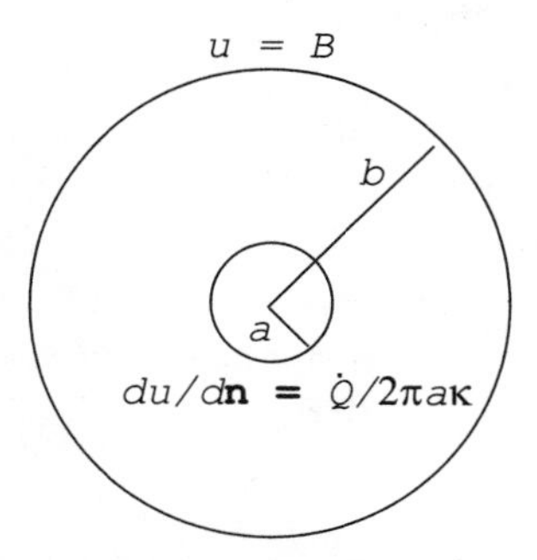

Figure 2.8 A cross section of the vertical exchanger of Figure 2.7. Far-earth temperature at distance $r = b$ is assumed to be unaffected.

§2.9 Forces Crushing a Capacitor (SSP 8)

A large parallel plate capacitor has its left plate $x = 0$ at voltage 0, while the right plate $x = d$ is at voltage v_0. See Figure 2.9. What is the force attempting to crush this capacitor?

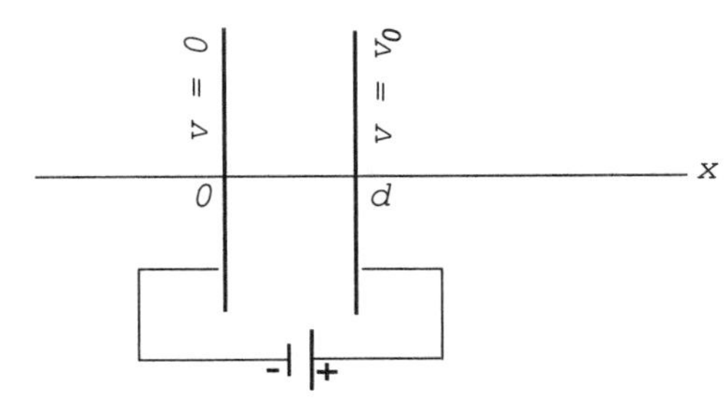

Figure 2.9 A parallel plate capacitor is charged with a potential difference of v_0 volts. Electrostatic forces are attempting to colapse the capacitor.

Intuitively, the voltage between the plates — well away from the edges — is more or less of the form $v = v(x)$. But then since static voltage is harmonic, it is by Exercise 2.7 a linear function of x. Applying the boundary conditions yields

$$v(x) = \frac{v_0}{d}x,$$

which is the work per Coulomb needed to move charge from the left plate to x. Hence

$$\nabla v = \frac{v_0}{d}\mathbf{i} = -E.$$

But by the Gauss theorem (Exercise 3.4), the magnitude of the electric field intensity is

$$|E| = \frac{\sigma}{\epsilon_0}$$

where σ is charge per unit area. Hence the magnitude of the force per unit area must be

$$|E|\sigma = \epsilon_0|E|^2 = \frac{v_0^2\epsilon_0}{d^2}.$$

§2.10 Neumann Problems on a Square (SSP 9)

Consider the unit square of Figure 2.10 with the Dirichlet condition $u = 0$ at three edges $x = 0, y = 0, y = 1$, but with the Neumann normal derivative condition $du/d\mathbf{n} = 1$ imposed on the right edge $x = 1$. Compare with SSP 1. As we derive later, the solution to this problem is

$$u = \frac{4}{\pi^2} \sum_{\substack{n=1 \\ n \text{ odd}}}^{\infty} \frac{\sinh n\pi x \, \sin n\pi y}{n^2 \cosh n\pi}.$$

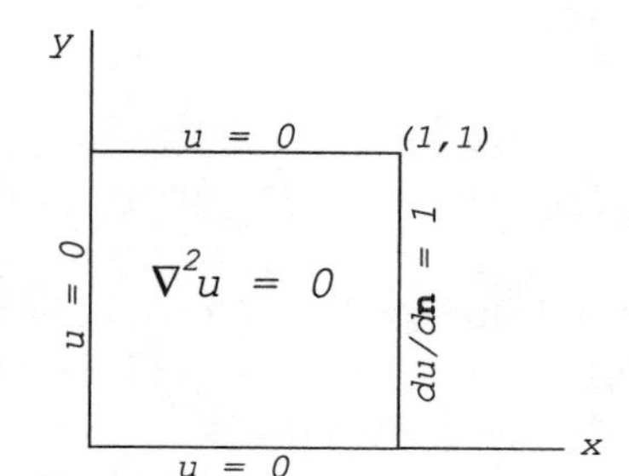

Figure 2.10 A steady mixed problem.

But given this basic solution $u = u(x, y)$, we may solve other related problems using superposition. For example, the steady problem of Figure 2.11 has solution

$$v(x, y) = 2\, u(y, x)$$

and so forth.

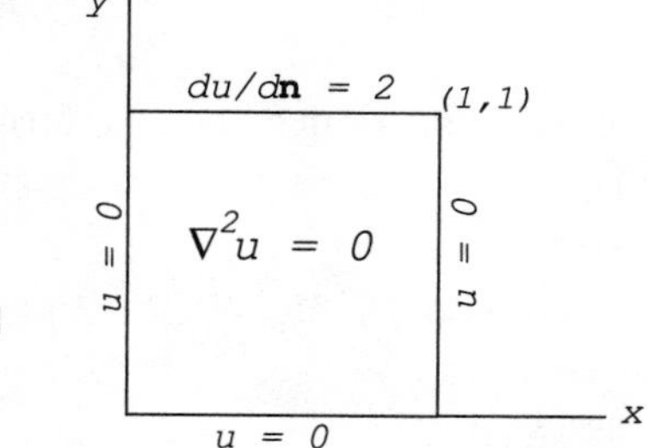

Figure 2.11

§2.11 Quarter Disks (SSP 10)

Consider the quarter unit disk with the mix of Dirichlet and Neumann conditions as shown in Figure 2.12.

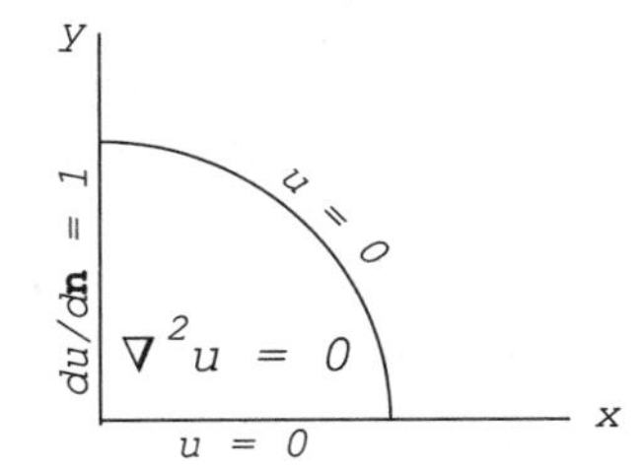

Figure 2.12 A steady mixed problem.

Analytically the description is

$$\nabla^2 u = 0, \qquad 0 < r < 1,\ 0 < \theta < \pi/2, \tag{2.16}$$

subject to

i) $u(r, 0) = 0,\ 0 < r < 1,$

ii) $u(1, \theta) = 0,\ 0 < \theta < \pi/2,$

iii) $u_\theta(r, \pi/2) = r,\ 0 < r < 1.$

Harmonic functions of form $u = c\theta$ will not solve this problem (exercise). However trial functions such as

$$u(r, \theta) = (r^\omega - r^{-\omega}) \sin \omega\theta$$

are harmonic and do satisfy the first two boundary conditions (Exercise 2.17). Can these trial solutions be superimposed in some way to satisfy the third boundary condition? For example, can we find some weighting function $f(\omega)$ so that

$$u = \int_0^\infty f(\omega)(r^\omega - r^{-\omega}) \sin \omega\theta \, d\omega \tag{2.17}$$

solves SSP 10? We shall study solutions of this type in Chapter 17.

§2.12 Robin Problem in a Hemisphere (SSP 11)

Consider the steady problem depicted in Figure 2.13. In symbols,

$$\nabla^2 u = 0, \quad 0 < \rho < 1,\ 0 \le \phi < \pi/2,\ 0 \le \theta \le 2\pi, \tag{2.18}$$

subject to

i) $\partial u/\partial \rho = -u$, for $\rho = 1$, and

ii) $u(\rho, \pi/2, \theta) = 0$, for $0 < \rho < 1$.

The mixed *Robin* condition i) translates to: *The rate heat is leaving is proportional to the interior temperature driving it out.* It is now intuitively clear that the steady state solution is the zero solution. More mathematically, $u \equiv 0$ is a solution and solutions are unique — see Exercise 3.8. We will solve the more interesting transient version of this problem later on.

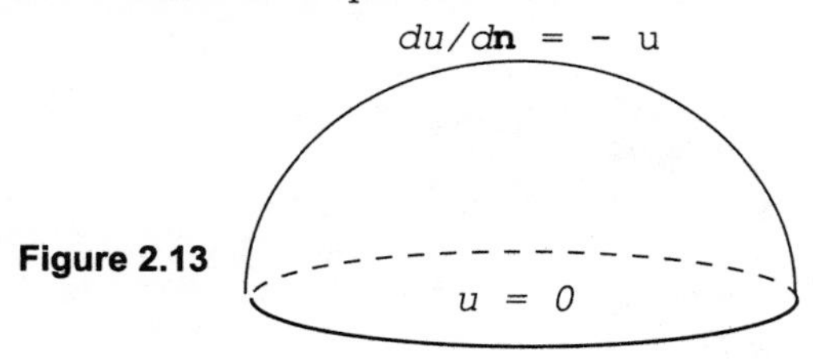

Figure 2.13

§2.13* Method of Conformal Mapping

A classical method for solving steady problems in the plane used extensively before the advent of the digital computer is the *Method of Conformal Mapping.* The method put simply is: Map the domain to a simpler domain where the solution is obvious, solve, map back. If the map is *conformal,* it will preserve harmonicity. In Figure 2.14a–e are various problems that yield to this method. Many useful worked examples appear in the superb text *Complex Variables and Applications* [Churchill and Brown] and in the classical aerospace engineering references [Abbot and von Doenhoff] and [Milne and Thompson]. Ironically, conformal mapping is enjoying a renaissance in finite element numerical algorithms.

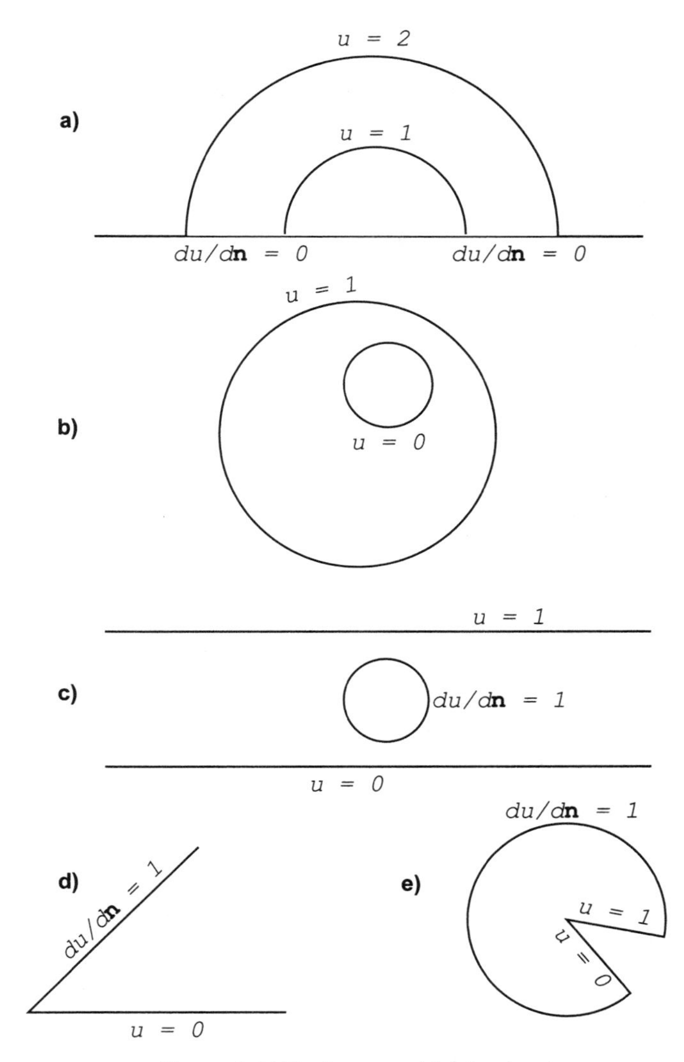

Figure 2.14 Regions on which Laplace's equation yields to conformal mapping.

Exercises

(Unless otherwise mentioned, r and θ will denote the radial and angular variables of polar coordinates.)

2.1 Show that $u = x^3 - 3xy^2$ is harmonic.

2.2 Show that $u = \theta$ is harmonic within any sector of opening less than 2π radians.

Hint: use $\theta = \arctan y/x$ and Exercise 2.1.

2.3 Does $u(2x, y)$ solve the steady problem on the rectangle shown in Figure 2.15 whenever $u(x, y)$ solves the steady problem on the square of Figure 2.1?

Answer: No. Why?

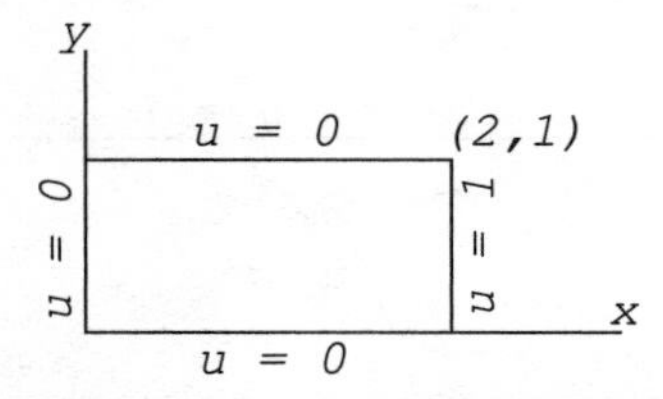

Figure 2.15 Steady-state temperatures (or voltages) on a rectangular region.

2.4 Solve the steady problem shown in Figure 2.16 using the solution $u(x, y)$ of SSP 1, Figure 2.1.

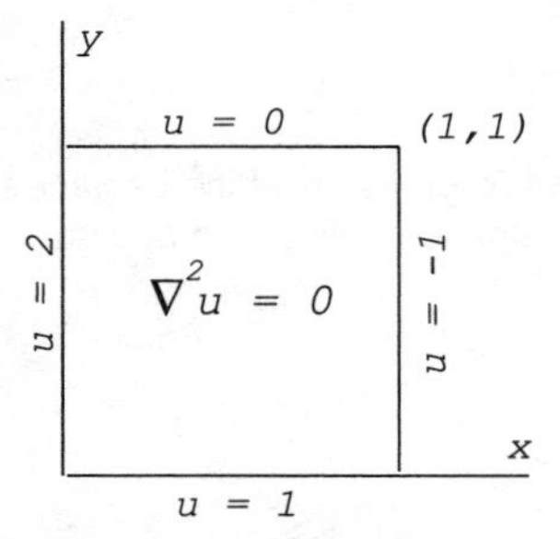

Figure 2.16 The steady problem of Exercise 2.4.

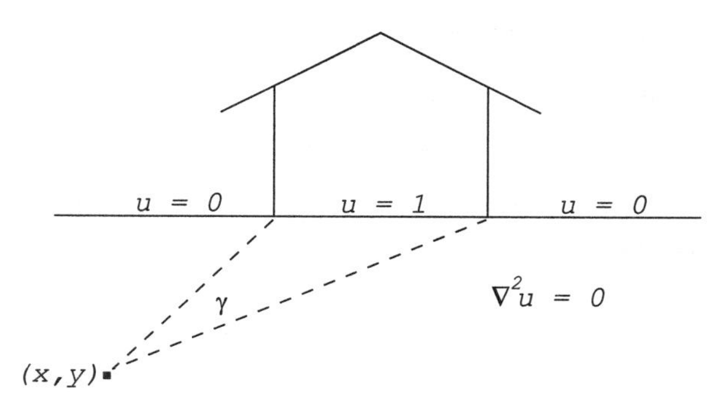

Figure 2.17 A crude attempt to model heat loss beneath heated buildings. (In reality, transient effects cannot be ignored.) A solution is $u = \gamma/\pi$, where γ is the included angle of the slab floor.

2.5 (Temperatures beneath heated buildings [Kusada et al.]). Solve the steady problem of Figure 2.17. Show that a solution is $u(x, y) = \gamma/\pi$ where γ is the included view angle of the heated portion from (x, y). Solve this problem two ways: first by superposition, then by Lévy's rule of thumb on Brownian motion.

2.6 By imposing the boundary conditions of Figure 2.5 on (2.7), verify Fourier's result (2.8).

2.7 Show that for functions of one variable $u = u(x)$, the harmonic functions are exactly the linear functions $u = Ax + B$.

2.8 Show in the plane that

$$\nabla^2 \log r = 0,$$

yet

$$\nabla^2 r = \frac{1}{r}.$$

2.9 Show that when $r = \sqrt{x^2 + y^2 + z^2}$,

$$\nabla \frac{1}{r} = -\frac{\mathbf{u}}{r^2}$$

where $\mathbf{u}$ is the unit radial vector $\mathbf{u} = (x, y, z)/r$. Show moreover that

$$\nabla^2 \frac{1}{r} = 0.$$

Conclude that the *electrostatic potential*

$$v = \frac{Q}{4\pi\epsilon_0 r}$$

(in energy per unit charge, i.e., volts) arising from the observed *electric field intensity*

$$E = \frac{1}{4\pi\epsilon_0}\frac{Q}{r^2}\mathbf{u}$$

(in force per unit charge, i.e., volts/m) induced by placing a charge of Q Coulombs at the origin is harmonic.

It is also an observed fact that when many discrete charges are present, the induced electric field intensity is the superposition (sum) of the individual fields. Conclude that electrostatic potential v arising from an arbitrary charge distribution is harmonic within charge-free regions.

2.10 Assume that a horizontal deeply buried pipe of radius $a = 1/2$ ft carries water at 180°F. Assume the surrounding earth of conductivity $\kappa = 0.5$ Btu-ft/h-°F-sqft is unaffected at a far-field distance of 8 ft, remaining at the yearly average air temperature of 50°F. Calculate the hourly loss of heat per foot of pipe.

2.11 Using Brownian motion, guess the bounded solution $u = u(x)$ to $\nabla^2 u = 0$ on the line segment $0 < x < 1$ subject to $u(0) = 1$ and $u(1) = 0$. Verify your answer analytically.

2.12 Again by Lévy's rule of thumb, guess the bounded harmonic function u on the unit disk $x^2 + y^2 < 1$ with boundary values

$$u = \begin{cases} 1 & \text{if} \quad 0 \le \theta \le \pi/2 \\ 0 & \text{if} \quad \pi/2 < \theta < 2\pi \end{cases} .$$

Verify your answer analytically.

2.13 The upper hemisphere of a charge-free spherical capacitor of radius R is at voltage V, the lower hemisphere at voltage 0, while the equator is an infinitesimally thin band of insulation. Find the voltage within.

2.14 The Laplacian at a point gives the proportional rate of increase of the error between the value at the point and the average value nearby.

Prove that when u is thrice differentiable,

$$a(r) = u(x_0, y_0) + \frac{r^2}{8\pi}\nabla^2 u(x_0, y_0) + O(r^4)$$

where $a(r)$ is the average value of $u(x, y)$ on the disk of radius r centered at (x_0, y_0).

Outline: Assume $x_0 = y_0 = u(x_0, y_0) = 0$. Expand $u(r\cos\theta, r\sin\theta)$ in a Taylor series in r. Average.

2.15 Find the change of variables and rescaling bringing SSP 5 to the dimensionless form (2.11). What is the solution to the problem as originally stated?

2.16 A heat exchanger of a closed-loop ground-coupled heat pump of radius $a = 1/6$ ft and length L ft is inserted vertically into wet sand of conductivity $\kappa = 2$ Btu/ft-h-^{O}F. The design heat load is $L\dot{Q} = 12000$ Btu/h. Assume that earth at a distance of $b = 8$ ft is unaffected and remains at the temperature $B = 52^{\text{O}}$F throughout the year. What length L insures that the heat pump never sees water below 37°F?

Answer: 246 ft. (This is overly conservative by a factor of 1.5. Transient analysis is warranted).

2.17 Show that indeed

$$u(r, \theta) = (r^{\omega} - r^{-\omega})\sin\omega\theta$$

is harmonic and satisfies the first two boundary conditions i) and ii) of SSP 10. Show via differentiating past the integral that the integral (2.17) also is harmonic and satisfies the same two boundary conditions for certain exponentially decaying $f(\omega)$.

2.18 Reformulate SSP 10 of §2.11 after making the substitution $v(r, \theta) = u(r, \theta) + (r/2)\sin 2\theta$.

2.19 Solve the steady problem of Figure 2.14a via conformal mapping.

Hint: See SSP 4.

2.20** Show via Leibniz's rule (§3.2) that the *Poisson integral*

$$u(x,y) = \frac{1}{2\pi}\int_0^{2\pi} b(\theta)\frac{1-x^2-y^2}{(x-\cos\theta)^2+(y-\sin\theta)^2}d\theta$$

solves the steady problem $\nabla^2 u = 0$ on the unit disk $x^2+y^2 < 1$ with boundary values $b(\theta)$. Do the case where $b = b(\theta)$ is continuous.

2.21 Write a routine to experimentally verify Lévy's solution (2.10).

Outline: Construct a series of random walks starting from (x, y), i.e., $(x_{n+1}, y_{n+1}) = (x_n, y_n) + \epsilon(\cos\theta_n, \sin\theta_n)$, where θ_n is chosen uniformly from $[0, 2\pi)$. The step length ϵ can either be taken to be constant but small, or normally distributed. Find the percentage of the trials that reach the upper boundary.

2.22 Find infinitely many solutions to Laplace's equation given the boundary conditions of Figure 2.3 in the case when $\theta_0 = \pi$.

2.23 Find unbounded solutions of Laplace's equation given the boundary conditions of Figure 2.4.

2.24 Why do deep earth temperatures equal the average annual temperature?

2.25 A possible explanation for gene clustering in bacterial chromosomes is suggested by the following model: Think of the essentially linear *E. Coli* cell as the interval $[0, 1]$. The enzyme surrounding a gene at $x = a$ is producing a reaction product at a constant rate; the product then diffuses throughout the cell. A second enzyme to the right at $x = b$ is absorbing the product at a rate proportional to its concentration at $x = b$. Show that the steady average concentration C is of the form

$$C = k(b^2 - a^2) + d.$$

Thus clustering yields a thermodynamic advantage [Svetic].

Chapter 3
The Heat Equation

§3.1 Flux

Consider a vector field F continuously differentiable everywhere on a *domain* (*region*) Ω of $\mathbf{R}^3$ — an open connected subset (see Exercises 1.36–1.40). Let S be an *orientable* surface within the solid Ω — a surface supporting a continuous unit normal vector $\mathbf{n}$. The *flux of* F *through* S (in the direction of $\mathbf{n}$) is the value of the surface integral

$$\text{flux} \quad = \int_S F \cdot \mathbf{n}\, d\sigma. \tag{3.1}$$

See Figure 3.1. This flux is said to have *intensity* (sometimes *density*) F.

When S is a closed surface enclosing the volume $V \subset \Omega$ and $\mathbf{n}$ is (as always) the outward pointing normal, the *Divergence Theorem* gives the flux as the integral of the divergence:

$$\int_S F \cdot \mathbf{n}\, d\sigma = \int_V \nabla \cdot F\, dx. \tag{3.2}$$

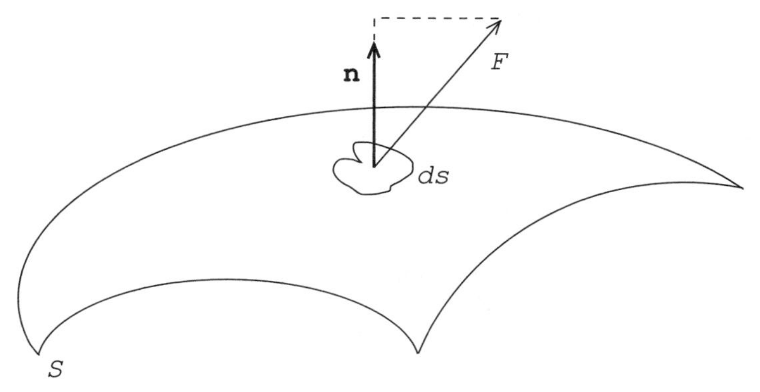

Figure 3.1 The flux of F through the surface S is a scalar quantity: the sum over all infinitesimal patches $d\sigma$ of the component of F in the normal direction $\mathbf{n}$ weighted by the area $d\sigma$ of the patch.

§3.2 The Heat Equation Derived

Suppose an isotropic material with thermal conductivity κ and temperature u occupies the region Ω. Then as can be inferred from observing simple insulating materials (Exercise 3.6), the flux of $-\kappa\nabla u$ is the thermal power flowing out through the surface S. The negative sign reflects that heat flows from higher to lower temperatures — against the gradient. To confuse the unwary, this thermal power flux is often called simply *heat flux*.

Therefore the thermal power leaving the region V enclosed by the surface S must be

$$\dot{Q}_{escaping} = -\int_S \kappa\nabla u \cdot \mathbf{n}\, d\sigma = -\int_V \nabla \cdot (\kappa\nabla u)\, dx. \tag{3.3}$$

On the other hand, relative to the reference temperature 0, the heat contained by the material within the volume V is

$$Q = \int_V c\rho u\, dx, \tag{3.4}$$

where c and ρ are the specific heat and density of the material. But recall the following powerful result.

Leibniz's Rule. Suppose the integral

$$G(t) = \int_X g(x,t)\, dx$$

exists for $a < t < b$, X a bounded set. Then differentiation of G on (a, b) can be performed under the integral

$$\frac{dG}{dt}(t) = \int_X \frac{\partial g(x,t)}{\partial t}\, dx$$

provided $\partial g/\partial t$ exists and is bounded on $(a, b) \times X$.

This rule is a manifestation of a theorem behind most interchange of limits: the *Lebesgue Dominated Convergence Theorem*. See Appendix A.

Leibniz's Rule applied to (3.4) yields

$$\dot{Q} = \int_V \frac{\partial c\rho u}{\partial t}\, dx. \tag{3.5}$$

Now perform a power balance. The thermal power $\dot{Q}$ within V must be accounted for in two ways — by escape through the boundary plus internal generation:

$$\int_V \frac{\partial c\rho u}{\partial t}\, dx = \int_V \nabla \cdot \kappa \nabla u \, dx + \int_V \dot{q}\, dx, \tag{3.6}$$

where $\dot{q}$ is the density of thermal power generation within the material Ω. But there is the easy

Fact. If $\int_E f(x)\, dx = \int_E g(x)\, dx$ for every subset E of Ω, then $f(x) = g(x)$ almost everywhere on Ω (with equality everywhere when both f and g are continuous).

Proof. Look at sets E where $f(x) - g(x)$ is positive.

Applying this Fact to (3.6) where the integrands are continuous yields the *heat equation*

$$\frac{\partial c\rho u}{\partial t} = \nabla \cdot \kappa \nabla u + \dot{q}. \tag{3.7}$$

For homogeneous isotropic time-invariant materials, i.e., where c, ρ, κ are constant, this becomes the more familiar

$$\frac{\partial u}{\partial t} = \alpha \nabla^2 u + \dot{q}/c\rho, \tag{3.8}$$

where

$$\alpha = \frac{\kappa}{c\rho} \tag{3.9}$$

is *diffusivity*.

Because of its success in modeling other diffusive phenomena, the heat equation is also called the *equation of diffusion*. The equation models the diffusion of spilled pollutants and voltages on certain badly designed telegraph cables (Exercises 1.31–1.34). We will see this equation also explain near surface earth temperatures, intersymbol distortion in submarine communications, RF currents on conductors, and other important phenomena.

§3.3 Conservation Theorems

The heat equation (3.8) is an example of a *conservation theorem*, a balance of the flux leaving a closed surface and its sources. The divergence theorem together with Leibniz's rule can uncover other valuable theorems of conservation. A simple example is fluid flow: Let $\mathbf{v}$ be the velocity field of a fluid of density ρ. As is traditional let $J = \rho\mathbf{v}$ denote the *current density.* The flux of J through S is the mass leaving through S per unit time and so, assuming fluid is neither created nor destroyed within S, conservation dictates that (Exercise 3.7)

$$\nabla \cdot J + \dot{\rho} = 0. \tag{3.10}$$

Or for a third (static) example, the celebrated result of K.F Gauss states that the flux of the electric field intensity is — within permittivity — the charge enclosed (Exercise 3.4), and so

$$\nabla \cdot E = \frac{\rho}{\epsilon_0}, \tag{3.11}$$

where ρ is the charge density. This is also known as the first of Maxwell's equations. In terms of electrostatic potential v,

$$\nabla^2 v = -\frac{\rho}{\epsilon_0}, \tag{3.12}$$

an instance of *Poisson's Equation.*

A strong form of conservation is where net flux of a field F leaving each closed surface S is 0, i.e., $\nabla \cdot F = 0$, or in terms of the potential u of F (if extant), *Laplace's Equation*:

$$\nabla^2 u = 0. \tag{3.13}$$

Thus steady state temperatures without internal generation, electrostatic potential in charge free regions, incompressible irrotational fluid displacement, all satisfy Laplace's equation, and are harmonic functions.

§3.4 Classical Uniqueness

Flux computations also plays a central role in two classical approaches for guaranteeing that well-posed PDEs have at most one solution. Our approach later in this book is non-classical and will not require the following results; they are included (in sketched form) because of their charm and usefulness elsewhere.

Uniqueness via Green's Identity

Suppose u_1 and u_2 are two bounded solutions of Laplace's equation (3.13) on the bounded domain Ω. Let $u = u_1 - u_2$. Suppose both u_1 and u_2 have identical limit values or flux limit values at almost every point of the boundary $S = \partial\Omega$, i.e.,

$$u\frac{du}{d\mathbf{n}} \equiv 0 \text{ on } \partial\Omega, \tag{3.14}$$

where as is traditional, the directional derivative in the outward normal direction is denoted by

$$\frac{du}{d\mathbf{n}} = \nabla u \circ \mathbf{n}.$$

Then assuming Ω is geometrically tame enough to support the divergence theorem, we obtain

$$0 = \int_{\partial\Omega} u\frac{du}{d\mathbf{n}} d\sigma = \int_{\Omega} \nabla \cdot u\nabla u \, dx$$

$$= \int_{\Omega} u\nabla^2 u \, dx + \int_{\Omega} (\nabla u)^2 \, dx = 0 + \int_{\Omega} |\nabla u|^2 \, dx,$$

and so $\nabla u \equiv 0$ on Ω and hence u is constant.

Examine carefully the conditions necessary for the above computation to obtain. I have suppressed details. The result is typical of Applied Mathematics, where to carefully state hypotheses is to drastically limit the scope of the widely applicable conclusion — facts are star-shaped bodies. The proof is more important than the result. Contrast with the example of Exercise 1.4, an (unbounded) harmonic function u on the unit disk with zero boundary values at almost every point. See Exercise 3.26.

A variation on the above argument will lead to a uniqueness theorem for transient diffusion problems. Here it is even more clear that we are employing a generalized integration by parts, an instance of *Green's First Identity* (Exercise 3.5). Let u be the difference of two solutions of the heat equation (3.8) and let

$$2\alpha Q(t) = \int_{\Omega} u^2 \, dx.$$

Then by Leibniz and the Divergence Theorem,

$$\dot{Q}(t) = \alpha^{-1} \int_\Omega u\dot{u}\, dx = \int_\Omega u\nabla^2 u\, dx = \int_{\partial\Omega} u\frac{du}{d\mathbf{n}}\, d\sigma - \int_\Omega (\nabla u)^2\, dx$$

$$= 0 - \int_\Omega |\nabla u|^2\, dx \le 0,$$

yet initially $Q(0) = 0$. But a non-negative function, initially 0 and decreasing with time, must be the zero function — a proof in the style of Lyapunov.

Uniqueness via the Maximum Principle

A harmonic function is the average of its nearby values. More precisely, behold

The Mean Value Theorem. Suppose u is harmonic on the domain Ω. Then at each point x_0 of Ω and every open ball B of positive radius r centered at x_0 with $\bar{B} \subset \Omega$,

$$u(x_0) = (\int_{\partial B} u\, d\sigma)/(\int_{\partial B} d\sigma) = (\int_B u\, dx)/(\int_B dx).$$

Proof Sketch. (Or see [Rudin]). Let $a(r)$ be the average value of u on the sphere $S = \partial B$, i.e.,

$$a(r) = \frac{\int_S u\, d\sigma}{\int_S d\sigma}.$$

Then its derivative by Leibniz is

$$a'(r) = \ \ldots\ = \frac{\int_S du/d\mathbf{n}\, d\sigma}{\int_S u\, d\sigma} = \frac{\int_B \nabla^2 u\, dx}{\int_S d\sigma} = 0.$$

Therefore the average value $a(r)$ is constant with r. But since u is continuous at x_0, its average value $a(r)$ must have limit $u(x_0)$ as $r \to 0$.

Corollary. (The Maximum Principle) A non-constant harmonic function has no interior extrema. Thus extrema (if any) must occur on the boundary.

Proof Sketch. A harmonic u cannot have (say) a maximum at the interior point x_0 and at the same time be the average of its nearby values unless constant nearby. All points of Ω where this maximum is achieved disconnect Ω unless u is constant everywhere on Ω. See Exercises 3.15 and 3.17 for versions of this result for Poisson and diffusion problems..

The Maximum Principle guarantees a modest uniqueness result for steady Dirichlet problems. As always, let u be the difference of two solutions, in this case solutions to Laplace's equation on a bounded domain Ω that are continuous on the closure $\bar{\Omega} = \Omega \cup \partial\Omega$ with identical boundary values. Then u is harmonic on Ω and continuous on $\bar{\Omega}$. But because $\bar{\Omega}$ is compact, u must possess maxima and minima (Exercise 3.5), which must by the Maximum Principle occur on the boundary — where u is identically 0.

§3.5 Nondimensional Variables

In many fields such as heat transfer, aerodynamics, filter design, it is standard practice to go over to *nondimensional (dimensionless, normalized) variables.* The advantages are many. We will consistently adopt this approach throughout this book.

For example, in the heat equation

$$\frac{\partial u}{\partial t} = \alpha \nabla^2 u \tag{3.15}$$

we could employ a change of variable $\theta = \alpha t$ to obtain the heat equation free of diffusivity

$$\frac{\partial u}{\partial \theta} = \nabla^2 u.$$

But there is even more advantage to normalizing both time and spatial variables simultaneously when possible. Consider again the 1-d heat equation (3.15) on (say) the line segment $0 < x < L$. Now make Fourier's change

$$\theta = \alpha t / L^2$$

and Biot's

$$\zeta = x/L$$

to obtain the *1-d heat equation in nondimensional form*

$$\frac{\partial u}{\partial \theta} = \frac{\partial^2 u}{\partial \zeta^2}, \quad 0 < \zeta < 1. \tag{3.16}$$

The exercises below provide practice in some of the other standard substitutions.

Exercises

3.1 Argue that the Möbius strip is non-orientable — it does not support a continuous unit normal.

3.2 Verify directly that the Divergence Theorem

$$\int_S F \cdot \mathbf{n}\, d\sigma = \int_V \nabla \cdot F\, dx.$$

holds for the vector field $F = x\mathbf{i} + y\mathbf{j} + z\mathbf{k}$ where S is the unit sphere $x^2 + y^2 + z^2 = 1$.

3.3 Show that the Divergence Theorem (3.2) fails for the field $F = \nabla(x^2+y^2+z^2)^{-1/2}$ when S is the unit sphere even though both the surface and volume integrals are extant. Why?

3.4 *Gauss' Law* states that the flux of the electric field intensity leaving a closed surface S is (within permittivity ϵ_0) the charge Q enclosed within. In symbols,

$$\int_S E \circ \mathbf{n}\, d\sigma = \frac{Q}{\epsilon_0}.$$

Deduce Gauss' law from *Coulomb's law*

$$E = \frac{1}{4\pi\epsilon_0}\frac{Q}{r^2}\mathbf{u},$$

where $r = \sqrt{x^2 + y^2 + z^2}$ and $\mathbf{u} = (x, y, z)/r$.

Outline: First do the case of a point charge Q at the center of a sphere by direct calculation. Then do an arbitrary surface S containing a point charge Q by thinking about cutting open an avocado. Make the leap of faith to the general result by superposition.

3.5 Prove *Green's First Identity* [O'Neill]:

$$\int_{\Omega} f\nabla^2 g\, dx = \int_{\partial\Omega} f\frac{dg}{d\mathbf{n}}\, d\sigma - \int_{\Omega} \nabla f \cdot \nabla g\, dx.$$

3.6 Think how simple homogeneous isotropic insulating materials conduct heat — exactly like Ohm's Law. A 2 inch thick sheet of extruded polystyrene with an R-value of 5 per inch experiences a flow perpendicular to the sheet of

$$\dot{Q} = (1/10)(T_1 - T_0)$$

Btu per hour per square foot, where $T_1 - T_0$ is the temperature change from one face to the other. Argue that for any conductive material, the thermal power flux intensity is $-\kappa\nabla u$.

Hint: Employ the Divergence Theorem.

3.7 Carefully ape the steps in the derivation of the Heat Equation (3.7) to derive (3.13), the *equation of continuity* for fluid flow.

3.8 Modify slightly the proof of classical uniqueness in §3.4 to apply to any mix of boundary conditions of type Dirichlet, Neumann, and the Robin mixed type $du/d\mathbf{n} = -cu$.

3.9 Show that the Mean Value Principle for harmonic functions in the one spatial variable x is a simple statement about similar triangles.

3.10 Fill in all details in the proof sketch of the Mean Value Principle for two variables.

Hint: use polar coordinates.

3.11 Find the change of variables carrying the wave equation (1.34) on the square $-L < x, y < L$ to nondimensional form

$$\frac{\partial^2 u}{\partial \bar{t}^2} = \frac{\partial^2 u}{\partial \bar{x}^2} + \frac{\partial^2 u}{\partial \bar{y}^2}, \quad 0 < \bar{x}, \bar{y} < 1.$$

3.12 What change of variables carries the heat equation (1.24) on the rectangle $-1 < x < 2,\ 1 < y < 3$ to nondimensional form? The lengths x, y are in feet, the diffusivity α in ft^2/s.

3.13 What change of variables carries the heat equation on a solid ball of center (x_0, y_0, z_0) and radius R to dimensionless form?

3.14 Suppose on the disc $x^2 + y^2 < 1$,

i) u is harmonic,

ii) u as well as both of its first partials are bounded, and

iii) the radial limit of $u du/d\mathbf{n}$ is 0 at almost every point of the boundary $x^2 + y^2 = 1$. Prove u is constant.

Hint: Use the proof of uniqueness via Green's First Identity on the smaller disc $x^2 + y^2 < r < 1$.

3.15 **(The Maximum Principle for Poisson)** Suppose u is a solution of Poisson's $\nabla^2 u = f$ on the bounded domain $\Omega \subset \mathbf{R}^n$ with u continuous on $\bar{\Omega}$ and $f(x) \geq 0$ on Ω. Then the maximum value of u on $\bar{\Omega}$ is achieved on the boundary $\partial\Omega$ (and possibly at interior points as well).

Outline: Suppose u has a maximum value at the interior point $x = 0$. Then the graph of u at 0 must be concave downward, thus $\nabla^2 u(0) \leq 0$ and hence $f(0) = 0$. Note that $v = \epsilon|x|^2$ solves $\nabla^2 v = 2n\epsilon$. The previous argument shows $u + v$ achieves its maximum value on the boundary $\partial\Omega$. Now let ϵ go to 0. See [Weinberger] for refinements of this Principle.

3.16 Formulate and prove a uniqueness result for Poisson problems.

3.17 **(The Maximum Principle for Diffusion)** Suppose u is a nonconstant solution of the heat equation $u_t = \nabla^2 u$ on the bounded domain Ω with u continuous on $\bar{\Omega}$. Then if u is maximum at (x_0, t_0), then either $t = 0$ or x_0 is on the boundary of Ω.

3.18 Formulate and prove a uniqueness result for diffusion problems.

3.19 Find the change of variables that brings Schrödinger's equation for the electron of the hydrogen atom (in spherical coordinates)

$$i\hbar \frac{\partial \psi}{\partial t} = -\frac{\hbar^2}{2m}\nabla^2 \psi - \frac{e^2}{4\pi\epsilon_0}\frac{1}{\rho}\psi$$

to the form

$$i\frac{\partial \psi}{\partial t} = -\nabla^2 \psi - \frac{1}{\rho}\psi.$$

Chapter 4
The Wave Equation

We first derive the wave equation, then solve the 1-d case via the method of characteristics and D'Alembert's method.

§4.1 The Wave Equation Derived

Let us think about compression waves moving through an elastic solid. Shear-free *stress* P is the result of applying the operator of elastic stiffness E to *strain*. Strain (within a sign) has displacement $\mathbf{u}$ as potential. More explicitly, if P is the force/area stressing an elastic material occupying the region Ω, then as a 3×3 matrix relation

$$P = E\,\nabla\mathbf{u},$$

where the column vector $\mathbf{u} = (u_1, u_2, u_3)^T$ at a point $x = (x_1, x_2, x_3)$ is the resulting displacement from equilibrium of the particle originally at that point. The ith column of $\nabla\mathbf{u}$ is the gradient of the ith coordinate of $\mathbf{u}$, etc. The diagonal matrix E has as entries the *moduli of elasticity* in the $\mathbf{i}, \mathbf{j}, \mathbf{k}$ directions, each in units of force per area — the spring constant of a unit length small tube of the fibers in that direction of the material per cross-sectional area. *Shear-free* means $\nabla \times \mathbf{u} = 0$, an idealization since any non-fluid will experience some shear when deformed. The vector flux of stress

$$\int_S P \cdot \mathbf{n}\, d\sigma = \int_V \nabla \cdot (E\,\nabla\mathbf{u})\, dx$$

is then the net force exerted by the volume V on the surrounding material. On the other hand, by Newton, this net force must be the inertial force, the net rate of change of momentum

$$\int_V \frac{\partial}{\partial t}(\rho \frac{\partial \mathbf{u}}{\partial t})\, dx,$$

where ρ is the density of the material at equilibrium. Think of a mass of quivering Jello^{TM}.

Equating these two net forces yields the (shear-free) *vector wave equation*

$$\frac{\partial}{\partial t}(\rho \frac{\partial \mathbf{u}}{\partial t}) = \nabla \cdot (E \, \nabla \mathbf{u}). \tag{4.1}$$

For common materials, tiny deformations produce immense forces, allowing us to neglect changes in density [Den Hartog]. Thus for common isotropic homogeneous materials, (4.1) is well approximated by the more familiar

$$\frac{\partial^2 \mathbf{u}}{\partial t^2} = c^2 \nabla^2 \mathbf{u} \tag{4.2}$$

where, as it turns out (Exercise 4.6), $c > 0$ is the speed of waves in this material.

The wave equation (4.1) well models longitudinal vibrations of beams, small normal vibrations of both strings and membranes, voltages on ideal transmission lines, acoustic waves, and electromagnetic propagation in an ideal dielectric. See Exercises 4.2–4.3.

§4.2 The Method of Characteristics

Note that if $f(\alpha)$ and $g(\beta)$ are everywhere twice-differentiable functions of real variables α and β, then it is routine to show (Exercise 4.6) that both

$$u(x,t) = f(x - ct) \tag{4.3a}$$

and

$$u(x,t) = g(x + ct) \tag{4.3b}$$

solve the *1-d wave equation*

$$\frac{\partial^2 u}{\partial t^2} = c^2 \frac{\partial^2 u}{\partial x^2} \tag{4.4}$$

for all x and t. Each such special solution is called a *traveling wave* since it moves intact, free of distortion at constant velocity c (Exercise 4.6). Note that the solution $u = f(x - ct)$ moves from left to right while $u = g(x+ct)$ moves from right to left. There is this astonishing converse.

Theorem on Characteristics. Suppose $u = u(x,t)$ is twice differentiable and solves the 1-d wave equation (4.4) on the rectangle $x_1 < x < x_2$, $t_1 < t < t_2$. Then there are twice differentiable functions $f(\alpha)$ and $g(\beta)$ such that

$$u(x,t) = f(x - ct) + g(x + ct). \tag{4.5}$$

Every solution of the 1-d wave equation is the sum of two traveling waves.

Proof. Make the change of variables

$$\alpha = x - ct \text{ and } \beta = x + ct, \tag{4.6}$$

bijectively mapping the rectangle $[x_1, x_2] \times [t_1, t_2]$ onto the inclined rectangle R bounded by the four lines $\alpha + \beta = 2x_i$ and $\beta - \alpha = 2ct_i$, $i = 1, 2$. See Figure 4.1(a) and Figure 4.1(b).

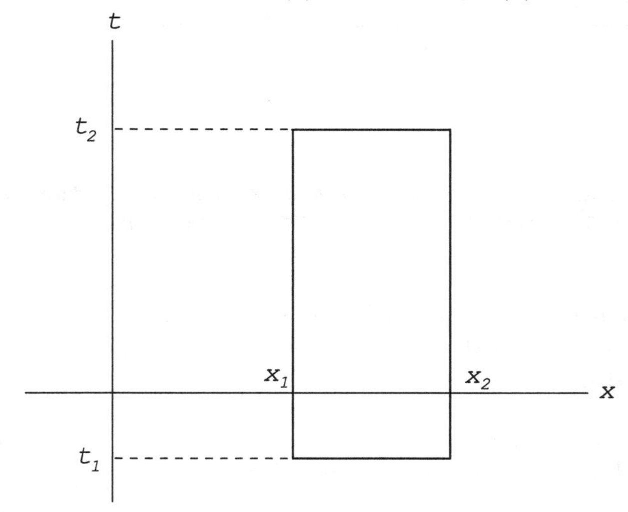

Figure 4.1(a) The domain of validity of the solution wave $u = u(x,t)$.

Set

$$v(\alpha, \beta) = u(x,t) = u(\alpha/2 + \beta/2, \beta/2c - \alpha/2c).$$

It is routine (Exercise 4.7) to show that on the interior of the inclined rectangle R,

$$\frac{\partial^2 v}{\partial \alpha \partial \beta} = 0, \tag{4.7}$$

and hence

$$v(\alpha, \beta) = f(\alpha) + g(\beta) \tag{4.8}$$

where f and g are determined within constants that sum to 0.

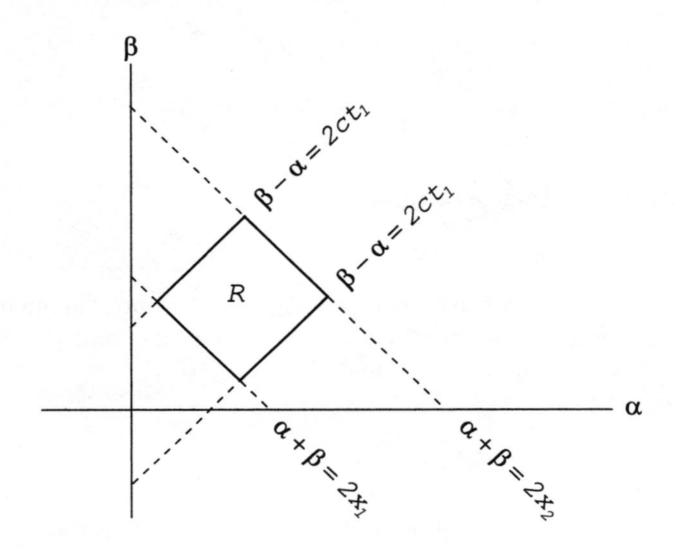

Figure 4.1(b) The change of variables $\alpha = x - ct$ and $\beta = x + ct$ transforms the domain of validity to an inclined rectangle R in the α-β plane.

Corollary. If $u = u(x, t)$ is a solution of (4.4) that is either valid for all t or for all x, then there is an extension of u to a solution valid for all x and t.

Proof. If the inclined rectangle R in the α-β plane is an infinite strip, f and g are everywhere defined. Extend u via (4.5).

Example 1. The function

$$u = \cos ct \, \sin x \tag{4.9}$$

is a solution of the wave equation $u_{tt} = c^2 u_{xx}$ that is the sum of two identical but oppositely traveling waves:

$$\cos ct \, \sin x = \frac{\sin(x - ct)}{2} + \frac{\sin(x + ct)}{2}.$$

Note that (4.9) is an example of a *periodic* solution

$$u(x, t+T) = u(x, t) \tag{4.10}$$

with period $T = 2\pi/c$. By the Corollary, a periodic solution can be extended to a solution for all x and t.

Solutions such as (4.9) are called *standing waves.* See Exercise 4.21.

Example 2. The function

$$u(x, t) = f(x - ct) + g(x + ct), \tag{4.11a}$$

where

$$f(\alpha) = \frac{1}{\alpha} \text{ and } g(\beta) = \frac{1}{\beta - 1}, \tag{4.11b}$$

solves $u_{tt} = c^2 u_{xx}$ on many distinct subdomains of the x-t plane — on rectangles that map to rectangles R contained within one of the 4 quadrants of the α-β plane formed by the lines $\alpha = 0$ and $\beta = 1$. For example, the solution on $0 < x < 1$, $-\infty < t < 0$ cannot be extended to a solution valid on a larger domain (Exercise 4.8).

Consult Weinberger's text for more detailed work with this method of characteristics.

§4.3 D'Alembert's Solution

Let us rederive the solution (4.5) of the 1-d wave equation more explicitly: Suppose we have in hand the initial conditions

$$u(x, 0) = \phi(x) \qquad \text{and} \qquad u_t(x, 0) = \psi(x).$$

From (4.5) at $t = 0$ we see that

$$f(x) + g(x) = \phi(x) \tag{4.12}$$

and

$$-cf'(x) + cg'(x) = \psi(x).$$

Integrating the second gives

$$g(x) - f(x) = \frac{1}{c} \int_0^x \psi(y)\, dy + g(0) - f(0). \tag{4.13}$$

Combining (4.12) and (4.13) yields (Exercise 4.14)

D'Alembert's Formula:

$$u(x,t) = \frac{\phi(x-ct) - \phi(x+ct)}{2} + \frac{1}{2c}\int_{x-ct}^{x+ct} \psi(y)\,dy. \tag{4.14}$$

Exercises

4.1 Consider the mass-spring system shown in Figure 4.2. Let $u_i(t)$ be the displacement from equilibrium of m_i and let k_i be the spring constant of the i-th spring. Show that a frictionless such system satisfies the matrix relation

$$M\ddot{\mathbf{u}} = -K\mathbf{u}$$

where M is the diagonal matrix of *mass* and where K is a tri-diagonal matrix of *stiffness*. What if each mass experiences (viscous) friction proportional to velocity? Argue that a beam is a distributed mass-spring system and hence its longitudinal vibrations satisfy the 1-d wave equation.

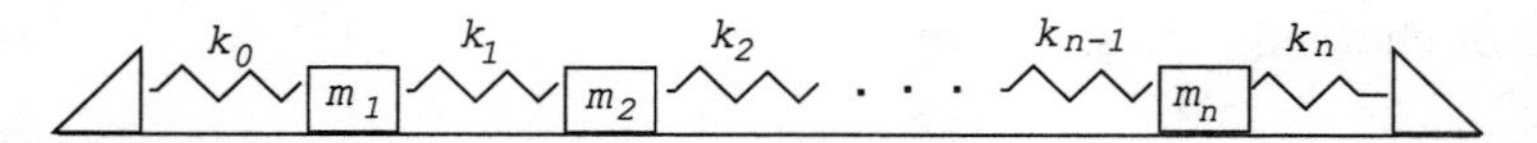

Figure 4.2 A mass-spring system.

4.2 Modify the derivation of the wave equation (4.2) for elastic solids to persuasively argue that the vertical displacements $u = u(x,y,t)$ of a vibrating membrane satisfy the *2-dimensional wave equation*

$$\frac{\partial^2 u}{\partial t^2} = c^2\left(\frac{\partial^2 u}{\partial x^2} + \frac{\partial^2 u}{\partial y^2}\right).$$

Hint: The membrane is not isotropic — the elasticity in the vertical direction is 0.

4.3 Modify the derivation of the wave equation (4.2) to make it plausible that vertical displacements of a vibrating string are modeled by the 1-dimensional wave equation.

4.4 Heaviside acerbically maintained that the ideal transmission line should consist of cascaded infinitesimal sections of distributed series inductance L (henrys per meter) and distributed shunt capacitance C (farads per meter). See Figure 4.3. Reconstruct Heaviside's deduction that voltage V and current I at time t and location x on such a transmission line satisfy the fundamental coupled system of PDEs known as the (*ideal*) *Telegrapher's Equations*

$$C\frac{\partial V}{\partial t} = -\frac{\partial I}{\partial x} \text{ and } L\frac{\partial I}{\partial t} = -\frac{\partial V}{\partial x}$$

and so voltage satisfies the 1-d wave equation

$$\frac{\partial^2 V}{\partial t^2} = \frac{1}{LC}\frac{\partial^2 V}{\partial x^2}.$$

Telegraph signals on such a line will be free of intersymbol distortion.

Hint: See Exercise 1.21. Recall that voltage v across and current i through an inductor of L henrys must obey the rule $Ldi/dt = v$.

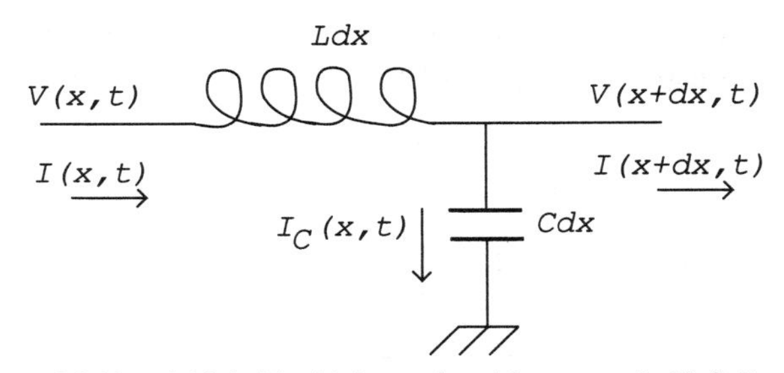

Figure 4.3 Heaviside's ideal telegraph cable: cascaded infinitesimal sections of distributed series inductance and shunt capacitance.

4.5 Find the change of variables carrying the wave equation $u_{tt} = c^2\nabla^2 u$ on the cube $-L < x, y, z < L$ to dimensionless coordinates

$$\frac{\partial^2 u}{\partial \bar{t}^2} = \frac{\partial^2 u}{\partial \bar{x}^2} + \frac{\partial^2 u}{\partial \bar{y}^2} + \frac{\partial^2 u}{\partial \bar{z}^2}, \qquad 0 < \bar{x}, \bar{y}, \bar{z} < 1.$$

4.6 Let $f(\alpha)$ be any twice differentiable function on the real line. Show both

$$u(x,t) = f(x - ct) \text{ and } u(x,t) = f(x + ct)$$

are solutions of the 1-d wave equation $u_{tt} = c^2 u_{xx}$. Show that the first moves from left-to-right, the second right-to-left, both at velocity $c > 0$.

4.7 Show every solution of the 1-d wave equation $u_{tt} = c^2 u_{xx}$ on the (possibly infinite) domain $x_1 < x < x_2, \ t_1 < t < t_2$ is the sum of two traveling waves, i.e., of the form

$$u(x,t) = f(x + ct) + g(x - ct).$$

Show also that f and g are determined within constants that sum to 0.

Outline: After setting $\alpha = x - ct, \ \beta = x + ct$, and $v(\alpha, \beta) = u(x,y)$, show $u_{\alpha\beta} = 0$, then deduce that $v(\alpha, \beta) = f(\alpha) + g(\beta)$.

4.8 Using (4.11), exhibit a solution $u = u(x,t)$ of the 1-d wave equation on a subdomain that possesses no extension to a larger domain.

4.9 **(Spherical characteristics)** Show that each radially symmetric solution $u = u(\rho, t)$ of the 3-d wave equation

$$\frac{\partial^2 u}{\partial t^2} = c^2 \nabla^2 u$$

is the sum of two traveling waves

$$u = \frac{f(\rho - ct) + g(\rho - ct)}{\rho}$$

and conversely. How are the magnitudes of these waves changing with time?

Outline: Put $\alpha = \rho - ct, \ \beta = \rho + ct$, and $v = (\alpha + \beta)u$. The wave equation then becomes $v_{\alpha\beta} = 0$.

4.10 What if any is the analogous result to spherical characteristics in the plane? After examining Exercises 4.7 and 4.9, make a conjecture. Test your conjecture.

4.11 Verify that like the wave equation, the 1-d *beam equation* $u_{tt} = -c^4 \, u_{xxxx}$ has sinusoidal traveling wave solutions $u = \sin(\omega t - \kappa x)$. Also verify that unlike the wave equation, these solutions *disperse* — they travel at a velocity $\nu = c\sqrt{\omega}$ that varies with frequency ω.

4.12 Solve $u_{tt} = c^2 u_{xx}$ subject to the initial conditions $u(x,0) = e^{-x}$ and $u_t(x,0) = \cos x$ using the method of characteristics.

4.13 In a recent tort action an overweight man sued his brother-in-law for reckless endangerment after falling (over land) while sliding down a cable strung from a tree to the opposite shore of his in-law's lake. The pudgy plantiff claimed that "the strap was jerked violently from my hands soon after I grabbed it and began to slide." Explain this violent jerk.

4.14 Carefully derive D'Alembert's formula (4.14).

4.15 Solve $u_{tt} = 4u_{xx}$ subject to $u(x,0) = \sin \pi x$ and $u_t(x,0) = 1$.

4.16 Consider the plucked string BVP 4 of §5.4. Show that the corner where the string is plucked splits into two traveling corners as the vibration evolves. Therefore a *classical solution* u to $u_{tt} = u_{xx}$ *cannot exist* since u_x is not continuous! See Figure 4.4.

Hint: Employ D'Alembert's solution

$$u(x,t) = \frac{f(x+t) + f(x-t)}{2}$$

where f is the initial shape extended to an odd periodic function. See Exercise 5.18.

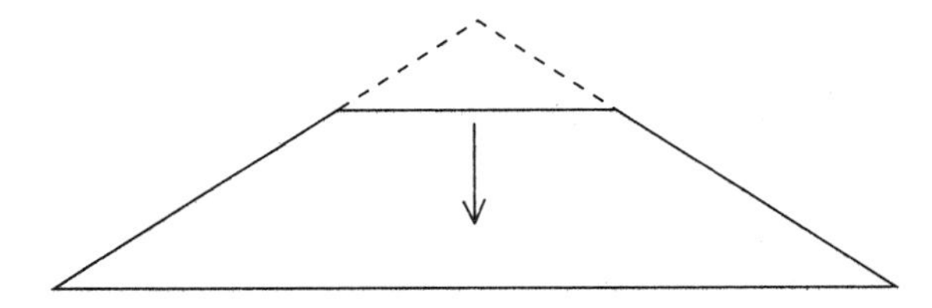

Figure 4.4 A string plucked at midstring. Two snapshots: the initial shape (dotted) and a later shape. The initial corner where initially plucked splits into two propagating corners.

4.17 Demurring from Kelvin's model of Figure 1.4, Heaviside insisted that an actual coaxial cable is more acurately modeled by the infinitesimal sections shown in Figure 4.5. Prove that the ideal telegrapher's model of Exercise 4.4 now becomes

$$CV_t = -I_x - GV$$

and

$$LI_t = -V_x - RI,$$

together giving

$$LCV_{tt} + (GL + RC)V_t = V_{xx} - RGV.$$

Finally show that signals u travel down such a cable without distortion, merely decaying in amplitude, i.e., $u = e^{-kt} f(x - ct)$, when and only when $RC = GL$. Unless the manufacturer meets this specification, signals will disperse.

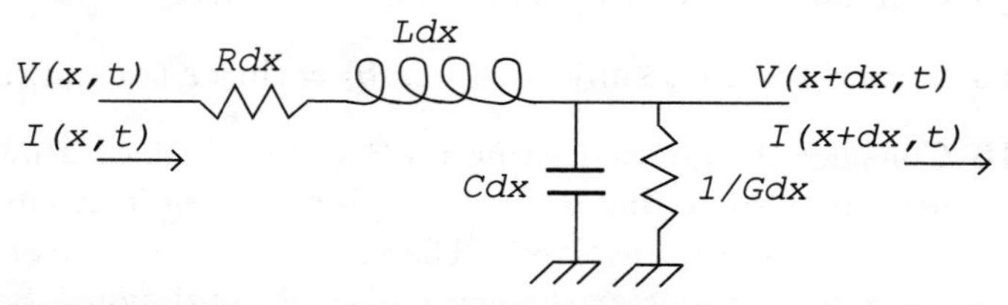

Figure 4.5 A real-world coaxial cable with distributed series resistance R and inductance L, and with distributed shunt capacitance C and conductance G.

The following 4 exercises practice notions that arise from measurable phenomenon of acoustic waves in ducts as well as voltages on transmission lines and within waveguides. For Exercises 4.18–4.21, suppose $u = u(x, t)$ is a periodic solution of the 1-d wave equation $u_{tt} = c^2 u_{xx}$ of period T, i.e.,

$$u(x, t) = u(x, t + T).$$

4.18 Show that for the *wavelength* $\lambda = cT$ and some constant k,

$$u(x + \lambda, t) = u(x, t) + 2k$$

for all x and t.

Hint: Since u is periodic, by the uniqueness portion of Exercise 4.7, $f(\alpha - \lambda) = f(\alpha) - k$ and $g(\beta + \lambda) = g(\beta) + k$.

4.19 We say $u = u(x,t)$ has *no mean flow* (*no DC offset*) if for all x,

$$\int_0^T u(x,t)dt = 0.$$

Prove that if u has no mean flow, then u is spatially periodic, i.e., $u(x+\lambda,t) = u(x,t)$, and is the sum of two uniquely determined periodic traveling waves $u(x,t) = f_0(x-ct) + g_0(x+ct)$ of no mean flow.

4.20 Let us say the *net energy flow past* x of the periodic $u = u(x,t)$ of no mean flow is

$$\int_0^T f_0(x-ct)^2 dt - \int_0^T g_0(x+ct)^2 dt.$$

Show the net energy flow past each x is identical.

Hint: Apply Leibniz's Rule.

4.21 A *standing wave* is a periodic solution of no mean flow with zero net energy flow.

Show every periodic solution u of no mean flow is the sum of a standing wave and traveling wave; the traveling wave carries the net energy flow of u in the correct direction.

Hint: $f_0(\alpha) + g_0(\beta) = \epsilon f_0(\alpha) + g_0(\beta) + (1-\epsilon) f_0(\alpha)$.

The traveling wave transports energy while the standing wave stores energy.

4.22 Write down a Schrödinger-like PDE satisfied by $u = e^{x-ct}$.

4.23 The *damped wave equation*

$$u_{tt} + 2\zeta u_t - c^2 u_{xx} = 0$$

has solutions of the form

$$u(x,t) = T(t)\sin\kappa x.$$

Find $T = T(t)$.

4.24 Prove that D'Alembert's formula (4.14) can be extended to the *driven wave equation*

$$u_{tt} = c^2 u_{xx} + f(x, t)$$

as

$$u(x, t) = \frac{\phi(x - ct) - \phi(x + ct)}{2} + \frac{1}{2c} \int_{x-ct}^{x+ct} \psi(y)\, dy$$

$$+ \frac{1}{2c} \int_0^t \int_{x-c(t-\tau)}^{x+c(t-\tau)} f(y, \tau)\, dy\, d\tau.$$

4.25 Rederive the characterization (4.5) of solutions to the 1-d wave equation (4.4) by means of the factorization

$$0 = (\frac{\partial^2}{\partial t^2} - c^2 \frac{\partial^2}{\partial x^2})u$$

$$= (\frac{\partial}{\partial t} - c\frac{\partial}{\partial x})(\frac{\partial}{\partial t} + c\frac{\partial}{\partial x})u.$$

Chapter 5

Separation of Variables

In this Chapter we work through a carefully selected list of standard problems that capture much of the core of the method of separation of variables. The approach will be intuitive and informal. Complete solutions are in some cases postponed to subsequent chapters. As we have seen, it is sufficient to consider the problems in nondimensional form. One good approach to learning this method is to rework the following problems repeatedly. We conclude with a disturbing critique of the method and certain tricks of the trade.

§5.1 Rod with Specified End Temperatures (BVP 1)

A homogeneous rod of length 1 with curved side insulated is initially and uniformly at temperature 1. Its ends are suddenly brought in contact with sinks at temperature 0. How will the temperature profile of the rod relax to 0? In symbols, we must solve

$$\frac{\partial u}{\partial t} = \frac{\partial^2 u}{\partial x^2}, \quad 0 < x < 1, \tag{5.1}$$

subject to

i) $u(x,0) = 1, \;\; 0 < x < 1,$

ii) $u(0,t) = 0, \;\; t \geq 0,$

iii) $u(1,t) = 0, \;\; t \geq 0.$

We now perform the steps of the powerful method of *separation of variables.* In brief, we find 'special solutions' satisfying only the boundary conditions, then superimpose these special solutions to realize the initial condition.

How special? Assume *variables separate*:

$$u(x,t) = T(t)X(x).$$

But then, if u is such a separated solution, the heat equation (5.1) becomes

$$\dot{T}(t)X(x) = T(t)X''(x)$$

where the dot is Newton's notation for differentiation with respect to time t and the prime is as usual differentiation with respect to the spatial variable x. Rearranging we obtain

$$\frac{\dot{T}(t)}{T(t)} = \frac{X''(x)}{X(x)}. \tag{5.2}$$

Since the left hand side of (5.2) is a function only of t while the right hand side only of x, the only conclusion possible is that both members are constant, i.e.,

$$\frac{\dot{T}(t)}{T(t)} = \frac{X''(x)}{X(x)} = \lambda. \tag{5.3}$$

The constant λ is called a *separation constant* or an *eigenvalue* of the Laplacian since $\nabla^2 X = X'' = \lambda X$.

For good and intuitive reasons to be explained later (Exercises 5.21 and 5.22), we will see that λ must be a negative real number (or see Exercise 5.1), i.e.,

$$\lambda = -\omega^2. \tag{5.4}$$

Thus we are led to the decoupled ordinary differential equations

$$\dot{T} + \omega^2 T = 0$$

$$X'' + \omega^2 X = 0$$

with solutions

$$T(t) = e^{-\omega^2 t}$$

and

$$X(x) = a \cos \omega x + b \sin \omega x.$$

Imposing boundary condition ii) on u hence X yields that $a = 0$. Imposing boundary condition iii) on u hence X yields that $\sin \omega = 0$ and so $\omega = n\pi$. Hence

$$T_n(t) = e^{-n^2\pi^2 t}$$
$$X_n(x) = \sin n\pi x$$
$$\lambda_n = -n^2\pi^2$$

yielding our 'special solutions'

$$u_n = e^{-n^2\pi^2 t} \sin n\pi x, \tag{5.5}$$

one for each $n = 1, 2, 3, \ldots$. But then by superposition, series of the form

$$u(x,t) = \sum_{n=1}^{\infty} c_n e^{-n^2\pi^2 t} \sin n\pi x \tag{5.6}$$

will also satisfy the heat equation (5.1) and boundary conditions ii) and iii). Only the initial condition i) remains to be satisfied.

Setting $t = 0$ and imposing on (5.6) the initial condition i) yields

$$1 = \sum_{n=1}^{\infty} c_n \sin n\pi x. \tag{5.7}$$

To compute these constants c_n we employ employ *Fourier's trick*, a generalization of a method associated with dot product — we multiply both sides of (5.7) by $\sin m\pi x$ and integrate term-by-term. More precisely, we use the *inner product* belonging naturally to the spatial domain of this problem, namely

$$\langle f, g \rangle = \int_0^1 f(y)g(y)\, dy. \tag{5.8}$$

With $\phi_m = \sin m\pi x$, we apply the linear functional $\langle \phi_m, \cdot \rangle$ to both sides of (5.7) to obtain

$$\langle \sin m\pi y, 1 \rangle \;=\; \sum_{n=1}^{\infty} c_n \langle \sin m\pi y, \sin n\pi y \rangle. \tag{5.9}$$

By direct calculation (Exercise 5.7) or, more theoretically, because the Laplacian with zero boundary conditions is self-adjoint (Exercise 5.23), the *eigenfunctions* $\sin n\pi x$ are *mutually orthogonal* (Exercise 5.24), i.e.,

$$\langle \sin m\pi y \;\; \sin n\pi y \rangle = \begin{cases} 0 & \text{if } m \neq n \\ \dfrac{1}{2} & \text{if } m = n. \end{cases} \tag{5.10}$$

Computing these integrals yields (Exercise 5.11)

$$c_n = \frac{\langle \sin n\pi y, 1 \rangle}{\langle \sin n\pi y, \sin n\pi y \rangle} = \frac{2}{n\pi}[1 - (-1)^n] \tag{5.11}$$

Thus it is plausible that the solution to the original problem is

$$u(x,t) = \frac{4}{\pi}\sum_{k=0}^{\infty} \mathrm{e}^{-(2k+1)^2\pi^2 t}\frac{\sin(2k+1)\pi x}{2k+1}. \tag{5.12}$$

Remark. Suppose the initial temperature of the rod is a more complicated function $u(x,0) = f(x)$ of location x. The calculations above remain valid until step (5.7), where, instead,

$$f(x) = \sum_{n=1}^{\infty} c_n \sin n\pi x, \tag{5.13}$$

thus yielding in the same way as in (5.9) the general constants

$$c_n = \frac{\langle \sin n\pi y, f(y)\rangle}{\langle \sin n\pi y, \sin n\pi y\rangle}. \tag{5.14}$$

But now think of this problem as an input–output problem, where the input is the initial condition $f(x)$, the output the solution $u = u(x,t)$. Set $\phi_n(x) = \sin n\pi x / \langle \sin n\pi x, \sin n\pi x\rangle = \sqrt{2}\sin n\pi x$. Then

$$u(x,t) = \sum_{n=1}^{\infty} e^{\lambda_n t}\langle \phi_n(y), f(y)\rangle \phi_n(x)$$

$$= \langle \sum_{n=1}^{\infty} e^{\lambda_n t}\phi_n(y)\phi_n(x), f(y)\rangle = \int_0^1 G(x,y,t) f(y)\, dy, \tag{5.15}$$

where G is the *Green's function* for this problem with zero boundary conditions where

$$G(x,y,t) = \sum_{n=1}^{\infty} e^{\lambda_n t}\phi_n(y)\phi_n(x)$$

$$= 2\sum_{n=1}^{\infty} e^{-n^2\pi^2 t}\sin n\pi x\, \sin n\pi y. \tag{5.16}$$

§5.2 Heat Lost from a Slab to Ambient (BVP 2)

Let us consider a mixed problem, where heat flux leaving the right end of a rod (or infinite slab) is driven by the gradient to ambient. More precisely, consider

$$\frac{\partial u}{\partial t} = \frac{\partial^2 u}{\partial x^2}, \quad 0 < x < 1, \tag{5.17}$$

subject to

i) $u(x, 0) = 1, \quad 0 < x < 1,$

ii) $u(0, t) = 0, \quad t \geq 0,$

iii) $u_x(1, t) = -u(1, t), \quad t \geq 0$ (a *Robin* condition).

The above might be a first attempt to model a *Trombe wall*, a massive masonry wall that absorbs energy from the sun during the day then discharges its heat to the interior space during the night [Duffie and Beckman].

Again assume a separable solution

$$u(x, t) = T(t)X(x)$$

leading to

$$\frac{\dot{T}(t)}{T(t)} = \frac{X''(x)}{X(x)} = \lambda = -\omega^2,$$

giving as before that

$$T(t) = e^{-\omega^2 t},$$

and after imposing ii),

$$X(x) = \sin \omega x.$$

But in this case, imposing condition iii) leads instead to the transcendental

$$-\omega \cos \omega = \sin \omega,$$

i.e.,

$$\tan \omega = -\omega. \tag{5.18}$$

A glance at a graph of $y = \tan\omega$ superimposed onto a graph of $y = -\omega$ (Figure 5.1) shows solutions

$$0 < \omega_1 < \omega_2 < \cdots$$

of (5.18) where, for large n,

$$\omega_n \approx \frac{2n-1}{2}\pi.$$

Thus it is plausible that the solution to our problem (5.17) is of the form

$$u(x,t) = \sum_1^\infty c_n e^{-\omega_n^2 t} \sin\omega_n x. \tag{5.19}$$

Imposing the initial condition i) leads to

$$1 = \sum_1^\infty c_n \sin\omega_n x.$$

Again, using the natural inner product (5.8),

$$\langle \sin\omega_m x,\ 1\rangle = \sum_{n=1}^\infty c_n \langle \sin\omega_m x,\ \sin\omega_n x\rangle.$$

Tedious computation (Exercise 5.8) obviated by later theory (Exercise 5.24) shows that again the functions $\sin\omega_n x$ are mutually orthogonal, and hence, after more computation (Exercise 5.9),

$$c_n = 2\frac{1-\cos\omega_n}{\omega_n(1+\cos^2\omega_n)}, \tag{5.20}$$

giving a plausible solution (5.19) to our original mixed problem (5.17).

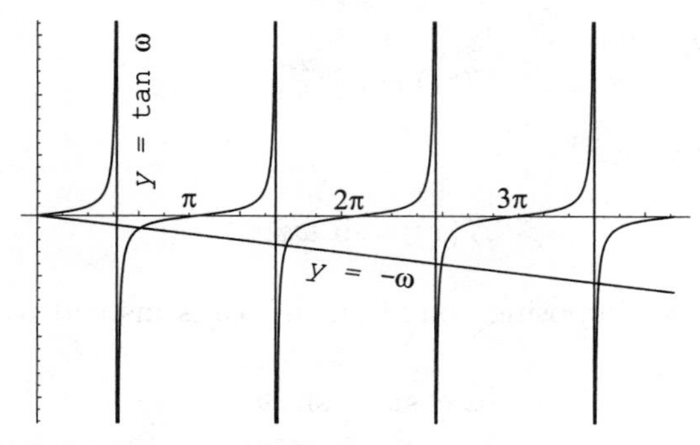

Figure 5.1 The graphs of $y = \tan\omega$ and $y = -\omega$ superimposed. The abscissas ω of the intersections yield the eigenvalues $\lambda = -\omega$.

5.3 The Fourier Ring Problem (BVP 3)

A long rod is bent and its ends welded to form a ring. Given an initial temperature profile around the ring, how will this temperature profile evolve over time? The problem in symbols becomes

$$\frac{\partial u}{\partial t} = \frac{\partial^2 u}{\partial x^2}, \quad -\pi < x < \pi, \tag{5.21}$$

subject to the initial profile

i) $u(x,0) = f(x)$,

and the spatially periodic boundary conditions

ii) $u(-\pi,t) = u(\pi,t)$,

iii) $u_x(-\pi,t) = u_x(\pi,t)$.

Assuming

$$u(x,t) = T(t)X(x)$$

yields

$$\frac{\dot{T}(t)}{T(t)} = \frac{X''(x)}{X(x)} = \lambda = -\omega^2,$$

and so

$$T(t) = e^{-\omega^2 t}$$

and

$$X = a\cos\omega x + b\sin\omega x.$$

Applying boundary conditions ii) and iii) yields that (exercise) $\omega_n = n$ and so

$$\begin{aligned} T_n(t) &= e^{-n^2 t}, \\ X_n(x) &= a_n\cos nx + b_n\sin nx, \end{aligned}$$

for $n = 0, 1, 2, \ldots$. Thus we are led to a solution of the Ring Problem

$$u(x,t) = \sum_{n=0}^{\infty} e^{-n^2 t}(a_n\cos nx + b_n\sin nx). \tag{5.22}$$

Imposing the initial condition i) and applying Fourier's trick yields the famous formulas of *Fourier Analysis*:

$$a_n = \frac{\langle f, \cos nx \rangle}{\langle \cos nx, \cos nx \rangle} = \frac{1}{\pi} \int_{-\pi}^{\pi} f(x) \cos nx \, dx \tag{5.23}$$

and

$$b_n = \frac{\langle f, \sin nx \rangle}{\langle \sin nx, \sin nx \rangle} = \frac{1}{\pi} \int_{-\pi}^{\pi} f(x) \sin nx \, dx \tag{5.24}$$

for $n > 0$, while a_0 is the average value of f, i.e.,

$$a_0 = \frac{\langle f, 1 \rangle}{\langle 1, 1 \rangle} = \frac{1}{2\pi} \int_{-\pi}^{\pi} f(x) \, dx, \tag{5.25}$$

and $b_0 = 0$ (Exercise 5.10). The Ring Problem has led us directly to *Fourier series*, the representation of functions

$$f(x) = \sum_{0}^{\infty} a_n \cos nx + b_n \sin nx$$

by trigonometric series.

5.4 The Plucked String (BVP 4)

Let us pluck a violin string and determine the vibrations produced. That is, consider the problem

$$\frac{\partial^2 u}{\partial t^2} = \frac{\partial^2 u}{\partial x^2}, \qquad 0 < x < 1, \tag{5.26}$$

subject to

i) $u(x, 0) = f(x)$,

ii) $u_t(x, 0) = 0, \quad 0 < x < 1$,

iii) $u(0, t) = 0, \quad t \geq 0$,

iv) $u(1, t) = 0, \quad t \geq 0$.

(As a Rule of Thumb, *well posed problems have as many conditions as derivatives.*)

Assume a solution of the form

$$u(x, t) = T(t)X(x)$$

and impose the wave equation (5.26) to obtain

$$\ddot{T} + \omega^2 T = 0$$

and

$$X'' + \omega^2 X = 0.$$

Imposing iii) and iv) as in BVP 1 yields that $\omega_n = n\pi$ and that $X_n(x) = \sin n\pi x$. Imposing initial condition ii) on T_n yields $T_n(t) = \cos n\pi t$ and hence our solution has form

$$u(x,t) = \sum_{n=1}^{\infty} c_n \cos n\pi t \, \sin n\pi x. \tag{5.27}$$

Physically the spatial eigenmodes (overtones) $X_n = \sin n\pi x$ are being modulated in amplitude harmonically by the temporal modes $T_n = \cos n\pi t$.

Suppose now that the string is plucked at its center, i.e.,

$$f(x) = \begin{cases} 2x & \text{if } \ 0 \le x \le 1/2 \\ 2 - 2x & \text{if } \ 1/2 \le x \le 1. \end{cases}.$$

That is, initially

$$f(x) = \sum_{1}^{\infty} c_n \sin n\pi x,$$

giving that (Exercise 5.11)

$$c_n = \frac{8\sin(n\pi/2)}{n^2\pi^2}. \tag{5.28}$$

5.5 A Circular Drum (BVP 5)

Let us sound a circular drum by carefully striking with the mallet at the very center so that vibration is radially symmetric. In symbols let us solve

$$\frac{\partial^2 u}{\partial t^2} = \nabla^2 u, \qquad 0 \le r < 1, \tag{5.29}$$

subject to

i) $u(r,0) = f(r), \ r < 1,$

ii) $u_t(r, 0) = 0$,

iii) $u(1, t) = 0, \quad t \geq 0$.

(Although this model appears to break the rule of thumb of *one condition per derivative*, there is a latent condition that will reveal itself during the calculation.)

Assume

$$u(r, t) = T(t)R(r),$$

giving that

$$\frac{\ddot{T}}{T} = \frac{\nabla^2 R}{R} = \lambda.$$

Again for theoretical reasons (Exercise 5.22) we are guaranteed that the eigenvalue λ is negative. Following a long notational tradition [Bessel] we set $\lambda = -\alpha^2$, an unfortunate conflict with an equally long tradition of α for diffusivity.

Recalling the form of the Laplacian in polar coordinates (1.31) we obtain the decoupled ODEs

$$\ddot{T} + \alpha^2 T = 0$$

and

$$\frac{d^2R}{dr^2} + \frac{1}{r}\frac{dR}{dr} = -\alpha^2 R.$$

Multiplying the second through by r^2, then setting

$$x = \alpha r \qquad \text{and} \qquad y = R(r),$$

we are lead directly (Exercise 5.6) to *Bessel's equation*

$$x^2y'' + xy' + x^2y = 0. \tag{5.30}$$

As discussed in great detail in Chapter 9, this ODE possesses one bounded and one unbounded fundamental solution. Limited to bounded solutions — our latent fourth condition — we see that within a constant multiple,

$$y = J_0(x),$$

a function known as the *Bessel function of order* 0 *of the first kind.* Think of $J_0(x)$ as a damped version of cosine — a graph is displayed in Figure 5.2.

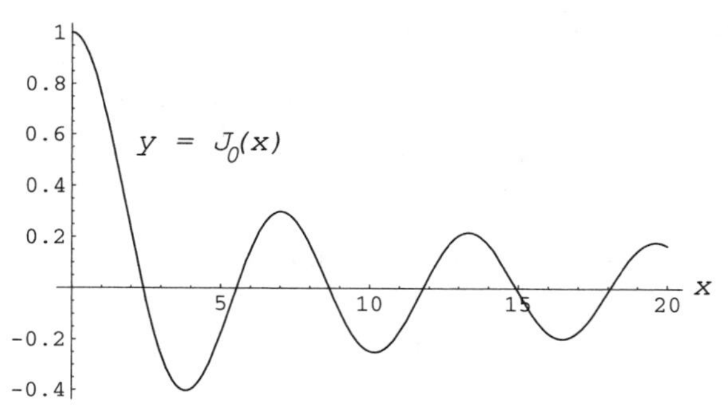

Figure 5.2 A graph of the Bessel function $J_0(x)$.

We shall see that a far reaching result of Rellich guarantees that $J_0(x)$ possesses positive zeros α_n increasing to ∞:

$$0 < \alpha_1 < \alpha_2 < \alpha_3 < \dots \ .$$

Table 5.1. The first five zeros of $J_0(x)$

α_1	α_2	α_3	α_4	α_5	α_6
2.40483	5.52008	8.65373	11.79153	14.93092	18.07106

Returning to our problem and resubstituting $r = x/\alpha$ we obtain the vibrations of a circular drum struck at the center:

$$u(r,t) = \sum_{n=1}^{\infty} c_n \ \cos \alpha_n t \ J_0(\alpha_n r) \tag{5.31}$$

where again by Exercise 5.24, the $J_0(\alpha_n r)$ are again mutually orthogonal with respect to the inner product belonging naturally to the geometry of the circular drumhead

$$\langle f, g \rangle = \int_0^1 f(r) g(r) r dr,$$

so that

$$c_n = \frac{\langle f(r), J_0(\alpha_n r) \rangle}{\langle J_0(\alpha_n r), J_0(\alpha_n r) \rangle} = \frac{\int_0^1 f(r) J_0(\alpha_n r) r \, dr}{\int_0^1 J_0(\alpha_n r)^2 r \, dr}. \tag{5.32}$$

§5.6 Quenching a Ball (BVP 6)

A solid ball of radius 1, initially and uniformly at temperature 1, is suddenly dropped into a bath at temperature 0. Let us discover how the interior temperatures relax to 0.

Intuitively the solution is radially symmetric:

$$\frac{\partial u}{\partial t} = \nabla^2 u, \quad 0 \le \rho < 1, \tag{5.33}$$

subject to

i) $u(\rho, 0) = 1, \quad 0 \le \rho < 1,$

ii) $u(1, t) = 0, \quad t \ge 0,$

together with the latent condition that we search for bounded solutions.

Assume

$$u(\rho, t) = T(t)R(\rho)$$

giving rise to

$$\frac{\dot{T}}{T} = \frac{\nabla^2 R}{R} = \lambda = -\omega^2.$$

Thus

$$T = e^{-\omega^2 t}$$

and, because of the form of the Laplacian in spherical coordinates (1.32),

$$R'' + \frac{2}{\rho}R' = -\omega^2 R, \tag{5.34}$$

where the prime denotes differentiation with respect to ρ.

The product rule and some algebra bring (5.34) to the form

$$(\rho R)'' = -\omega^2 \rho R. \tag{5.35}$$

That is, $y = \rho R$ solves $y'' + \omega^2 y = 0$. Hence

$$\rho R = a \cos \omega\rho + b \sin \omega\rho,$$

i.e.,

$$R = a\frac{\cos \omega\rho}{\rho} + b\frac{\sin \omega\rho}{\rho}.$$

Insisting on bounded solutions means $a = 0$. Imposing the zero surface condition ii) yields $\sin\omega = 0$, i.e., $\omega = n\pi, \quad n = 1, 2, \ldots$. Therefore the temperatures within the cooling ball are

$$u(\rho, t) = \sum_{1}^{\infty} c_n e^{-n^2\pi^2 t} \cdot \frac{\sin n\pi\rho}{\rho}, \tag{5.36}$$

a sum of damped *sinc* functions. Initially

$$1 = \sum_{1}^{\infty} c_n \frac{\sin n\pi\rho}{\rho}.$$

But the spatial eigenmodes $R_n = \sin(n\pi\rho)/\rho$ are (as usual) mutually orthogonal under the inner product that belongs to the geometry of the sphere:

$$\langle f, g\rangle = \int_0^1 f(\rho)g(\rho)\rho^2 d\rho, \tag{5.37}$$

that is,

$$\langle R_m, R_n\rangle = 0 \ \text{ for } \ m \neq n \tag{5.38}$$

and consequently (exercise)

$$c_n = \frac{2(-1)^{n+1}}{n\pi}.$$

§5.7 Square Regions Revisited (SSP 1)

Let us solve a steady state problem with separation of variables, namely the first steady problem of this book (Figure 2.1):

$$\nabla^2 u = 0, \quad 0 < x, y < 1, \tag{5.39}$$

subject to

i) $u(0, y) = 0, \quad 0 \leq y \leq 1,$

ii) $u(1, y) = 1, \quad 0 < y < 1,$

iii) $u(x, 0) = 0, \quad 0 \leq x \leq 1,$

iv) $u(x, 1) = 0, \quad 0 \leq x \leq 1.$

Allow x to play the role of t in a transient problem:

$$\frac{\partial^2 u}{\partial x^2} = -\frac{\partial^2 u}{\partial y^2}.$$

We then proceed as before, finding special separated solutions to i), iii), and iv), then combine these special solutions to satisfy the troublesome condition ii).

Assume a solution

$$u(x,y) = X(x)Y(y)$$

and impose Laplace's equation to obtain

$$-\frac{d^2X}{dx^2}/X = \frac{d^2Y}{dy^2}/Y = \lambda = -\omega^2.$$

The sign choice $\lambda = -\omega^2$ follows from the boundary conditions $Y(0) = 0 = Y(1)$. Hence

$$X = a\cosh \omega x + b \sinh \omega x$$

and

$$Y = c\cos \omega y + d\sin \omega y.$$

Applying iii) and iv) to Y and i) to X yields

$$X_n = \sinh n\pi x,$$

$$Y_n = \sin n\pi y,$$

resulting in the solution

$$u(x,y) = \sum_{n=1}^{\infty} c_n \sinh n\pi x \, \sin n\pi y. \tag{5.40}$$

Imposing the remaining condition ii) gives

$$1 = \sum_{n=1}^{\infty} c_n \sinh n\pi \, \sin n\pi y \tag{5.41}$$

and so again aping BVP 1,

$$c_n = \begin{cases} \frac{4}{n\pi \sinh n\pi} & \text{if } n \text{ odd} \\ 0 & \text{if } n \text{ even} \end{cases}. \tag{5.42}$$

§5.8 A Critique of the Method

During the solution frenzy above we ignored many serious mathematical (and physical) questions. For example examine the steps taken during the solution of BVP 1. Although we obtain solutions of the heat equation of the form

$$u_n = T_n(t)X_n(x),$$

are we sure that an infinite sum of such solutions

$$u(x,t) = \sum_1^\infty c_n T_n(t) X_n(x) \tag{5.43}$$

is also a solution? There are of course famous examples of convergent series that cannot be differentiated term-by-term — see Exercise 5.14. Second, what guarantees that all initial shapes (conditions) $f(x)$ can be realized using the spatial eigenmodes

$$f(x) = \sum_1^\infty c_n X_n(x)? \tag{5.44}$$

See Exercise 5.15. Third, in what sense does the above series (5.44) of spatial eigenmodes equal the initial shape $f(x)$? Certainly not in the usual sense, for each x. See Exercise 5.16. Fourth, even if we suppose all physically meaningful initial shapes $f(x)$ can be written in terms of the spatial eigenmodes

$$f(x) = \sum_1^\infty c_n X_n(x),$$

are we justified in multiplying by some X_m and then integrating term-by-term? Sometimes it is not possible to integrate term-by-term — see Exercise 5.17. Fifth, are these representations unique? Could there be two sets of c_n yielding the same initial shape $f(x)$? This can indeed happen in naive models — see Chapter 14.

Last and most disturbing, Exercise 5.18 reveals that the solution to the vibrating string BVP 4, when plucked at its center, is not smooth and thus cannot be a solution in the classical sense — there is no classical solution! To escape this mathematical morass we must leave 19th century Mathematics.

From our new viewpoint, a spatial temperature profile or a vibrational shape will be a point of an infinite dimensional space — a *Hilbert space.* A solution to a transient problem will begin at an initial shape, then evolve along a trajectory in this Hilbert space. These undulating shapes, snapshots at time t, will be written with time-varying coordinates with respect to time invariant axes. The axes are the eigenfunctions belonging to the spatial differential operator arising naturally from the problem.

These theoretical underpinnings are sketched in the next Chapter. But we must wait until Chapter 19 for the modern Sobolev methods that substantiate all these calculations. In the meantime we will develop our problem solving skills.

§5.9 Tricks of the Trade

Often problems are first posed in a form that will not yield to separation of variables. There are however standard tricks for reposing such problems more favorably. What follows is a short list of several common difficulties and their resolution in general terms. Concrete examples will appear in later sections.

Difficulty I. Nonhomogeneous models.

Consider a problem of the nonhomogeneous form

$$\dot{u} = Au + F \tag{5.45}$$

with the *driving* or *forcing* function $F = F(x,t)$. Physically think of F as an exogenous influence like distributed heat sources within a material, distributed actuator action, current flow, distributed acoustic noise, etc.

Approach 1. Find some solution $b(x,t)$ of (5.45) ignoring the initial condition (and if possible with 0 boundary conditions). By setting $v = u - b$ the problem becomes a homogeneous problem about v subject to new conditions.

Approach 2. Solve the homogeneous problem ignoring the driving function, then use the eigenmodes obtained to solve the nonhomogeneous problem. In our (by now) generic symbol set, express

$$F = \sum_{n=1}^{\infty} f_n(t)\, \phi_n,$$

assume

$$u = \sum_{n=1}^{\infty} T_n(t)\, \phi_n,$$

then solve the decoupled ODEs

$$\dot{T}_n - \lambda_n T_n = f_n$$

subject to the initial conditions $T_n(0)$ determined from

$$u(x, 0) = \sum_{n=1}^{\infty} T_n(0)\, \phi_n.$$

Difficulty II. Nonhomogeneous boundary conditions.

Merely subtract off a cleverly chosen correction $b(x)$:

$$v(x,t) = u(x,t) - b(x)$$

so that v possesses homogeneous boundary conditions. The trade-off is that homogeneous problems

$$\dot{u} = Au$$

become nonhomogeneous

$$\dot{v} = Av + F.$$

One now proceeds as in (I) above. Often simple polynomials $b(x)$ suffice. Often $b(x)$ is the steady state of the original problem.

Difficulty III. Time-varying boundary conditions.

Attempt as in (II) to subtract off a cleverly chosen $b(x,t)$ so that $v = u - b$ has time-invariant homogeneous boundary conditions.

Or introduce the new state variable $v = \dot{u}$ and transform the problem into a problem about v rather than u; I have found this very useful but ad hoc. More systematically, go to *state (phase) space* — see below.

Difficulty IV. High order time derivatives.

Problems of type

$$u^{(m)} = Au$$

can be solved as in (I) by using the eigenmodes of A to yield high order ODEs

$$T_n^{(m)} - \lambda_n T_n = 0.$$

Alternatively, employ a reduction to first order by going over to *state space*: introduce the *states*

$$u_i = u^{(i-1)}, \quad i = 1, 2, \ldots, m.$$

The problem now is recast in the equivalent first order vector form

$$\left|\begin{matrix} u_1 \\ u_2 \\ \cdot \\ \cdot \\ \cdot \\ u_m \end{matrix}\right|^{\cdot} = \begin{pmatrix} 0 & 1 & 0 & 0 & \cdots & 0 \\ 0 & 0 & 1 & 0 & \cdots & 0 \\ & & & \cdot & & \\ & & & \cdot & & \\ & & & \cdot & & \\ A & 0 & 0 & 0 & \cdots & 0 \end{pmatrix} \left|\begin{matrix} u_1 \\ u_2 \\ \cdot \\ \cdot \\ \cdot \\ u_m \end{matrix}\right|.$$

Difficulty V. The spatial operator is not densely defined.

But often all physically meaningful initial shapes are representable in terms of the eigenfunctions. This no difficulty at all — work in the subspace spanned by the eigenfunctions.

Difficulty VI. Mixed spatial partials.

Elliptic operators such as say

$$A = \frac{\partial^2}{\partial x^2} + 2\frac{\partial^2}{\partial x \partial y} + 4\frac{\partial^2}{\partial y^2}$$

can be brought by a change of variables and change of inner product to the Laplacian via *Sylvester's Law of Inertia*. See Exercise 14.16.

Difficulty VII. Mixed temporal and spatial partials.

A balance of forces in vibrating systems often leads to generic BVPs of type

$$M\ddot{u} + C\dot{u} + Au = F$$

where M, C, and A are called *mass, (viscous) damping* and *stiffness* respectively — spatial differential operators for distributed systems, matrices for lumped. The term $C\dot{u}$ is often a mixed partial as in the *Euler-Bernoulli beam* model (§8.6)

$$\frac{\partial^2 u}{\partial t^2} - c\frac{\partial^3 u}{\partial t \partial x^2} + \frac{\partial^4 u}{\partial x^4} = F,$$

where damping

$$C = -c\frac{\partial^2}{\partial x^2}.$$

Other times damping C and stiffness A 'commute,' hence C preserves the eigenspaces of A, hence almost always giving that the eigenmodes ϕ_n of A are also eigenmodes of C, i.e., $C\phi_n = \mu_n \phi_n$. *Damping is modal.* The approach is now clear: use the eigenmodes of A to write the solution, thus obtaining the decoupled ODEs

$$\ddot{T}_n + \mu_n \, \dot{T}_n + \lambda_n \, T_n = f_n.$$

Unfortunately it often happens that C does not share the eigenmodes of A — *damping is not modal.* The classical approach is to then employ *nonorthogonal expansions via the adjoint problem.* However as has been pointed out recently, this is equivalent to going over to a more physically natural inner product where orthogonal expansions obtain. More in Chapter 14.

An encyclopedia of such tricks of the trade is Weinberger's 1965 text. Willard Miller has thought deeply about Separation of Variables and its role in Mechanics.

Exercises

5.1 In (5.4) of BVP 1, taking $\lambda = \omega^2 > 0$ leads to a conflict with the boundary conditions.

5.2 Modify BVP 1 by replacing condition iii) by $u(1,t) = 1$. Demonstrate that an attempt to separate variables now fails. Next, after applying the Rule of Thumb: *subtract off the steady state*, successfully solve the modified problem by separation of variables.

5.3 Code and graph snapshots of the rod temperatures of BVP 1. Compare the results of a finite difference model to the results obtained by truncating the series solution (5.11).

5.4 Find the first 20 eigenvalues $\lambda_n = -\omega_n^2$ of BVP 2 to 6 places.

Ans: $\lambda = -4.115858,\ -24.139342,\ -63.659107,\ \ldots$.

Hint: Employ the Newton-Raphson algorithm to solve

$$\omega \cos \omega + \sin \omega = 0.$$

5.5 Why are violins bowed (a continuous microscopic plucking) approximately 1/8th of the way from the bridge? Why is the resulting sound much richer than when the string is bowed in the center?

5.6 Make the substitution $x = \alpha r$ and $y = R(r)$ and reduce

$$\frac{d^2R}{dr^2} + \frac{1}{r}\frac{dR}{dr} = -\alpha^2 R$$

to Bessel's equation

$$x^2 y'' + xy' + x^2 y = 0.$$

5.7 Verify equations (5.10) and (5.11).

5.8 Prove that the eigenfunctions $\phi_n = \sin \omega_n x$ of BVP 2 are indeed mutually orthogonal.

5.9 Verify (5.20).

5.10 Verify equations (5.23)–(5.25).

5.11 Verify (5.28).

5.12 By following the steps of SSP 1, solve the steady problem on the rectangle of Figure 5.3.

Figure 5.3

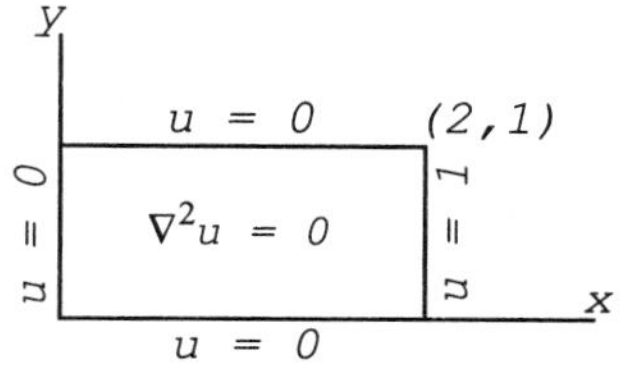

5.13 Find the sound made by a square drum $0 \le x, y \le 1$.

Answer: (See Figure 15.9)

$$u(x, y, t) = \sum_{m,n=1}^{\infty} c_{mn} \cos \pi t\sqrt{m^2 + n^2}\ \sin m\pi x\ \sin n\pi y.$$

5.14 Given that for $-\pi < \theta < \pi$,

$$\frac{\theta}{2} = \sum_{n=1}^{\infty} (-1)^{n+1} \frac{\sin n\theta}{n},$$

show that the derivative of this function cannot be obtained by differentiating the series term-by-term.

5.15 Show there are no c_n for which

$$1 = \sum_{1}^{\infty} c_n \sin n\pi x$$

on $-1 \le x < 1$ in any reasonable sense.

5.16 In BVP 1 we found c_n for which

$$1 = \sum_{1}^{\infty} c_n \sin n\pi x.$$

on $0 \le x \le 1$. But in what sense? Try $x = 0$.

5.17 Let $F_n(x) = n$ for $0 < x < 1/n$, 0 otherwise. Take $F_0 \equiv 0$. Show that the series

$$\sum_{n=1}^{\infty}(F_n(x) - F_{n-1}(x))$$

converges at each point x to 0 yet the series cannot be integrated term-by-term over the interval $[0, 1]$.

5.18 Make your best guess of future shapes of the plucked string of BVP 4 as determined in (5.27) and (5.28). Verify your guess by solving the problem via finite divided differences

$$\frac{u_i^{j+1} - 2u_i^j + u_i^{j-1}}{\Delta t^2} = \frac{u_{i+1}^j - 2u_i^j + u_{i-1}^j}{\Delta x^2}$$

or by truncating the series solution (5.27) and graphing. You will be astonished.

5.19 In 1966 Mark Kac asked, "Can one hear the shape of a drum?" Must two drums of different shapes have different eigenvalues? [Protter]. Argue that this is an important question. (This question was resolved negatively in 1992 [Gordon, Webb, Wolpert]).

Hint: Think of petroleum exploration, anti-submarine warfare, or curing composite materials.

5.20 As a rule of thumb, the natural vibrational frequencies of one (spatial) dimensional bodies are more or less equally spaced, as in the Plucked String BVP 4. In contrast, two or more dimensional bodies have vibrational frequencies that become more densely packed in the higher octaves. Illustrate this by plotting against f the number $N = N(f)$ of natural vibrational frequencies less than f of the square drum of Exercise 5.13. What is the asymptotic value of N?

Answer: $N \sim \pi f^2/2$.

Hint: The number of integral *lattice* points (m, n) within the circle of radius r is approximately the area of the circle.

Argue that active sound suppression within heating ducts is a far easier problem than sound suppression within an automobile or airplane fuselage.

5.21* Prove that the Laplacian $A = \nabla^2$ subject to zero boundary conditions on a bounded domain Ω is a *negative definite* operator, i.e., for $\phi \neq 0$ satisfying the boundary conditions,

$$\langle A\phi, \phi \rangle < 0.$$

Outline: Integrate by parts via Green's first identity [O'Neill]:

$$\int_\Omega f \nabla^2 g \, dx = \int_{\partial\Omega} f \nabla g \cdot \mathbf{n} \, ds - \int_\Omega \nabla f \cdot \nabla g \, dx.$$

5.22 Prove that the eigenvalues of a negative definite operator are negative.

5.23*In fact, prove that the Laplacian $A = \nabla^2$ subject to either zero or zero flux conditions on the boundary of a bounded domain Ω is a *self-adjoint* (*Hermitian*) operator with respect to the *Hilbert* inner product $\langle .,. \rangle$, i.e.,

$$\langle A\phi, \psi \rangle = \langle \phi, A\psi \rangle$$

for all ϕ, ψ satisfying the boundary conditions, where

$$\langle \phi, \psi \rangle = \int_\Omega \phi\psi \, dx.$$

Outline: Integrate twice by parts via Green's first identity.

Note that $\langle A\phi, \psi \rangle = [\phi, \psi]$ where

$$[\phi, \psi] = -\int_\Omega \nabla\phi \cdot \nabla\psi \, dx,$$

the *Dirichlet* inner product.

5.24 Prove that eigenfunctions belonging to distinct eigenvalues of a self-adjoint operator are orthogonal.

Outline: If $A\phi = \lambda\phi$ and $A\psi = \mu\psi$, then $\lambda\langle\phi, \psi\rangle = \mu\langle\phi, \psi\rangle$.

5.25 Modify BVP 1 (the rod) to include a constant source term g:

$$\frac{\partial u}{\partial t} = \frac{\partial^2 u}{\partial x^2} + g, \quad 0 < x < 1,$$

arising say from an embedded nichrome heating element. Solve twice, using both approaches suggested in (I) above: first by subtracting off the steady state, then by using the eigenfunctions of the homogeneous problem.

Answer:

$$u(x,t) = \frac{gx(1-x)}{2} + \sum_{n=1}^{\infty} c_n\, e^{-n^2\pi^2 t} \sin n\pi x$$

where

$$c_n = 2(1-(-1)^n)(\frac{1}{n\pi} - \frac{g}{n^3\pi^3}).$$

5.26 Modify BVP 1 by replacing the initial condition by $u(x,0)=0$ and the second boundary condition $u(1,t) = 0$ by the time varying condition $u(1,t) = \sin t$. Solve.

Answer:

$$u = x\sin t - \sum_{n=1}^{\infty} \frac{2(-1)^n n\pi}{n^4\pi^4+1} \mathrm{e}^{-n^2\pi^2 t} \sin n\pi x$$

$$+ \cos t \sum_{n=1}^{\infty} \frac{2(-1)^n n\pi}{n^4\pi^4+1} \sin n\pi x + \sin t \sum_{n=1}^{\infty} \frac{2(-1)^n}{n\pi(n^4\pi^4+1)} \sin n\pi x.$$

Notice how transients die away, leaving a periodic trajectory.

5.27 Show for the vibrational model

$$\frac{\partial^2 u}{\partial t^2} + c\frac{\partial^2 u}{\partial t \partial x} - \frac{\partial^2 u}{\partial x^2} = F, \quad 0 < x < 1,$$

subject to the boundary conditions $u(0,t) = 0 = u(1,t)$, that damping is not modal.

Chapter 6

Hilbert Space

Power and clarity can be ours by taking a slightly more abstract view. One of the most useful structures of modern Mathematics is named for the leading mathematician of the 20th century, David Hilbert. The ideas of a norm, inner product, orthogonality, Hilbert space, differential operators and their eigenvalues make precise the steps of separation of variables and the very notion of a solution to a boundary value problem.

§6.1 Norms on $\mathbf{R}^n$

A *norm* on a real vector space V is a mapping $v \mapsto \|v\|$ from V to the nonnegative reals with the three properties

$$\|v\| \geq 0 \text{ with equality exactly when } v = 0, \tag{6.1a}$$

$$\|cv\| = |c| \cdot \|v\|, \tag{6.1b}$$

$$\|u + v\| \leq \|u\| + \|v\|, \tag{6.1c}$$

for all (real) scalars c and vectors u and v in V.

For example, on the n-tuples $\mathbf{x} = (x_1, x_2, \ldots, x_n)$ of $\mathbf{R}^n$, there are three commonly applied norms (Exercise 6.1):

$$\|\mathbf{x}\|_\infty = \max_{i=1,\ldots,n} |x_i|, \tag{6.2}$$

$$\|\mathbf{x}\|_1 = \sum_{i=1}^{n} |x_i|, \tag{6.3}$$

$$\|\mathbf{x}\|_2 = \Big(\sum_{i=1}^{n} x_i^2\Big)^{1/2}. \tag{6.4}$$

The last is, of course, the familiar distance formula for the length of the vector x. There are many other possible norms (Exercise 6.1), making the following result all the more astonishing.

Theorem A. All norms on a real vector space V of dimension n are equivalent, i.e., all yield the identical convergent sequences. In fact, given any two norms, $\|\cdot\|$ and $\|\cdot\|'$ on V, each will *dominate* the other, i.e., there are positive constants c and c' such that

$$c\|v\| \le \|v\|' \le c'\|v\|$$

for all v in V.

Proof. See [MacCluer, 2000].

§6.2 Hilbert Spaces

The familiar distance formula norm (6.4) is not only physically appealing but is the most mathematically tractable since it arises from an inner product.

Definition. An *inner product* on a real vector space V is a symmetric, positive definite bilinear form $\langle\cdot,\cdot\rangle : V \times V \to \mathbf{R}$, i.e.,

$$\langle u, v\rangle = \langle v, u\rangle, \tag{6.5a}$$

$$\langle v, v\rangle \ge 0 \text{ with equality exactly when } v = 0, \tag{6.5b}$$

$$\langle u + v, w\rangle = \langle u, w\rangle + \langle v, w\rangle, \tag{6.5c}$$

$$\langle cu, v\rangle = c\langle u, v\rangle, \tag{6.5d}$$

for all real scalars c and vectors u, v, w in V.

The norm arising from this inner product is

$$\|v\| = \sqrt{\langle v, v\rangle}. \tag{6.6}$$

To see that this is indeed a norm, we must check the three axioms of (6.1), where only the third presents any difficulty. The crucial missing fact is the single most useful inequality of mathematics.

The Cauchy–Schwarz Inequality. For any inner product,

$$\langle u, v\rangle \le \|u\| \cdot \|v\| \tag{6.7}$$

with equality exactly when u and v are scalar multiples of one another.

Proof. Expand out $\langle u-cv, u-cv\rangle \ge 0$ and set $c = \|u\|/\|v\|$ (Exercise 6.5).

A real vector space X with an inner product under which X is *complete*, i.e., all Cauchy convergent sequences converge, is called a **Hilbert space**, the most important mathematical object of the twentieth century.

A sequence of vectors $\phi_1, \phi_2, \phi_3, \ldots$ from a Hilbert space X is an *orthogonal* sequence if the terms are mutually orthogonal, i.e.,

$$\langle \phi_i, \phi_j \rangle = 0 \text{ if } i \neq j.$$

A sequence $\phi_1, \phi_2, \phi_3, \ldots$ is *orthonormal* if

$$\langle \phi_i, \phi_j \rangle = \begin{cases} 1 & \text{if } i = j \\ 0 & \text{otherwise.} \end{cases} \tag{6.8}$$

An orthonormal sequence $\phi_1, \phi_2, \phi_3, \ldots$ is an *orthonormal basis* for the Hilbert space X if every element f of X can be written as a norm convergent series

$$f = \sum_n c_n \phi_n. \tag{6.9}$$

We restrict our attention to *separable* Hilbert spaces, spaces that possess a finite or countably infinite orthonormal basis. These are the spaces that arise physically in quantum mechanics, heat transfer, and distributed vibrations.

Example 1. $\mathbf{R}^n$ is a Hilbert space under the ordinary dot product

$$\langle \mathbf{x}, \mathbf{y} \rangle = \sum_{i=1}^{n} x_i y_i \tag{6.10a}$$

with norm

$$\|\mathbf{x}\| = \Big(\sum_{i=1}^{n} x_i^2\Big)^{1/2}. \tag{6.10b}$$

Example 2. Take Example 1 to the limit: Consider all ∞-tuples of real numbers

$$\mathbf{x} = (x_1, x_2, x_3, \ldots) \tag{6.11a}$$

that are *square summable,* i.e.,

$$\sum_{i=1}^{\infty} x_i^2 < \infty. \tag{6.11b}$$

All such sequences form a complete normed vector space arising from the inner product

$$\langle \mathbf{x}, \mathbf{y} \rangle = \sum_{i=1}^{\infty} x_i y_i \tag{6.11c}$$

under the norm

$$\|\mathbf{x}\| = \Big(\sum_{i=1}^{\infty} x_i^2\Big)^{1/2} \tag{6.11d}$$

(Exercises 6.11 and 6.12). This Hilbert space is sometimes called $\mathbf{R}^\infty$ but more often $\mathbf{l}^2$ or ℓ^2, pronounced "little ell two." The obvious orthonormal basis is the *natural basis*

$$\mathbf{e}_i = (0, 0, 0, \ldots, 1 \ (i\text{th position}), 0, 0, \ldots).$$

Example 3. Let Ω be a bounded domain, an open connected subset of $\mathbf{R}^n$. The set

$$X = L^2(\Omega) \tag{6.12a}$$

of all square-integrable, real-valued functions f, i.e., functions f with

$$\int_\Omega f(x)^2 \, dx < \infty, \tag{6.12b}$$

forms a separable Hilbert space under the inner product

$$\langle f, g \rangle = \int_\Omega f(x) g(x) \, dx \tag{6.12c}$$

and L^2 norm

$$\|f\| = \Big(\int_\Omega f(x)^2 \, dx\Big)^{1/2}. \tag{6.12d}$$

Since $X = L^2(\Omega)$ possesses a countably infinite orthonormal basis, each element of X is determined by an infinite-tuple of its components with respect to this orthonormal basis. Thus X is structurally indistinguishable from ℓ^2. The verification of all this requires several deep analytic results (see [Rudin], [Halmos], or [Young]).

Thus a complete list of separable real Hilbert spaces is

$$\mathbf{R}, \mathbf{R}^2, \mathbf{R}^3, \mathbf{R}^4, \ldots, \ell^2.$$

§6.3 Orthogonal Complements

The central geometric notion of Hilbert spaces is *perpendicularity* or *orthogonality*. To repeat, two elements f and g are *orthogonal* if $\langle f, g \rangle = 0$. There is a corresponding notion for sets.

Definition. Let E be a subset of the Hilbert space X. The (closed) subspace $E^\perp$ of all f with $\langle f, e \rangle = 0$ for all e in E is called the *orthogonal complement* of the set E. (See Exercises 6.13 and 6.14).

For example, the z-axis is the orthogonal complement of the xy-plane in 3-space, or more interestingly, of the set consisting only of $\mathbf{i}$ and $\mathbf{j}$.

§6.4 The Gram-Schmidt Process

Often we are confronted with a sequence β_n of elements from a Hilbert space X — although not mutually orthogonal — whose *span* is *dense* in X, i.e., where the subspace of all finite linear combinations

$$\text{span}\{\beta_j\} = \{\beta\,;\beta = \sum_j c_j\beta_j\}$$

comes arbitrarily close to any given element f in X, or equivalently (Exercise 6.15),

$$(\text{span}\{\beta_j\})^\perp = \mathbf{0}.$$

From this *approximate spanning set* we build an orthonormal basis as follows.

The Gram-Schmidt Process: Suppose we have already found an orthonormal sequence $\phi_1, \phi_2, \ldots, \phi_{n-1}$ with the same span as $\beta_1, \beta_2, \ldots, \beta_{n-1}$.

Step 1. Set

$$\gamma_n = \beta_n - \sum_{j=1}^{n-1} \langle \beta_n, \phi_j \rangle \phi_j$$

thus subtracting off the components of β_n in the previous ϕ_j directions.

Step 2. Normalize

$$\phi_n = \gamma_n / \|\gamma_n\|.$$

Step 3. Loop back to Step 1 with $n+1$ replacing n.

It may happen in Step 1 that $\gamma_n = 0$, indicating that β_n depends on the previous β_j and thus can be discarded. Several Exercises below provide drill in this algorithm [Chaney and Kincaid]. For further reading see [Halmos, 1957] and [Young].

§6.5 Differential Operators

The spatial derivatives of a linear PDE on the spatial domain Ω will be thought of as (linear) operators on the Hilbert space $L^2(\Omega)$.

Example 1. Let the Laplacian $A\phi = \phi''$ on $X = L^2(0,1)$ be subject to the Dirichlet conditions $\phi(0) = 0 = \phi(1)$. This is the spatial operator of BVPs 1 and 4. Note that by integrating twice by parts,

$$\langle A\phi, \psi \rangle = \int_0^1 \phi'' \psi \, dx$$

$$= \phi'\psi|_0^1 - \langle \phi', \psi' \rangle = \phi'\psi|_0^1 - \phi\psi'|_0^1 + \langle \phi, \psi'' \rangle. \tag{6.13}$$

By the boundary conditions on the ϕ, we see for the second term on the right of (6.13) that

$$\phi\psi'|_0^1 = \phi(1)\psi'(1) - \phi(0)\psi'(0) = 0.$$

If we were to impose the identical conditions on ψ, the first term of the right side of (6.13) would also vanish

$$\phi'\psi|_0^1 = 0.$$

Thus when ψ satisfies the identical boundary conditions as ϕ,

$$\langle A\phi, \psi \rangle = \langle \phi, A\psi \rangle. \tag{6.14}$$

Such an operator A that commutes across the inner product is called *self-adjoint*, or in the case of complex Hilbert spaces, *Hermitian*. In spite of the abuse to come, I will often use "Hermitian" even when the operator A of (6.14) is operating on a real Hilbert space.

Note that by taking $\psi = \phi$ in (6.13), we see that after one integration by parts,

$$\langle A\phi, \phi \rangle = -\langle \phi', \phi' \rangle, \tag{6.15}$$

and so

$$\langle A\phi, \phi \rangle \le 0, \tag{6.16}$$

i.e., A is *negative semi-definite.* If $\langle A\phi, \phi\rangle = -\langle \phi', \phi' \rangle = 0$, then $\phi' \equiv 0$, and hence ϕ is a constant function; but ϕ enjoys 0 boundary conditions and is therefore the zero function. In short,

$$\langle A\phi, \phi \rangle < 0 \tag{6.17}$$

when ϕ is not the zero function. So in fact, the operator A is *negative definite,* and thus any of its eigenvalues must be negative (Exercise 5.22).

Example 2. Let $S\phi = \phi'$ on $X = L^2(0,1)$ subject to the *periodic boundary conditions* $\phi(0) = \phi(1)$. Again by parts

$$\langle S\phi, \psi \rangle = \langle \phi', \psi \rangle = \phi\psi|_0^1 - \langle \phi, \psi' \rangle,$$

and so if we impose the identical periodic conditions on ψ,

$$\phi\psi|_0^1 = \phi(1)\psi(1) - \phi(0)\psi(0) = 0,$$

giving that

$$\langle S\phi, \psi \rangle = \langle \phi, -S\psi \rangle,$$

i.e., S is *skew-symmetric.* Hence S cannot possess nonzero real eigenvalues (Exercise 6.30).

Example 3. Let $A\phi = \phi''$ on $X = L^2(0,1)$ subject to the Dirichlet condition $\phi(0) = 0$ on the left, but the Robin condition $\phi'(1) = -\phi(1)$ on the right. (This is the spatial operator of BVP 2.) Similar to the computations of Example 1, A is Hermitian and negative definite (Exercise 6.31).

It may be a comfort to know that most of the common practical BVPs have Hermitian (self-adjoint) spatial differential operators.

§6.6 Resolvents

What is the domain of a differential operator A? Consider for example, the operator $A\phi = \phi''$ on $X = L^2(0,1)$ subject to zero boundary conditions (Example 1 of §6.5). Certainly the domain of A should contain all continuously differentiable functions on $(0,1)$ that are continuous on the closure $[0,1]$ and vanish at the endpoints. But as it turns out, we are forced to include within the domain a surprisingly wide class of ill-behaved functions — see Exercise 6.29. There are deep issues here that were resolved in the mid-twentieth century by the Russian school headed by Sobolev. See Chapter 19.

These delicate theoretical issues can be avoided for a given differential operator A by a simple trick — explicitly exhibit a continuous (bounded) *resolvent* $R(\lambda)$ of A at some scalar λ, i.e., an integral operator $R(\lambda)$ that inverts $\lambda I - A$ on at least the obvious ϕ in the domain of A. Then *define* the domain of A to be the range of $R(\lambda)$. The operator A will then be said to be *well conditioned* by $R(\lambda)$. It is easy to see that A is Hermitian exactly when its resolvent $R(\lambda)$ is Hermitian at some real λ (Exercise 6.56).

Example 1 (reprise). Let the Laplacian $A\phi = \phi''$ on $X = L^2(0,1)$ be subject to $\phi(0) = 0 = \phi(1)$. Our goal is to invert the relation

$$(\lambda - A)\phi = f, \tag{6.31}$$

i.e.,

$$R(\lambda)f = \phi. \tag{6.32}$$

Since A is negative definite we should expect A to be invertible, i.e., for the resolvent to exist at $\lambda = 0$. The following calculations are typical.

Using the handy *Volterra* integral operator V where

$$(Vf)x = \int_0^x f(y)\,dy, \tag{6.33}$$

and its constant-function-valued companion

$$\xi f = \int_0^1 f(y)\,dy, \tag{6.34}$$

we see that if $A\phi = \phi'' = -f$, then $-Vf = V\phi'' = \phi' - \phi'(0)$. Integrating once again, $-V^2 f = \phi - \phi'(0)x$. But $-\xi V f = \xi\phi' - \phi'(0)\xi 1 = -\phi'(0)$. Thus

$$R(0)f = (\xi V - V^2)f. \tag{6.35}$$

More explicitly, the resolvent of A at $\lambda = 0$ is

$$(R(0)f)x = \int_0^1 \int_0^y f(z)\,dz\,dy - \int_0^x \int_0^y f(z)\,dz\,dy. \qquad (6.36)$$

Example 2 (reprise). Let $S\phi = \phi'$ on $X = L^2(0,1)$ be subject to the periodic conditions $\phi(0) = \phi(1)$. Since constant functions are eigenfunctions belonging to $\lambda = 0$, we must invert $\lambda - S$ at some nonzero λ, say at $\lambda = 1$. So set $(1 - S)\phi = f$. This is the familiar ODE

$$\phi' - \phi = -f.$$

After multiplying through by the integrating factor e^{-x} and integrating we have the general solution

$$\phi = ce^x - e^x V e^{-x} f.$$

Imposing the periodic boundary conditions gives (Exercise 6.62) the resolvent at $\lambda = 1$:

$$(R(1)f)x = \frac{e^x}{1 - e^{-1}} \int_0^1 e^{-y} f(y)\,dy - e^x \int_0^x e^{-y} f(y)\,dy. \qquad (6.37)$$

These resolvent integral operators

$$(Rf)x = \int_\Omega \kappa(x,y) f(y)\,dy, \qquad (6.38)$$

are *integral operators*. The function κ is called the *kernel* of the operator R, and, when $R = (\lambda - A)^{-1}$, the *Green's function* of the problem $(\lambda - A)u = f$.

For example, the Volterra operator of (6.33)

$$(Vf)x = \int_0^x f(y)\,dy$$

is the operator

$$(Vf)x = \int_0^1 \kappa(x,y) f(y)\,dy$$

with kernel $\kappa(x,y) = 1$ if $y < x$, 0 otherwise.

A bounded operator R is *compact* if for every (norm) bounded sequence f_n, the image sequence $\phi_n = Rf_n$ possesses a convergent subsequence. Compact operators share many of the desirable properties of operators on finite dimensional spaces.

The Hilbert-Schmidt Principle. When the kernel κ of a integral operator (6.38) is absolutely square integrable, i.e.,

$$\int_\Omega \int_\Omega \kappa(x,y)^2 \, dx \, dy < \infty,$$

the operator is compact.

Such (compact) integral operators with square-integrable kernels κ are called *Hilbert-Schmidt* operators. Thus for example, the Volterra operator (6.33), and the spatial operators of Examples 1–2 are Hilbert-Schmidt operators (Exercise 6.55).

And now a workhorse of mathematics:

The Spectral Theorem. The eigenvectors ϕ_n of a compact Hermitian operator R form a complete orthonormal basis for X. The eigenspace E_μ spanned by the eigenvectors belonging to a nonzero eigenvalue μ is finite dimensional. The eigenvalues $\mu_n \to 0$.

Thus for example, we are assured that the eigenfunctions of BVP 1 and 2 are complete since we have explicitly found their resolvents and shown them compact and Hermitian.

Summary: For the majority of practical boundary value problems, one can verify a separation of variables solution by computing explicitly the resolvent $R(\lambda)$ at some convenient λ and checking that $R(\lambda)$ is Hermitian and compact by the Hilbert-Schmidt criterion above. The spectral theorem then guarantees that the spatial eigenfunctions form an orthogonal basis for your proposed solution. The remaining details are discussed in the next section.

§6.7* Justifying Separation of Variables

I will now sketch the theory that underlies the method of separation of variables. Consider (say) the heat transfer problem

$$u_t = \nabla^2 u \text{ on the bounded domain } \Omega, \qquad (6.39)$$

subject to zero boundary conditions and satisfying the initial condition $u(x,0) = f(x)$. The Laplacian $A = \nabla^2$ is thought of as an operator on the Hilbert space $X = L^2(\Omega)$, a mapping from a dense subset of X into X. But there is a deep result of Rellich.

Rellich's Theorem. The Laplacian subject to zero boundary conditions on a bounded domain Ω has a compact resolvent.

(See Chapter 19.) Rellich guarantees that $A = \nabla^2$ possesses a compact resolvent $R = -A^{-1}$ that is defined and continuous everywhere on X, with the domain of A as range. As we have seen in Exercises 5.21–5.24, R is self-adjoint and positive definite. Because the resolvent R is compact and self-adjoint, by the spectral theorem its eigenfunctions $\phi_1, \phi_2, \ldots$, (the eigenfunctions of A), form a complete orthogonal basis of $X = L^2(\Omega)$. The corresponding (positive) eigenvalues of R increase monotonically $-1/\lambda_n \to 0$.

A *strong* solution $u = u(.,t)$ to the problem (6.38) is a moving spatial temperature profile in the infinite dimensional space X that satisfies

$$\dot{u} = Au, \tag{6.40}$$

where $\dot{u}$ is the limit of the usual difference quotient, but where the limit is in the norm sense in X. This solution u begins at the initial shape f, then evolves along a trajectory in this Hilbert space. These undulating shapes, snapshots at time t, can be written with time-varying coordinates with respect to the time invariant eigenfunction axes ϕ_n:

$$u(\cdot, t) = \sum_{n=1}^{\infty} q_n(t)\phi_n. \tag{6.41}$$

It is easy to show (Exercise 6.41) that if $\dot{u}$ exists at t, then

$$\dot{u} = \sum_{n=1}^{\infty} \dot{q}_n\phi_n. \tag{6.42}$$

Because R is a bounded operator, it will distribute over the convergent sum (6.41), from which it easily follows (Exercise 6.42) that A itself will distribute, i.e.,

$$Au = A\sum_{n=1}^{\infty} q_n \, \phi_n = \sum_{n=1}^{\infty} q_n \, A\phi_n = \sum_{n=1}^{\infty} \lambda_n q_n \, \phi_n, \tag{6.43}$$

and so $\dot{q}_n = \lambda_n q_n$. In other words,

the solution is a superposition of separated solutions.

Each square-integrable initial shape f can be realized in the form

$$f = \sum_{n=1}^{\infty} c_n \phi_n \tag{6.44}$$

since the eigenfunctions form a complete orthogonal basis of X. This *strongly* convergent sum — convergent in the norm sense — is *a fortiori* convergent in the *weak* sense, i.e., for every ψ in X,

$$\langle \psi, \sum_{k=1}^{N} c_k \phi_k \rangle \to \langle \psi, f \rangle,$$

(Exercise 6.7) giving when $\psi = \phi_n$ the universal formula for the coefficients of any orthogonal Fourier expansion (6.18):

$$c_n = \frac{\langle \phi_n, f \rangle}{\langle \phi_n, \phi_n \rangle}. \tag{6.45}$$

But does a strong solution exist? Consider the solutions to the temporal ODEs $\dot{q_n}(t) = \lambda_n q_n(t)$ subject to initial value $q_n(0) = \langle f, \phi_n \rangle / \langle \phi_n, \phi_n \rangle$, namely

$$q_n(t) = \frac{\langle f, \phi_n \rangle}{\langle \phi_n, \phi_n \rangle} e^{\lambda_n t}. \tag{6.46}$$

However there is the following classic estimate.

Weyl's Theorem. The eigenvalues of the Laplacian with zero boundary conditions on the bounded domain $\Omega \subset \mathbf{R}^n$ grow by the asymptotic rule

$$\lambda_k \sim -4\pi^2 (\frac{k}{\mu(B)\mu(\Omega)})^{2/n} \tag{6.46}$$

where $\mu(B)$ and $\mu(\Omega)$ are the volumes of the unit ball B in $\mathbf{R}^n$ and Ω . [Protter]

Thus the strong convergence of

$$v = \sum_{n=1}^{\infty} \lambda_n q_n(t) \phi_n \tag{6.47}$$

is guaranteed and hence (6.41) will be a strong solution with limiting value f as $t \to 0^+$.

Combining Rellich and Weyl's results yields a useful general principle which I will refer to as

The Rellich-Weyl Principle. All diffusion problems

$$\dot{u} = \nabla^2 u$$

with zero boundary conditions on a bounded domain Ω are uniquely solvable for any given square-integrable initial shape f by orthogonal separation of variables.

Exercises

6.1 Show that each of (6.2)–(6.4) are indeed norms on $\mathbf{R}^n$. Find infinitely many more such norms.

6.2 Show that the series

$$\sum_{n=1}^{\infty} \frac{\sin n\pi x}{n}$$

is norm convergent in $L^2(0,1)$.

Hint: Show Cauchy convergence.

6.3* Find a sequence of f_n in $L^2(0,1)$ with $||f_n||_2 \to 0$ which converges nowhere pointwise.

Hint: Think of radar pulses moving from left to right on an oscilloscope display with screen-wrap, all of amplitude 1 but whose durations decrease to 0. Such signals will converge nowhere pointwise yet their energy content will die away.

6.4 Numerically estimate $||\mathrm{e}^{-x^2}||$ in $L^2(0,1)$.

Answer: 0.773398.

6.5 Carry out the details in the proof of Cauchy's inequality (6.7). Deduce the triangle inequality (6.1c).

6.6 Prove from the axioms (6.5) for a Hilbert space that $||f|| = \sqrt{(f,f)}$ possesses all three of the defining properties of a norm listed in (6.1).

6.7 Carefully prove that *strong convergence implies weak convergence,* then deduce the fundamental formula (6.19) that explicitly gives the coefficients of any orthogonal expansion.

6.8 In quantum mechanics it is necessary to employ complex scalars for Hilbert spaces. This requires one minor modification of the axioms (6.5):

$$\langle f, g\rangle = \langle g, f\rangle \text{ is replaced by } \langle f, g\rangle = \overline{\langle g, f\rangle}$$

where the over bar is complex conjugation. With this modification prove $\|f\| = \sqrt{\langle f, f\rangle}$ is still a norm.

6.9 For Ω a domain of $\mathbf{R}^n$, show carefully that

$$\langle f, g\rangle = \int_\Omega f(x)g(x)\,dx$$

is an inner product on $L^2(\Omega)$.

6.10 How would the inner product of Exercise 6.9 be modified for the complex Hilbert space of all absolutely square-integrable complex-valued functions on Ω?

6.11 Since Cauchy's inequality holds in each $\mathbf{R}^n$, show that the infinite series (6.11c) is absolutely convergent.

6.12 Deduce from Exercise 6.11 that ℓ^2 is closed under addition and is in fact a vector space.

6.13 Prove that for any subset E — subspace or not — its orthogonal complement $E^\perp$ will be a closed subspace.

6.14 Prove that $(E^\perp)^\perp$ is the smallest closed subspace containing E.

6.15 Show that the span of a collection of β_j is dense in X exactly when $\langle g, \beta_j\rangle = 0$ for all j implies $g = 0$.

6.16 Prove that for any orthonormal collection $\{\phi_n\}_n$, the series

$$\sum_n c_n\phi_n \text{ converges in norm when and only when } \sum_n c_n^2 < \infty.$$

6.17 (Bessel's inequality). Let $\{\phi_n\}_n$ be any orthonormal collection and f any element. Prove

$$\|f\|^2 \geq \sum_n \langle f, \phi_n \rangle^2.$$

At the same time, also prove this fundamental *regression* result:

Let $f_a = \sum_n x_n \phi_n$ be an approximation of f. Then the least approximation error $\|f - f_a\|$ will occur when each $x_n = \langle f, \phi_n \rangle$.

Hint: Expand out $\langle f - \sum_n x_n \phi_n, f - \sum_n x_n \phi_n \rangle$ and complete the square.

6.18 (Riemann-Lebesgue Lemma). Let $\{\phi_n\}_{n=1}^{\infty}$ be any orthonormal sequence and f an arbitrary element. Prove $\langle f, \phi_n \rangle \to 0$. This translates to *finite energy signals are asymptotically band limited.* See §7.

Hint: Apply the N-th Term Test.

6.19 Perform the Gram-Schmidt process on the three vectors $\mathbf{u} = (1, 1, 1), \mathbf{v} = (1, -1, 1), \mathbf{w} = (1, 1, -1)$ of $\mathbf{R}^3$.

6.20 Perform the Gram-Schmidt process on $1, x, x^2$ of $X = L^2(-1, 1)$ under the usual inner product

$$\langle f, g \rangle = \int_{-1}^{1} f(x) g(x)\, dx.$$

6.21 The square-wave signal $v(t) = \operatorname{sgn} \sin \pi t$ (volts) appearing across a 50 Ohm resistive termination is observed for 2 seconds. Find the best approximation v_a in energy (Joules) to the signal $v(t)$ of harmonic form

$$v_a(t) = a_0 + a_1 \cos \pi t + b_1 \sin \pi t + a_2 \cos 2\pi t + b_2 \sin 2\pi t$$

$$+ a_3 \cos 3\pi t + b_3 \sin 3\pi t.$$

Hint: Energy is the time-integral of power $P = VI = V^2/50$ and hence proportional to the square of the L^2 norm of voltage V. Employ Bessel's theorem on regression (Exercise 6.17).

6.22* Prove the **Lemma on Product Domains:** if $\{\phi_n\}_{n=1}^{\infty}$ is an orthonormal basis of $L^2(\Omega)$ and if $\{\psi_n\}_{n=1}^{\infty}$ is an orthonormal basis of $L^2(\Lambda)$, then $\{\phi_m\psi_n\}_{m,n=1}^{\infty}$ is an orthonormal basis of $L^2(\Omega \times \Lambda)$.

For example, $\sin m\pi x \, \sin n\pi y$ is an orthonormal basis for L^2 of the unit square $0 < x, y < 1$, since by Rellich's Theorem, the $\sin n\pi x$ form an orthogonal basis for $L^2(0, 1)$.

Hint: Employ Fubini's Theorem (Appendix A).

6.23* Prove **Gram's theorem on regression**: Given any $n + 1$ elements $f, \phi_1, \phi_2, \phi_3, \ldots, \phi_n$ of the Hilbert space X, the least error

$$\|f - \sum_{i=1}^{n} x_i\phi_i\|$$

will occur exactly when the x_i solve the system of n linear equations

$$\sum_{i=1}^{n} x_i\langle\phi_i, \phi_j\rangle = \langle f, \phi_j\rangle, \quad j = 1, 2, 3, \ldots, n.$$

Hint: Show that the closest approach to f in the subspace spanned by the ϕ_i is the perpendicular projection of f into this subspace.

Deduce Bessel's theorem on regression (Exercise 6.17).

6.24 Find the resolvent at $\lambda = 0$ of $A\phi = \phi''$ on $L^2(0, 1)$ subject to $\phi(0) = 0 = \phi'(0)$.

6.25 Find the resolvent at $\lambda = 1$ of the operator of BVP 3, the Ring Problem.

6.26* Not every reasonable operator has eigenvalues. Prove that the Volterra integral operator V given by

$$(Vf)(x) = \int_0^x f(t)\, dt$$

is a bounded operator on $X = L^2(0, 1)$ that possesses no eigenvalues λ.

Outline: Use Hölder's inequality to show that V is defined on all of X. Next show V is bounded (continuous): $\|Vf\| \leq \|f\|$. Lastly show that $Vf = \lambda f$ is impossible.

6.27 (Lyapunov) Let A be negative semi-definite. Then any trajectory of the system

$$\dot{u} = Au$$

is non-increasing in norm. Thus trajectories that initiate within a ball remain within that ball. The system is *stable in the Lyapunov sense.*

6.28 More strongly than in Exercise 6.27, if $\langle A\phi, \phi\rangle \leq -c\langle \phi, \phi\rangle$ with $c > 0$, then all trajectories exponentially decay to zero. In control language, the system is *exponentially stable.*

6.29 Consider the Laplacian $A\phi = \phi''$ on $X = L^2(0,1)$ subject to zero boundary conditions $\phi(0) = 0 = \phi(1)$. Show that the domain of A contains the function $\phi(x) = 1/4 - (x - 1/2)^2$ if $x \geq 1/2$ and 0 otherwise, by taking in (6.36) its step function preimage $f(x) = 2$ when $x \geq 1/2$ and 0 otherwise. Thus *the domain of A contains functions without 2nd derivatives everywhere!*

6.30 Show that the skew symmetric operator $S\phi = \phi'$ subject to the periodic boundary conditions $S(0) = S(1)$, (Example 2 of §6.2), cannot possess non-zero real eigenvalues in the real Hilbert space $X = L^2(0,1)$. Show that if complex scalars are admitted, all eigenvalues of S are purely imaginary. Find them.

6.31 Show that the spatial operator of BVP 2 (Example 3 of §6.5) is Hermitian and negative definite.

6.32 Find the resolvent of the operator

$$A = \begin{pmatrix} 1 & 0 \\ 0 & 1 \end{pmatrix}.$$

Answer:

$$R(\lambda) = \frac{1}{\lambda - 1}\begin{pmatrix} 1 & 0 \\ 0 & 1 \end{pmatrix}.$$

6.33 Find the resolvent of the operator

$$A = \begin{pmatrix} 1 & 0 \\ 0 & 0 \end{pmatrix}.$$

Answer:

$$R(\lambda) = \frac{1}{\lambda - 1}\begin{pmatrix} 1 & 0 \\ 0 & 0 \end{pmatrix} + \frac{1}{\lambda}\begin{pmatrix} 0 & 0 \\ 0 & 1 \end{pmatrix}.$$

6.34 Find the resolvent of the operator

$$A = \begin{pmatrix} 0 & 1 \\ 0 & 0 \end{pmatrix}.$$

Answer:

$$R(\lambda) = \frac{1}{\lambda}\begin{pmatrix} 1 & 0 \\ 0 & 1 \end{pmatrix} + \frac{1}{\lambda^2}\begin{pmatrix} 0 & 1 \\ 0 & 0 \end{pmatrix}.$$

6.35 Find the resolvent of the operator

$$A = \begin{pmatrix} 0 & -1 \\ 1 & 0 \end{pmatrix}.$$

Answer:

$$R(\lambda) = \frac{\lambda}{\lambda^2+1}\begin{pmatrix} 1 & 0 \\ 0 & 1 \end{pmatrix} + \frac{1}{\lambda^2+1}\begin{pmatrix} 0 & -1 \\ 1 & 0 \end{pmatrix}.$$

6.36 Find the partial fraction expansion of the resolvent of the operator

$$A = \begin{pmatrix} 1 & 4 \\ 3 & 2 \end{pmatrix}.$$

6.37 Compute the resolvent of the spatial operator of BVP 2, i.e., of the operator given in Example 3, §6.5, above.

6.38* For any well-conditioned A with an orthonormal basis of eigenvectors ϕ_n belonging to the eigenvalues λ_n, show the resolvent of A is

$$R(\lambda)f = \sum_{n=1}^{\infty} \frac{\langle f, \phi_n\rangle \phi_n}{\lambda - \lambda_n}$$

for all λ well away from the eigenvalues λ_n.

6.39 Less generally than the previous exercise, if in addition the resolvent is a Hilbert-Schmidt operator

$$(R(\lambda)f)x = \int_\Omega \kappa_\lambda(x, y) f(y)\, dy$$

where $\kappa_\lambda \in L^2(\Omega \times \Omega)$, show that (in the norm sense)

$$\kappa_\lambda(x, y) = \sum_{n=1}^{\infty} \frac{\phi_n(x)\phi_n(y)}{\lambda - \lambda_n}$$

and

$$\|\kappa_\lambda\|^2 = \sum_{n=1}^{\infty} \frac{1}{|\lambda - \lambda_n|^2}.$$

6.40 Verify (6.37).

6.41 Establish (6.42) — that strong time differentiation distributes over such sums.

6.42 Establish (6.43) — that the Laplacian distributes over such sums.

Hint: The domain of $A = \nabla^2$ is the range of its resolvent R.

6.43 Find a differential operator that is positive semi-definite yet not Hermitian.

6.44 Find a Hermitian differential operator that is not semi-definite.

Hint: Try $A\phi = \phi'' + 2\pi^2\phi$ with $\phi(0) = 0 = \phi(1)$. Compute $\langle A\phi, \phi\rangle$ first for $\phi = \sin \pi x$ then for $\phi = \sin 2\pi x$.

6.45 Show that the operator $A\phi = \phi^{(4)}$ subject to $\phi(0) = \phi'(0) = 0 = \phi(1) = \phi'(1)$ is Hermitian positive definite. This operator arises in the Bernoulli-Euler model of transverse vibrations of a clamped-clamped beam — see §8.6.

6.46 Often in mechanical vibrational problems given by $\ddot{u} = Au$, the quantity $-\langle A\phi, \phi\rangle/2$ is (after an adjustment for units) potential energy. Show this holds for the simple undamped mass-spring system $m\ddot{x} = -kx$, then for the longitudinal vibrations of a beam.

6.47 Show that for two differentiable trajectories u and v,

$$\dot{\langle u, v\rangle} = \langle \dot{u}, v\rangle + \langle u, \dot{v}\rangle.$$

6.48 For a dynamical system governed by $\ddot{u} = Au$ with A Hermitian, call

$$E = \langle \dot{u}, \dot{u}\rangle/2 - \langle Au, u\rangle/2$$

total energy. Show energy is conserved along each trajectory. You may assume $(Au)^{\cdot} = A\dot{u}$.

6.49 Prove that the system of BVP 1 is exponentially stable. That is, if $u = u(t)$ is the temperature profile of the rod with initial profile f, then

$$\|u(t)\| \leq e^{\lambda_1 t}\|f\|$$

where $\lambda_1 = -\pi^2$ is the *dominant* eigenvalue. Thus temperature relaxes exponentially to 0 (in the L^2 sense). How general is this phenomenon?

6.50 Prove that an orthogonal projection of $\mathbf{R}^n$ is Hermitian positive semi-definite.

6.51 Find the eigenfunctions and the corresponding eigenvalues of $A\phi = \phi'' - p\phi'$, $(p > 0)$ subject to $\phi'(0) = 0 = \phi(1)$.

Hint: There are three distinct cases: $0 < p < 2$, $p = 2$, $p > 2$.

6.52 Prove that the eigenvalues of a Hermitian operator on a complex Hilbert space are necessarily real.

6.53 Let A be the "under-conditioned" operator $A\phi = \phi''$ on $X = L^2(0,1)$ subject to the (too few) conditions $\phi(0) = 1$. Show that a resolvent $R(\lambda)$ cannot exist for any λ. (This operator breaks the rule *one condition per derivative.*)

6.54 Let A be the "over-conditioned" operator $A\phi = \phi'$ on $X = L^2(0,1)$ subject to the (too many) conditions $\phi(0) = 0 = \phi(1)$. Show that an everywhere defined resolvent cannot be found.

6.55 Show that the (Hermitian) resolvents of the spatial operators of BVP 1 and 2 are both compact.

6.56 Show that a operator A is Hermitian exactly when its resolvent $R(\lambda)$ is Hermitian at some real λ.

6.57 Show that for finite dimensional Hilbert spaces $X = \mathbf{R}^n$, a linear first order problem $\dot{u} = Au$ subject to $u(0) = f$, (where A is an $n \times n$ matrix of constants), has one and only one solution, namely

$$u(t) = \mathrm{e}^{At} f,$$

where

$$\mathrm{e}^{tA} = I + \frac{tA}{1!} + \frac{t^2A^2}{2!} + \frac{t^3A^3}{3!} + \cdots .$$

6.58* Argue that it is reasonable when variables separate to write the solution to $\dot{u} = Au$, $u(0) = f$ in the fundamental form

$$u(t) = e^{At} f,$$

since after all, with respect to the basis ϕ_n of eigenmodes, A is the diagonal matrix

$$A = \begin{pmatrix} \lambda_1 & 0 & 0 & 0 & \cdots \\ 0 & \lambda_2 & 0 & 0 & \cdots \\ 0 & 0 & \lambda_3 & 0 & \cdots \\ & & & & \ddots \end{pmatrix}$$

and the solution $u(t)$ is obtained by applying the matrix

$$\begin{pmatrix} e^{\lambda_1 t} & 0 & 0 & 0 & \cdots \\ 0 & e^{\lambda_2 t} & 0 & 0 & \cdots \\ 0 & 0 & e^{\lambda_3 t} & 0 & \cdots \\ & & & & \ddots \end{pmatrix}$$

to the initial profile $u_0 = f$.

6.59 Suppose Ω is a bounded solid with time and temperature invariant isotropic material properties. Show that the net **entropy** of Ω, when at absolute temperature profile u, is (within an additive constant)

$$S(u) = \int_\Omega c\rho \log u \, dx.$$

Show further that when Ω is insulated away from the outside world, $S(u(t))$ is nondecreasing along each conduction trajectory, i.e.,

$$\frac{d}{dt} S(u(t)) \geq 0.$$

Thus in conduction problems, *the Second Law is a consequence of the heat equation.*

Outline: From (3.5),

$$S(u(t)) = \int \frac{dQ}{T} = \int_0^t \int_\Omega \frac{c\rho u_t}{u} dx\, dt.$$

To see that entropy is nondecreasing, apply Leibniz's rule, the heat equation, then Green's First Identity.

6.60 Prove that in any inner product space, the elements $\phi_1, \phi_2, \ldots, \phi_n$ are independent if and only if the Gram matrix $(\langle \phi_i, \phi_j \rangle)$ is nonsingular.

6.61 Stronger yet, prove that in any inner product space, the elements $\phi_1, \phi_2, \ldots, \phi_n$ are independent if and only if the Gram matrix $(\langle \phi_i, \phi_j \rangle)$ is positive definite.

6.62 Establish the formula (6.37) for the resolvent of the skew symmetric operator S of Exercise 6.30.

6.63 Show that a Hilbert-Schmidt operator (6.38) is self-adjoint exactly when its kernel κ is symmetric, i.e., when $\kappa(x, y) = \kappa(y, x)$.

6.64 Prove that a linear operator R on a Hilbert space X is continuous if and only if it is *bounded,* i.e., for some $b > 0$,

$$\|Rf\| \leq b\|f\|$$

for all f in X. The infimum $\|R\|$ of all such b is called the *norm* of the operator R.

6.65 Deduce from the spectral theorem that a compact Hermitian operator R has an eigenvalue λ at the *norm radius* $|\lambda| = \|R\|$. Conversely, deduce the spectral theorem from this fact.

6.66 Suppose A is a well-conditioned Hermitian negative definite operator with a complete orthonormal basis of eigenvectors $\phi_1, \phi_2, \ldots$ belonging to the eigenvalues $\lambda_1 \geq \lambda_2 \geq \cdots$. Show that the *Rayleigh quotient* is maximized at $\phi = \phi_1$ with maximum value λ_1, i.e.,

$$\sup_{\phi \in dom A} \frac{\langle A\phi, \phi \rangle}{\langle \phi, \phi \rangle} = \lambda_1.$$

Chapter 7
Trigonometric Expansions

In this chapter we see that every periodic signal of finite energy can be written in terms of its component frequencies — as its Fourier series. This expansion is in the Hilbert space sense, not in the classical sense. We will learn and practice the beautiful pointwise convergence results of Fourier and Dini, then touch on the beginnings of signal processing.

§7.1 Sine, Cosine, and Fourier Series

Rellich's Theorem (Chapter 19) guarantees that the eigenfunctions

$$\phi_n(x) = \sin n\pi x, \quad n = 1, 2, 3, \dots\ , \tag{7.1}$$

of BVP 1 form a complete orthogonal basis of $X = L^2(0,1)$ with respect to the ordinary inner product

$$\langle f, g \rangle = \int_0^1 f(x)g(x)dx. \tag{7.2}$$

A simple rescaling yields

Result A. The functions

$$\phi_n = \sin nx, \quad n = 1, 2, 3, \dots, \tag{7.3}$$

form a complete orthogonal basis of $X = L^2(0, \pi)$ under the inner product

$$\langle f, g \rangle = \int_0^\pi f(x)g(x)dx. \tag{7.4}$$

In fact, each square-integrable f has the (unique) *sine series* expansion

$$f(x) = \sum_{n=1}^{\infty} c_n \sin nx \tag{7.5}$$

where

$$c_n = \frac{\langle f, \phi_n \rangle}{\langle \phi_n, \phi_n \rangle} = \frac{2}{\pi} \int_0^\pi f(x) \sin nx \, dx. \tag{7.6}$$

The equality in (7.5) is in the norm sense.

Result B. The functions

$$\psi_n = \cos nx, \quad n = 0, 1, 2, 3, \ldots, \tag{7.7}$$

also form a complete orthogonal basis for $X = L^2(0, \pi)$. Each square-integrable f has the (unique) *cosine series* expansion

$$f(x) = d_0 + \sum_{n=1}^{\infty} d_n \cos nx, \tag{7.8}$$

where

$$d_0 = \frac{\langle f, \psi_0 \rangle}{\langle \psi_0, \psi_0 \rangle} = \frac{1}{\pi} \int_0^\pi f(x) \, dx \tag{7.9a}$$

and

$$d_n = \frac{\langle f, \psi_n \rangle}{\langle \psi_n, \psi_n \rangle} = \frac{2}{\pi} \int_0^\pi f(x) \cos nx \, dx. \tag{7.9b}$$

Proof. The ψ_n are certainly mutually orthogonal (Exercise 7.1). If they are not complete, some f lies in their orthogonal complement, i.e., $\langle f, \psi_n \rangle = 0$ for $n = 0, 1, 2, 3, \ldots$. But then by integrating by parts (Exercise 7.2), for $\phi_n = \sin nx, \;\; n = 1, 2, 3, \ldots,$

$$\langle Vf, \phi_n \rangle = \frac{(-1)^{n+1}}{n} \langle f, \psi_0 \rangle + \frac{1}{n} \langle f, \psi_n \rangle = 0, \tag{7.10}$$

where V is the Volterra operator

$$(Vf)x = \int_0^x f(y) dy. \tag{7.11}$$

But since the $\phi_n = \sin nx$ are complete,

$$Vf = 0,$$

(Exercise 6.15) and hence $f = 0$ since V has no eigenvalues (Exercise 6.26).

Theorem. (Joseph Fourier) The trigonometric functions

$$1;\ \cos x,\ \sin x;\ \cos 2x,\ \sin 2x;\ \cos 3x,\ \sin 3x;\ \ldots \tag{7.12}$$

form a complete orthogonal basis for $X = L^2[-\pi, \pi]$ under the inner product

$$\langle f, g\rangle = \int_{-\pi}^{\pi} f(x)g(x)\,dx. \tag{7.13}$$

More explicitly, every square integrable function f of X possesses a unique expansion into its norm convergent *Fourier series*

$$f(x) = a_0 + \sum_{n=1}^{\infty}(a_n \cos nx + b_n \sin nx), \tag{7.14}$$

where

$$a_0 = \frac{1}{2\pi}\int_{-\pi}^{\pi} f(x)\,dx, \tag{7.15}$$

and where for $n > 0$,

$$a_n = \frac{1}{\pi}\int_{-\pi}^{\pi} f(x)\cos nx\,dx, \tag{7.16}$$

and

$$b_n = \frac{1}{\pi}\int_{-\pi}^{\pi} f(x)\sin nx\,dx. \tag{7.17}$$

Proof. Each f is the unique sum of an even and an odd function

$$f(x) = \frac{f(x) + f(-x)}{2} + \frac{f(x) - f(-x)}{2}. \tag{7.18}$$

Once the even part of f is written as a cosine series and the odd part as a sine series on the half interval $[0, \pi]$, the equalities will extend to the whole interval $[-\pi, \pi]$ (Exercise 7.3).

Example 1. Consider the square wave

$$f(x) = \begin{cases} 0 & \text{if} \quad -\pi \le x < 0 \\ 1 & \text{if} \quad 0 \le x \le \pi \end{cases} \tag{7.19}$$

shown in Figure 7.1. Note that

$$g(x) = f(x) - \frac{1}{2}$$

is odd and hence its Fourier series on $[-\pi, \pi]$ is free of even (cosine) terms giving That (in the norm sense)

$$f(x) = \frac{1}{2} + \sum_{1}^{\infty} b_n \sin nx. \tag{7.20}$$

(Note this not true pointwise at $x = 0$.) Moreover

$$\begin{aligned} b_n &= \frac{1}{\pi}\int_{-\pi}^{\pi} f(x)\sin nx\, dx \\ &= \frac{1}{\pi}\int_{0}^{\pi} \sin nx\, dx = \frac{1-(-1)^n}{n\pi}. \end{aligned} \tag{7.21}$$

Thus $b_{2k} = 0$. But this should have been clear without computation: the odd function f when extended periodically has an odd frequency of 1 cycle per period 2π and hence must consist solely of odd frequencies. (Why? Exercise 7.9). In summary,

$$f(x) = \frac{1}{2} + \frac{2}{\pi} \sum_{n=1, n \text{ odd}}^{\infty} \frac{\sin nx}{n}. \tag{7.22}$$

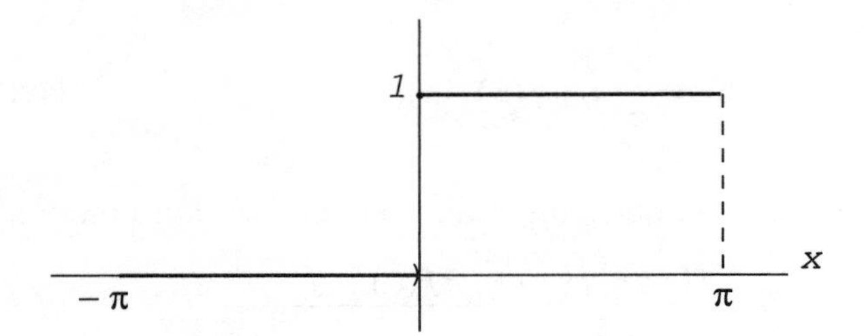

Figure 7.1 The square wave *f(x)*.

Alternatively, recall from BVP 1 that

$$1 = \frac{4}{\pi} \sum_{n=1, n \text{ odd}}^{\infty} \frac{\sin nx}{n} \tag{7.23}$$

on the half interval $[0, \pi]$. Thus when considered on the full interval, the sum of (7.23) must be the limited *Walsh* function

$$h(x) = \begin{cases} -1 & \text{if} \quad -\pi < x < 0 \\ 1 & \text{if} \quad 0 < x < \pi. \end{cases}$$

Merely add 1 and divide by 2 to obtain the expansion (7.22) for f.

Simple rescaling gives general Fourier series expansions on any interval.

Corollary. For f in $X = L^2[-L, L]$,

$$f(x) = a_0 + \sum_{n=1}^{\infty} (a_n \cos \frac{\pi n x}{L} + b_n \sin \frac{\pi n x}{L}) \tag{7.24}$$

where

$$a_0 = \frac{1}{2L} \int_{-L}^{L} f(x)\, dx, \tag{7.25}$$

and for $n > 0$,

$$a_n = \frac{1}{L} \int_{-L}^{L} f(x) \cos \frac{\pi n x}{L}\, dx, \tag{7.26}$$

and

$$b_n = \frac{1}{L} \int_{-L}^{L} f(x) \sin \frac{\pi n x}{L}\, dx \tag{7.27}$$

§7.2 Pointwise Convergence

Let us briefly visit the ill-behaved but interesting classical notion of *pointwise* convergence — a far less practical notion than convergence with respect to energy norms. See Chapters 14, 18, and 19.

For any absolutely integrable function f we may define the *Fourier series* of f as the series

$$a_0 + \sum_{n=1}^{\infty} (a_n \cos nx + b_n \sin nx) \tag{7.28}$$

using the standard formulae (7.15)–(7.18) for the coefficients a_n and b_n.

We say the series (7.28) *converges pointwise* at $x = x_0$ when the sequence of (scalar) partial sums

$$S_N(x_0) = a_0 + \sum_{n=1}^{N} (a_n \cos nx_0 + b_n \sin nx_0)$$

converges.

Kolomogorov has shown that there absolutely integrable functions f whose Fourier series converge nowhere pointwise! The evidence seems to suggest that the Fourier series of a randomly chosen such function f converges pointwise somewhere with probability 0.

On a more positive note, there is Carleson's celebrated 1966 proof showing that the Fourier series of an L^2 square-integrable function f must converge almost everywhere pointwise to f [Katznelson, Carleson].

The following is a far easier yet useful pointwise convergence theorem from the 19th century.

Dini's Criterion. Suppose the difference quotient

$$Q(x) = \frac{f(x) - f(x_0)}{x - x_0}$$

is (absolutely) integrable on $[-\pi, \pi]$. Then the Fourier series of f converges at $x = x_0$ to $f(x_0)$.

Thus the Fourier series of f converges to f at the very least wherever f is differentiable. In contrast $Q(x)$ will not be integrable at step discontinuities.

(The following proof is most important. It details an inversion of the transform $f \longmapsto a_n, b_n$. Variations of this proof justify the inversion formulae for the Mellin, Fourier, Laplace, and other transforms — central to topics as diverse from the distribution of prime numbers to image enhancement).

Proof. Consider the partial sum

$$S_N(x) = a_0 + \sum_{n=1}^{N} (a_n \cos nx + b_n \sin nx). \tag{7.28}$$

Then (Exercise 7.4) this partial sum is given by a convolution

$$S_N(x) = \frac{1}{2\pi} \int_{-\pi}^{\pi} f(y) D_N(x - y) dy \tag{7.29}$$

where

$$D_N(x) = 1 + 2\cos x + 2\cos 2x + 2\cos 3x + \ldots + 2\cos Nx \quad (7.30)$$

is known as the *Dirichlet kernel.* But (Exercise 7.5)

$$D_N(x) = \frac{\sin(2N+1)x/2}{\sin x/2}. \quad (7.31)$$

Moreover it is clear from integrating (7.30) term-by-term that

$$\frac{1}{2\pi}\int_{-\pi}^{\pi} D_N(x)\,dx = 1. \quad (7.32)$$

Hence from (7.29), (7.32), and then (7.31),

$$S_N(x_0) - f(x_0) = \frac{1}{2\pi}\int_{-\pi}^{\pi} [f(x) - f(x_0)]D_N(x_0 - x)\,dx$$

$$= \frac{1}{\pi}\int_{-\pi}^{\pi} \frac{f(x) - f(x_0)}{x - x_0}\,\frac{(x - x_0)/2}{\sin(x - x_0)/2}\sin((N + 1/2)(x - x_0))\,dx. \quad (7.33)$$

By assumption,

$$Q(x) = \frac{f(x) - f(x_0)}{x - x_0},$$

and hence

$$F(x) = \frac{f(x) - f(x_0)}{x - x_0}\,\frac{(x - x_0)/2}{\sin(x - x_0)/2}$$

is absolutely integrable. The desired result $S_N(x_0) \to f(x_0)$ now follows from the following deep Principle discussed further in the next subsection.

The Riemann-Lebesgue Lemma. For any absolutely integrable F,

$$\lim_{\omega\to\infty}\int_{-\pi}^{\pi} F(x)\cos\omega x\,dx = \lim_{\omega\to\infty}\int_{-\pi}^{\pi} F(x)\sin\omega x\,dx = 0. \quad (7.34)$$

Corollary. If both right and left limits $f(x_0^+)$ and $f(x_0^-)$ exist and if both right and left difference quotients

$$Q_+(x) = \frac{f(x) - f(x_0^+)}{x - x_0} \text{ for } x > x_0, \text{ 0 otherwise,}$$

$$Q_-(x) = \frac{f(x) - f(x_0^-)}{x - x_0} \text{ for } x < x_0, \text{ 0 otherwise,}$$

are absolutely integrable, then the Fourier series of f converges at $x = x_0$ to the average

$$\frac{f(x_0^+) + f(x_0^-)}{2}$$

of the right and left hand limiting values of f.

Proof. Exercise 7.6.

Example 2. Reconsider Example 1, the square wave f of Figure 7.1. The right and left hand difference quotients at 0 are identically 0 and hence *a fortiori* integrable. The Fourier series of f must then by the Corollary converge at $x = 0$ to the average of 0 and 1 (which is obvious by direct substitution in (7.22)).

Example 3. Consider $f(x) = x^2$ on $-\pi \leq x \leq \pi$. Because f is even, its representation requires no sine terms and so

$$f(x) = \sum_{n=0}^{\infty} a_n \cos nx, \tag{7.35}$$

where a_0 is the average value

$$a_0 = \frac{1}{2\pi} \int_{-\pi}^{\pi} x^2 \, dx = \frac{\pi^2}{3},$$

and where for $n > 0$ (Exercise 7.7),

$$a_n = \frac{1}{\pi} \int_{-\pi}^{\pi} x^2 \cos nx \, dx = \frac{2}{\pi} \int_{0}^{\pi} x^2 \cos nx \, dx = \frac{4(-1)^n}{n^2}. \tag{7.36}$$

Thus $f(x) = x^2$ has Fourier series

$$x^2 = \frac{\pi^2}{3} + 4 \sum_{n=1}^{\infty} \frac{(-1)^n}{n^2} \cos nx \tag{7.37}$$

where the equality is in the L^2 sense. But by Dini, the series must converge pointwise at both $x = 0$ and $x = \pi$ giving the famous

$$\frac{\pi^2}{12} = 1 - \frac{1}{2^2} + \frac{1}{3^2} - \frac{1}{4^2} + \dots \tag{7.38a}$$

and

$$\frac{\pi^2}{6} = 1 + \frac{1}{2^2} + \frac{1}{3^2} + \frac{1}{4^2} + \dots \; . \tag{7.38b}$$

7.3 Signal Processing

Suppose we observe some time-varying signal $f(t)$ on channel. Suppose this signal is *periodic*, i.e., $f(t+T) = f(t)$ for some T and all t. See Figure 7.2. By rescaling we may as well assume the period $T = 2\pi$. It is physically reasonable to assume that f is a signal of *finite energy* per period, i.e., f is square-integrable over one period, a function of $L^2(-\pi, \pi)$. But then in the L^2 sense, f has and equals its Fourier series

$$f(t) = a_0 + \sum_{n=1}^{\infty} (a_n \cos nt + b_n \sin nt). \tag{7.39}$$

Let us change from the *quadrature* form (7.39) to a form that explicitly displays amplitude, frequency, and phase:

$$f(t) = c_0 + \sum_{n=1}^{\infty} c_n \sin(nt + \theta_n) \tag{7.40}$$

where $c_0 = a_0$ and $c_n = \sqrt{a_n^2 + b_n^2}$ (Exercise 7.8).

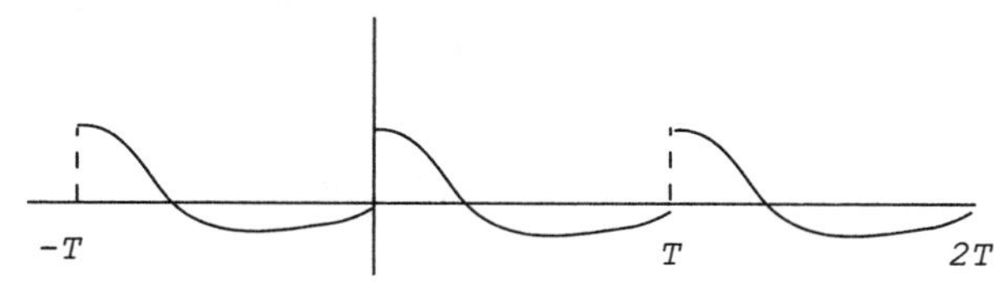

Figure 7.2 A periodic signal *f(t)* of period *T*.

As we have seen for any orthogonal basis, total energy is the sum total of the energies in all the frequencies

$$\|f\|^2 = c_0^2 + \frac{1}{2} \sum_{n=1}^{\infty} c_n^2. \tag{7.41}$$

This spectral power density distribution is observable! If a periodic voltage $f(t)$ is applied against ground at one end of a long wire antenna, a selective radio receiver will detect the individual frequency components $\sin(nt + \theta_n)$ occurring with amplitude c_n. *A signal when broadcast breaks into its Fourier series.*

Filtering and enhancement of signals is achieved by a manipulation of the values of the Fourier coefficients a_n and b_n — by hardware or nowadays by software via the *Fast Fourier Transform.* One standard reference of this subject is Papoulis's *enceinte* book *Signal Analysis.*

The mathematical power of Dini's Criterion for pointwise convergence of Fourier series stems from the **Riemann-Lebesgue Lemma** given in (7.34). We have already seen in Exercise 6.18 a more abstract version: in any Hilbert space X, if $\phi_1, \phi_2, \phi_3, \ldots$ is an infinite orthonormal sequence and f any element, then

$$\lim_{n\to\infty} \langle f, \phi_n \rangle = 0, \tag{7.42}$$

an easy consequence of Bessel's inequality. In our case, here in $X = L^2(-\pi, \pi)$, (7.42) translates to

$$\lim_{n\to\infty} a_n = \lim_{n\to\infty} b_n = 0. \tag{7.43}$$

But this is not quite what is needed in Dini's Criterion. Instead we need the 'smear' version (7.34) at non-integral frequencies for absolutely integrable functions f, a larger class than L^2. We assume this stronger version as an unproved Principle [Rudin]. See Exercise 7.13. This Principle, put as an epigram, translates to

absolutely integrable signals are asymptotically bandlimited.

Exercises

7.1 Show that the functions $\psi_n = \cos nx$ are mutually orthogonal in $L^2(0, \pi)$ under the usual inner product (7.4).

7.2 Establish (7.10) by integrating by parts.

7.3 Prove that if the equality

$$f = a_0 + \sum_{n=1}^{\infty} a_n \cos nx$$

holds (norm-wise) in $L^2(0, \pi)$ for the even function f, then the equality also holds in $L^2(-\pi, \pi)$. State and prove the analogous statement for odd functions.

7.4 Prove (7.29), that each partial sum S_N of the Fourier series of f is given by convolution of f with the corresponding Dirichlet kernel D_N.

7.5 Prove (7.31), the explicit formula for the Dirichlet kernel.

Hint: Sum $1 + z + z^2 + \ldots + z^N$ with $z = e^{ix}$.

7.6 Prove the Corollary to Dini's Criterion on convergence to the average of left and right hand limits.

7.7 Prove (7.36) and (7.38).

7.8 Show that the standard *quadrature* form of a Fourier series (7.39) can be re-written in the *amplitude-frequency-phase* form (7.40).

7.9 Argue without computation that the Fourier series of the odd function f of Example 1 will have only odd frequency components.

7.10 Find the Fourier series of the sawtooth wave $f(x) = x, \quad -\pi \leq x \leq \pi$.

7.11 Deduce from Dini's Criterion that the Fourier series of an everywhere differentiable periodic function f must converge at each point x to the value $f(x)$.

7.12 If f is twice continuously differentiable, its Fourier series converges in the L^∞ sense, i.e., the series converges uniformly to f.

Hint: Integrate twice by Parts.

7.13 Prove the Riemann-Lebesgue Lemma (7.34) for the special case that F is piecewise continuously differentiable.

Hint: Integrate by parts.

7.14 Prove

$$\sum_{n=1}^{\infty} \frac{\sin n}{n} = \frac{\pi - 1}{2} = \sum_{n=1}^{\infty} (\frac{\sin n}{n})^2.$$

Outline: Obtain the sine series

$$\frac{\pi - x}{2} = \sum_{n=1}^{\infty} \frac{\sin nx}{n},$$

then apply Dini's Criterion. Integrate term-by-term from 0 to 2.

7.15 What is the resulting spectral power density when a single tone amplitude modulates a carrier? That is, find the Fourier series of

$$f(t) = (1 + \delta \sin \alpha t) \sin \omega t.$$

Assume $\alpha N = \omega$ for N a very large integer.

7.16 Attempt to discover the power spectral density when a single tone phase modulates a carrier. That is, find the Fourier series of

$$f(t) = \sin(t\omega - \delta \sin \alpha t).$$

Assume $\omega = N\alpha$ where N is a huge integer.

Answer: (See §9.)

$$f(t) = \sum_{1}^{\infty} (J_{N-n}(\delta) - J_{N+n}(\delta)) \sin n\alpha t.$$

7.17 Numerically investigate *Gibb's phenomenon* by graphing partial sums $S_N(x)$ of (7.22), the Fourier series for the step function $f(x)$ of Figure 7.1. Study the oscillatory nature of $S_N(x)$ near $x = 0$.

7.18 Show

$$\frac{\pi^2}{8} = 1 + \frac{1}{3^2} + \frac{1}{5^2} + \cdots.$$

7.19 Does the Fourier series of $f(x) = |x|^{1/2}$ converge at $x = 0$?

7.20 Does the Fourier series of $f(x) = |x|^{-1/2}$ converge at $x = 0$?

Chapter 8

Rectangular Problems

Let us pose and solve several problems whose underlying spatial domain Ω is rectangular, thus yielding trigonometric eigenmodes.

§8.1 Quenching a Block (BVP 7)

A homogeneous isotropic rectangular block, initially at a uniform temperature, is dropped into bath. Solve for the interior temperatures as they relax to the temperature of the bath.

After rescaling the problem becomes

$$\frac{\partial u}{\partial t} = \nabla^2 u, \tag{8.1}$$

$$0 < x < a, \quad 0 < y < b, \quad 0 < z < c,$$

subject to

i) $u(0, y, z, t) = 0 = u(a, y, z, t)$,

ii) $u(x, 0, z, t) = 0 = u(x, b, z, t)$, and

iii) $u(x, y, 0, t) = 0 = u(x, y, c, t)$ when $t > 0$, where initially

iv) $u(x, y, z, 0) = 1$.

Assume separation, that $u(x, y, z, t) = T(t)X(x)Y(y)Z(z)$, and hence

$$\frac{\dot{T}}{T} = \frac{X''}{X} + \frac{Y''}{Y} + \frac{Z''}{Z} = \lambda \tag{8.2}$$

where the prime $'$ denotes, in a common abuse of notation, the partial derivative with respect to the appropriate single variable. Applying the zero boundary conditions i)–iii) yields

$$X_k = \sin k\pi x/a,$$

$$Y_l = \sin l\pi y/b,$$

$$Z_m = \sin m\pi z/c,$$

$$T_{klm} = e^{-((k/a)^2+(l/b)^2+(m/c)^2)\pi^2 t}$$

with general solution u

$$= \sum_{k,l,m=1}^{\infty} d_{klm} e^{-((k/a)^2+(l/b)^2+(m/c)^2)\pi^2 t} \sin k\pi x/a \ \sin l\pi y/b \ \sin m\pi z/c. \tag{8.3}$$

Imposing the initial condition (iv), after multiplying both sides of (8.3) by one term and integrating, we obtain in the usual way that

$$d_{klm} = \frac{8}{klm\pi^3}(1-(-1)^k)(1-(-1)^l)(1-(-1)^m),$$

and thus the expansion requires only odd frequency spatial modes.

Examine the underpinnings of the above computation. By Rellich's Theorem (§6.7) we know that the spatial eigenmodes of this Laplacian constitute a complete orthogonal basis. The separated $X_k Y_l Z_m$ are indeed eigenmodes of ∇^2. By the Lemma on Product Domains (Exercise 6.22), an orthogonal basis for a spatial domain Ω that is itself a cartesian product can be obtained by multiplying orthogonal bases for the spatial factors. Thus the separated eigenmodes $X_k Y_l Z_m$ form the complete list of the eigenfunctions of the Laplacian ∇^2 on this rectangular domain with zero boundary conditions. Moreover the sum

$$\sum_{k,l,m=1}^{\infty} d_{klm} \lambda_{klm} e^{\lambda_{klm} t} \tag{8.4}$$

where $\lambda_{klm} = -\pi^2((k/a)^2 + (l/b)^2 + (m/c)^2)$ is continuous for $t > 0$. Therefore by the methods of §6.7, (8.3) is the (unique) strong solution of our problem (8.1).

§8.2 Deep Earth Temperatures (BVP 8)

At what depth can a submarine detect radio transmissions? To what depth within a conductor do rapidly varying voltages at the surface penetrate? Why is deep subsurface earth temperature equal to the average yearly air temperature? Are these all instances of the same mathematical phenomenon? Let us first tackle the conceptually simple problem of deep earth temperatures.

Assume that below some far depth $x = L$ within the earth, vertical heat transfer ceases. (We will later resolve this problem more persuasively without this assumption, which is after all a portion of the desired conclusion). Assume also — after rescaling — that air temperature varies sinusoidally over one year about the average value of 0. We will see in Exercise 8.7 that daily variations can be neglected.

We now have two superimposed one dimensional diffusion problems: a rod initially at 0 with left end temperatures varying sinusoidally about 0, and a rod initially at some temperature profile with the left end brought suddenly and held to 0; the right end in both cases is insulated. But the second problem decays to a zero steady state. We have thus obtained some justification for the observed fact that deep ground temperature equals annual average air temperature.

But how do surface variations effect subsurface temperatures? This is the problem posed as Exercise 8.2.

Presumably the transients from the initiation of the present postglacial climate have died away. Then as deduced in Exercise 8.5, the periodic steady state temperature $u(x, t)$ at depth x as one might intuit is of the form

$$u(x, t) = X(x) \sin(\omega t - \theta(x)) \tag{8.5}$$

when driven by a sinusoidal surface temperature $u(0, t) = \sin \omega t$. It remains to determine how the amplitude $X(x)$ of the subsurface variations decrease with depth x.

Because we are dealing with a *periodic steady state* problem, let us employ a trick associated with the name Helmholtz. Rewrite (8.5) in *phasor* (*quadrature*) form

$$\begin{aligned} \rho(x, t) &= u(x, t + \pi/2\omega) + iu(x, t) \\ &= e^{i(\omega t - \theta(x))} X(x) = e^{i\omega t} Y(x), \end{aligned} \tag{8.6}$$

keeping in mind that our temperature u is the imaginary part of ρ and that the real part of ρ is merely u anticipated by $\pi/2\omega$ time units. By this standard trick we have decoupled time from space.

Imposing the heat equation on both the real and imaginary parts of ρ, we see that

$$\dot{\rho} = \alpha\rho'',$$

i.e.,

$$\alpha Y'' = i\omega Y \tag{8.7}$$

where α is the diffusivity of the soil. Thus

$$Y = c_1 e^{-x\zeta\sqrt{\omega/\alpha}} + c_2 e^{x\zeta\sqrt{\omega/\alpha}} \tag{8.8}$$

where ζ is the principal square root of i, *viz.*, $\zeta = (1+i)/\sqrt{2}$. On physical grounds we would infer that $c_2 \approx 0$ since temperature variations fall off with depth. Thus since $Y(0) = X(0) = 1$, $c_1 \approx 1$. Therefore

$$Y(x) \approx e^{-x\sqrt{\omega/2\alpha}} e^{-ix\sqrt{\omega/2\alpha}} \tag{8.9}$$

yielding the famous *skin effect* formulae

$$X(x) \approx e^{-x\sqrt{\omega/2\alpha}} \text{ (amplitude)} \tag{8.10a}$$

and

$$\theta(x) \approx x\sqrt{\omega/2\alpha} \text{ (phase lag).} \tag{8.10b}$$

More carefully, imposing the boundary conditions $\rho(0,t) = e^{i\omega t}$ and $\rho_x(L,t) = 0$, we can explicitly compute c_1 and c_2 (Exercise 8.6) to see that

$$|Y(x) - e^{-x\sqrt{\omega/2\alpha}} e^{-ix\sqrt{\omega/2\alpha}}| < \frac{2}{e^{L\sqrt{\omega/2\alpha}} - 1}. \tag{8.11}$$

But in this problem (unlike the next) L can be chosen arbitrarily large. Therefore the above skin effect formulae (8.10) are not only approximate but exact for subsurface temperatures.

It is traditional to define the *depth of penetration* as

$$\delta = \frac{\sqrt{2\alpha}}{\sqrt{\omega}} = \frac{\sqrt{\alpha}}{\sqrt{\pi f}} \tag{8.12}$$

where $f = \omega/2\pi$ is the frequency of the surface temperature. This depth of penetration $x = \delta$ is where the influence at the surface has diminished to $e^{-1} \approx 0.37$ of its original amplitude. Note that higher frequencies penetrate to lesser depth, inversely with the square root. Daily fluctuations are negligible — see Exercise 8.7.

Small deviations from these results were observed beneath the permafrost of Alaska [Lachenbruch and Marshall]. Transient analysis has added to the mounting evidence of a global warming trend [Lewis].

§8.3 Current within a Flat Conductor (BVP 9)

Suppose a sinusoidally varying voltage of frequency f is applied at one end of a flat solid conductor with a rectangular cross-section, say at one end of a printed circuit board trace, or say the silver-plated flat copper ribbon used in building the inductors in high power radio frequency amplifiers. Assume all initial transients have died out. Disregard effects at the edges of this ribbon conductor.

By manipulation of the four laws of Maxwell (Exercise 8.8), the induced (vector) electric field intensity E within the conductor must satisfy

$$\mu\epsilon\frac{\partial^2 E}{\partial t^2} + \mu\sigma\frac{\partial E}{\partial t} = \nabla^2 E - \nabla\nabla \cdot E. \tag{8.13}$$

By the very definition of a conductor, there is no free charge ($\nabla \cdot \epsilon E = 0$) and the permittivity $\epsilon \approx 0$. (Consult the classic *Fields and Waves in Modern Radio* by Ramo and Whinnery). Thus (8.13) becomes

$$\mu\sigma\frac{\partial E}{\partial t} = \nabla^2 E. \tag{8.14}$$

Moreover because conductivity σ is so high, there is negligible tilt of the electric intensity E towards the midplane of the conductor — E is essentially parallel to the surface and points down the length of the conductor. Let then $u(x, y, t)$ be the component of E pointing lengthwise down the conductor at some fixed distance y from the source, where x is depth measured down from one surface. By symmetry, u_x must vanish at the midplane of the conductor. From (8.14),

$$\mu\sigma\frac{\partial u}{\partial t} = \nabla^2 u. \tag{8.15}$$

Because the wavelength is typically far longer than the thickness of the conductor, the Laplacian term $\partial^2 u/\partial y^2$ can be neglected. So as in ground temperatures (8.11), the electric intensity falls off with depth according to the approximate exponential rule

$$\textit{relative amplitude at depth } x \approx e^{-x\sqrt{\pi f \mu \sigma}} \tag{8.16}$$

good within the error given in (8.11), with depth of penetration

$$\delta = \frac{1}{\sqrt{\pi f \mu \sigma}}. \tag{8.17}$$

Because current density $J = \sigma E$, the penetration of current also obeys the rules (8.16) and (8.17).

For example, the current of a 100 MHz signal will penetrate only 0.0066 mm into a flat copper conductor! (Exercise 8.9).

Because of this *skin effect*, radio transmissions to submarines are sent at the extremely low frequency of 75 Hz and even so, are detectable only above 100 meter depths. Intersymbol distortion problems limit signaling to one bit per second throughput. See Exercise 1.31.

§8.4 A Trapped Quantum Particle (BVP 10)

Because of a fundamental uncertainty in the very act of taking measurements, the dynamics of a small particle of mass m roaming somewhere within the spatial domain Ω is governed by *Schroedinger's Equation*

$$i\hbar\frac{\partial \psi}{\partial t} = -\frac{\hbar^2}{2m}\nabla^2\psi + V\psi \qquad (8.18),$$

where ψ is the *wave function* of the particle, ψ a function of $L^2(\Omega)$ of norm 1. An observation (measurement) of a property of the particle is taken by an instrument whose mathematical analog is a Hermitian operator, say A. The resulting measurement on average is then the expected value $\langle A\psi, \psi\rangle$. For example the expected position of the particle is $q = \langle x\psi, \psi\rangle$ while kinetic and potential energy equal $-\frac{\hbar^2}{2m}\langle\nabla^2\psi, \psi\rangle$ and $\langle V\psi, \psi\rangle$ respectively. See Appendix B.

In dimensionless variables Schroedinger's Equation becomes

$$i\frac{\partial \psi}{\partial t} = -\nabla^2\psi + V\psi. \tag{8.19}$$

The instrument V that observes potential energy is always given as simple multiplication by the classical potential function $a(x)$: $V\psi = a\psi$.

Suppose our particle is confined to lie somewhere within a bounded domain Ω because of a potential $a(x)$ that is exceedingly large outside Ω; assume the wave function ψ vanishes outside Ω. Suppose further that the potential function $a(x)$ vanishes within Ω. Think of a particle within an otherwise empty cube. We have a 'trapped quantum particle' whose energy is all kinetic.

But then Schrodinger's Equation (8.19) simplifies to

$$\frac{\partial\psi}{\partial t} = i\nabla^2\psi, \tag{8.20}$$

where the Laplacian has zero boundary conditions. Therefore by Rellich's Theorem the wave function has the orthogonal expansion

$$\psi = \sum_{n=1}^{\infty} c_n e^{i\lambda_n t}\psi_n(x) \tag{8.21}$$

where $\nabla^2\psi_n = \lambda_n\psi_n$, subscripted so that

$$0 < -\lambda_1 \leq -\lambda_2 \leq -\lambda_3 \leq \dots \ .$$

The $E_n = -\lambda_n$ are the only allowable and possible energy states of the particle. This has been observed — substantiated by experiment. The normalized spatial eigenmodes ψ_n are called *stationary states.* Stationary states belonging to the dominant eigenvalue λ_1 (least energy state $E_1 = -\lambda_1$) are the *ground states.* See the delightful *Quantum Mechanics* by P.C.W. Davies, [Gillespie], or the classic [Pauline and Wilson].

§8.5 Quantum Tunneling (BVP 11)

A simple *p-n* junction diode overdoped with impurities exhibits an odd property: as forward bias voltage is applied, current will flow, at first increasing monotonically with voltage, but then as the applied voltage is increased even further, current flow will *decrease.* See Figure 8.1. This region of *negative resistance* has been exploited in microwave oscillator circuits where such a diode is the active element in a cavity resonator.

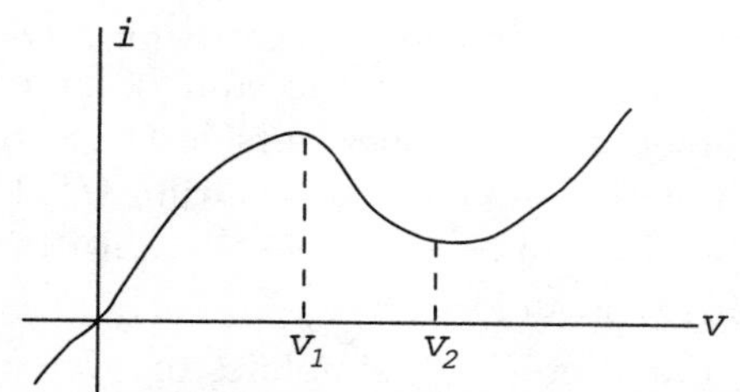

Figure 8.1 A tunnel diode exhibits *negative resistance* on certain intervals of forward bias $v_1 < v < v_2$.

The diode current is actually the superposition of three currents: *excess*, *diffusion*, and *tunneling (Esaki)* current as depicted in Figure 8.2. The ordinary diffusion and excess currents will flow once a threshold is reached — the *barrier potential* created when the majority carriers combine at the interface of the p- and n-materials. Explaining the tunnel current requires a foray into Quantum Mechanics.

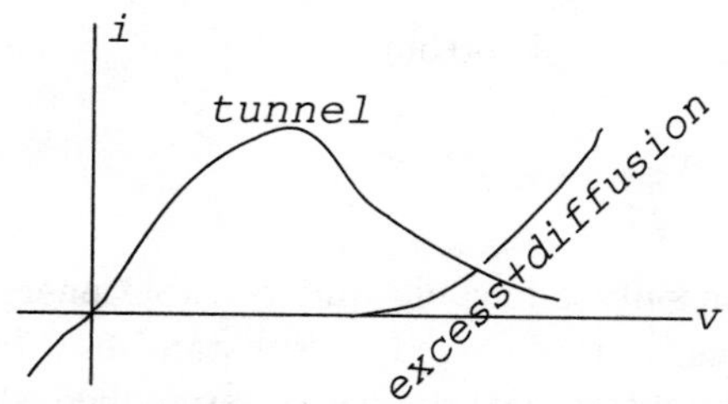

Figure 8.2 The curent through a tunnel diode is the sum of three currents: *tunnel, excess,* and *diffusion.*

Model the diode as the line segment $[-\pi, \pi]$ with the n-material lying between $-\pi \leq x < 0$ and the p-material at $0 < x \leq \pi$. As the *first* approximation let us assume the forward bias induces a net potential that is a vertical step up at $x = 0$. That is, potential energy is observed by an instrument V that is multiplication by

$$a(x) = \begin{cases} \infty & \text{if} \quad x \leq -\pi \\ 0 & \text{if} \quad -\pi < x < 0 \\ a_0 & \text{if} \quad 0 \leq x < \pi \\ \infty & \text{if} \quad \pi \leq x \end{cases} . \tag{8.22}$$

See Figure 8.3.

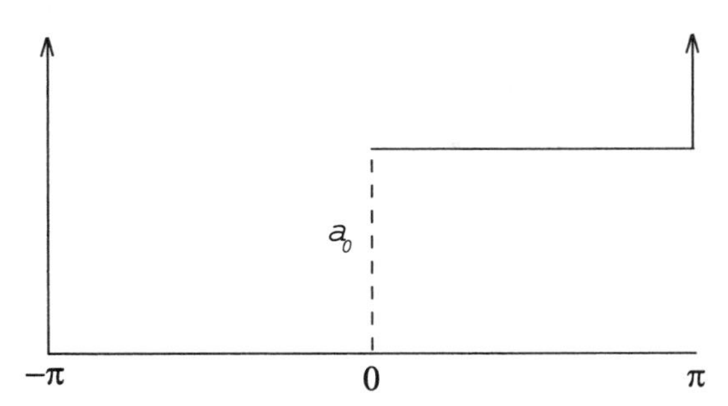

Figure 8.3 In the first approximation, forward biasing a tunnel diode induces a step increase in potential at the interface of the *n*- and *p*-regions.

Classical mechanics would rule that an electron with kinetic energy $E < a_0$ in the n-region $(-\pi, 0)$ would not be able to surmount the potential barrier at $x = 0$ to enter the p-region $(0, \pi)$. But as discussed in BVP 10, the electron obeys instead Schroedinger's

$$i\frac{\partial \psi}{\partial t} = -\nabla^2\psi + V\psi.$$

Let us solve for the wave function ψ, state by stationary state ψ_n.

We search for a stationary state ψ of energy E, i.e., an eigenfunction ψ of the Schroedinger operator

$$H = -\nabla^2 + V$$

belonging to the value E,

$$-\nabla^2\psi + V\psi = E\psi,$$

which translates in this case to

$$\psi'' + (E - a)\psi = 0, \qquad a = a_0 \text{ or } 0. \tag{8.23}$$

Solve (8.23) in the n- and p-region separately to obtain

$$\psi = \begin{cases} \sin\sqrt{E}(x+\pi) & \text{for} \quad -\pi < x < 0 \\ b\sinh\sqrt{a_0 - E}(x-\pi) & \text{for} \quad 0 < x < \pi \end{cases}. \tag{8.24}$$

Keep in mind that the wave function ψ must eventually be normalized to norm 1.

Once the resolvent of H is found to be compact and Hermitian at $\lambda = 0$ and in fact a Hilbert-Schmidt operator (Exercise 8.11), it is guaranteed that stationary states ψ as well as their derivatives ψ' are (absolutely) continuous. Physically speaking, probability current and momentum are conserved.

Insisting that the stationary state ψ and its derivative ψ' be continuous across $x = 0$ leads (Exercise 8.12) to the eigenvalue (*characteristic*) equation

$$-\frac{\tan \pi\sqrt{E}}{\sqrt{E}} = \frac{\tanh \pi\sqrt{a_0 - E}}{\sqrt{a_0 - E}}, \tag{8.25}$$

and the concomitant value of the fitting constant

$$b = -\frac{\sin \pi\sqrt{E}}{\sinh \pi\sqrt{a_0 - E}}. \tag{8.26}$$

In the usual way, find the real positive solutions E of the eigenvalue equation (8.25) by superimposing the graphs of the left- and righthand side as functions of E. For small values of barrier height, say $a_0 = 0.4$, there will be no solutions $E < a_0$ of (8.25) — see Figure 8.4a. For larger heights, say $a_0 = 3$, there will be several solutions $E < a_0$ — see Figure 8.4b.

In summary, for low barrier heights, no current will flow unless entering kinetic energy exceeds the barrier potential. But counter to intuition, for higher potential barriers, the first few lowest stationary states will *tunnel* into the p-region even though their kinetic energy is classically insufficient to climb the barrier.

For example, if $a_0 = 3$, the two lowest stationary states will tunnel. Their wave functions are shown in Figure 8.5.

The actual real-world tunnel diode is far more physically and mathematically complex. As the majority carriers combine at the junction, a narrow *depletion region* $(-\epsilon, \epsilon)$ is created where the potential ramps up rather than rising vertically. Moreover current flow is determined by the acceptors available across the barrier at the same energy. As forward bias is increased, the depletion region widens and the net barrier potential drops until tunneling is extinguished and all states must leap. Solving for and matching the wave functions in the three regions — especially the depletion region with its ramp potential — requires computations of horrendous complexity. See [Zener] or [Wentzel-Kramers-Brillouin]. To sample this complexity, attempt Exercise 8.14.

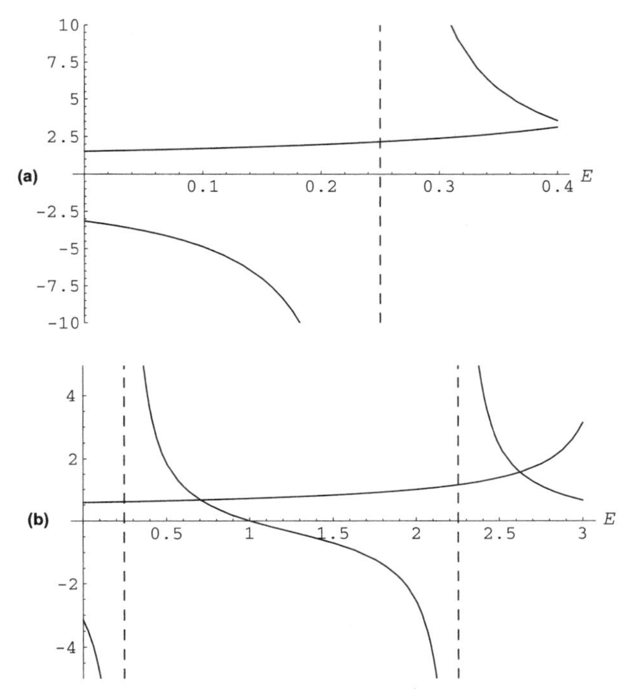

Figure 8.4 The graphs of the right- and left-hand side of the eigenvalue equation (8.25) are superimposed. For the low barrier potential $a_0 = 0.4$, (Figure 8.4a) no tunneling will occur: there is no solution $E < a_0$. For the higher barrier potential (Figure 8.4b) $a_0 = 3$, tunneling will occur: there are two solutions $E < a_0$.

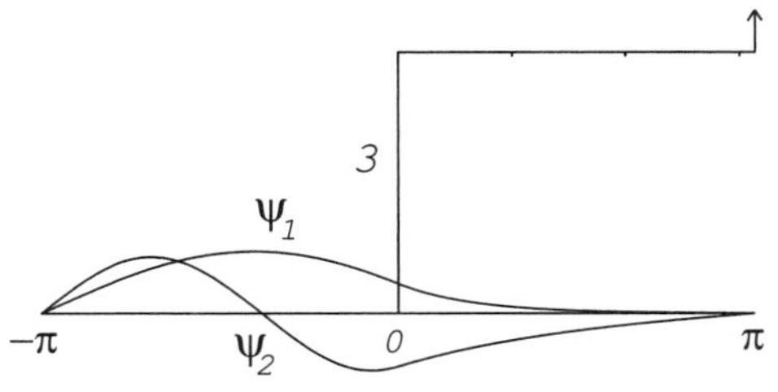

Figure 8.5 For a barrier height of $a_0 = 3$, the first two stationary states ψ_1 and ψ_2 will tunnel. They are shown with the step barrier potential superimposed. Their pre-tunneling kinetic energies E_i are *0.704* and *2.624* resp.

§8.6 Transverse Vibrations of the Beam (BVP 12)

In the first approximation a long thin steel beam can be thought of as a system of infinitesimal springs and masses as in Figure 4.2. Or think of a composite bundle of parallel elastic fibers. We saw in Exercise 4.1 that the restoring force of each fiber is proportional to $\partial^2 u/\partial x^2$ and hence longitudinal displacement satisfies the wave equation. But what about *transverse* vibrations, vibrations normal to the unbent beam?

Any bending will stretch the fibers on the outside of the bend while compressing the fibers on the inside of the bend, leaving an intermediate *neutral axis* of unstressed fibers. But there are shear forces on the 'glue' holding together the horizontal fibers, rotational inertia, etc. A complete model of a beam is difficult to obtain — see [Meirovich]. For small deflections, the **Bernoulli-Euler model** has been found accurate for many applications: if $u(x,t)$ is vertical deflection from rest of a horizontal beam,

$$m(x)\frac{\partial^2 u}{\partial t^2} + \frac{\partial^2}{\partial x^2}(EI(x)\frac{\partial^2 u}{\partial x^2}) = f(x,t) \qquad (8.27)$$

where $m(x)$ is the mass density per unit length, where $EI(x)$ is the flexural rigidity density, and where $f(x,t)$ is an external deflecting force.

Three of the most useful boundary conditions at the left end $x = 0$ are

a) *clamped (fixed)*: $u(0,t) = u_x(0,t) = 0$,

b) *pinned (hinged)*: $u(0,t) = u_{xx}(0,t) = 0$, or

c) *free (floating)*: $u_{xx}(0,t) = u_{xxx}(0,t) = 0$.

Analogous conditions can be imposed at the right end as well. Consult [Meirovitch] for the physical intuition behind these boundary conditions.

Fact. The operator $\nabla^4 = \partial^4/\partial x^4$ on $L^2(0,1)$ is self-adjoint, positive semidefinite, with compact resolvent for each of the nine combinations of boundary conditions a)–c) above.

Proof. Exercise 8.15.

Let us solve one of the 9 standard problems: *find all possible vibrations of a freely vibrating cantilevered homogeneous beam* (Figure 8.6).

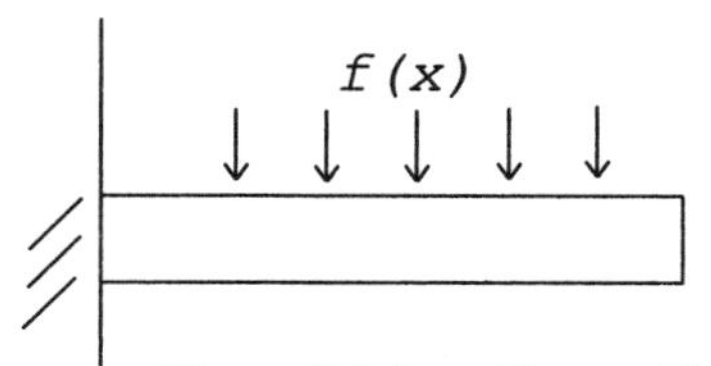

Figure 8.6 A cantilevered beam welded to a vertical wall.

The nondimensional model is

$$\frac{\partial^2 u}{\partial t^2} = -\frac{\partial^4 u}{\partial x^4} \tag{8.28}$$

on $0 < x < 1$ subject to

i) $u(0,t) = 0 = u_x(0,t)$,

ii) $u_{xx}(1,t) = 0 = u_{xxx}(1,t)$,

iii) $u(x,0) = f(x)$,

iv) $u_t(x,0) = 0$,

which is, as the Rule of Thumb dictates, one condition for each of the 6 derivatives taken in the model.

We search for eigenfunctions of the differential operator

$$A\phi = \phi'''' \tag{8.29}$$

and find that

$$\phi = a\cos\beta x + b\sin\beta x + c\cosh\beta x + d\sinh\beta x \tag{8.30}$$

with eigenvalue $\lambda = \beta^4$.

The clamped conditions at the left end require (Exercise 8.16) the eigenfunctions to be of the form

$$\phi = \cosh\beta x - \cos\beta x + \mu(\sinh\beta x - \sin\beta x). \tag{8.31}$$

Imposing the right end free conditions yields (Exercise 8.17) the characteristic equation

$$\cos\beta = \frac{-1}{\cosh\beta} \tag{8.32}$$

together with the ancillary

$$\mu = -\frac{\cosh\beta + \cos\beta}{\sinh\beta + \sin\beta}. \tag{8.33}$$

Superimposing the graphs of the left- and righthand sides of (8.32) yields an increasing sequence of positive β_n of asymptotic value $\beta_n \sim n\pi/2$, n odd.

The fundamental vibrational frequencies of the clamped-free beam are therefore $\omega_n = \beta_n^2$, $n = 1, 2, 3, \ldots$, with general orthogonally decoupled displacement

$$u = \sum_{n=1}^{\infty}(a_n \cos\omega_n t + b_n \sin\omega_n t)\phi_n, \tag{8.34}$$

where the spatial eigenfunctions ϕ_n are given by (8.31–8.33). Imposing the initial conditions iii)–iv) gives

$$a_n = \langle f, \phi_n\rangle / \langle \phi_m, \phi_n\rangle \text{ and } b_n = 0.$$

§8.7 Rectangular Waveguides (BVP 13)

At microwave frequencies it is no longer advisable to use two-wire transmission lines because of the ohmic loss in the two conductors, the loss in the dielectric used to separate them, and radiation from the line itself. (See Exercise 8.21). Instead radio frequency energy is transported from place to place in *waveguides*, piped like a fluid, within (usually) rectangular metal tubing.

Within the waveguide, the Equations of Maxwell become (Exercise 8.8)

$$\begin{gathered} \nabla\cdot E = 0 \qquad \nabla\cdot H = 0 \\ \nabla\times E = -\mu\frac{\partial H}{\partial t} \qquad \nabla\times H = \epsilon\frac{\partial E}{\partial t} \end{gathered} \tag{8.35}$$

with the corollaries

$$\nabla^2 E = \mu\epsilon\frac{\partial^2 E}{\partial t^2} \tag{8.36a}$$

and

$$\nabla^2 H = \mu\epsilon\frac{\partial^2 H}{\partial t^2}. \tag{8.36b}$$

Because of the high conductivity of the waveguide walls, the electric field intensity E will ideally have no component tangent to the waveguide walls. Hence by Stokes (Exercise 8.31) there is no flux of the magnetic field intensity H leaving through the walls. In short, *E and H have tangential and normal zero boundary conditions respectively.*

Suppose we are presented with a rectangular waveguide of horizontal width a and height b as in Figure 8.7. Let us assume $b < a$. (Almost universally in practice $b \approx a/2$.) As we now deduce, the waveguide is incapable of carrying microwaves of free space wavelength λ longer than $\lambda_c = 2a$. *The longest dimension a determines the cutoff frequency: $f_c = 1/2a\sqrt{\mu\epsilon}$.* However the waveguide is capable of supporting a bewildering array of higher frequencies and propagation modes [Ramo and Whinnery]. *Whatever Maxwell allows, will occur.*

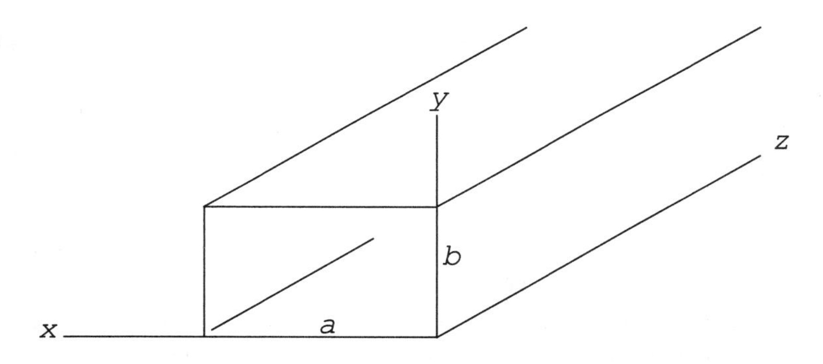

Figure 8.7 A rectangular waveguide extending to infinity.

Suppose the waveguide is carrying a sinusoidal signal of frequency $f = \omega/2\pi$. Let us assume that the waveguide is infinite in extent or so well matched at the termination that there is only outward propagating energy. Assume transients have long since dissipated. Then it is intuitively clear that a snapshot will reveal E spatially repeats periodically down the waveguide. Subsequent snapshots will reveal that this pattern is traveling intact down the tube with some velocity ν. Thus the electric field intensity is of the form

$$E = \sin(\omega t - \gamma z)\ [u(x,y)\mathbf{i} + v(x,y)\mathbf{j} + w(x,y)\mathbf{k}\,] \tag{8.37}$$

with $\gamma = \omega/\nu$.

Because each component satisfies the wave equation (8.36a), in particular (Exercise 8.22)

$$\nabla^2 w = -(\mu\epsilon\omega^2 - \gamma^2)w. \tag{8.38}$$

Hence $-\mu\epsilon\omega^2 + \gamma^2$ is an eigenvalue of the Laplacian on the rectangular cross section of the waveguide $0 < x < a$, $0 < y < b$, with zero boundary conditions (since the electric field has zero tangential boundary conditions). But the eigenfunctions of this problem (8.38) are

$$\phi_{mn} = \sin m\pi x/a \ \sin n\pi y/b,$$

with dominant mode $m = n = 1$ giving that

$$\mu\epsilon\Omega^2 - \gamma^2 \geq \frac{\pi^2}{a^2} + \frac{\pi^2}{b^2},$$

and hence

$$\mu\epsilon\omega^2 > \frac{\pi^2}{b^2}. \tag{8.39}$$

But $2\pi f = \omega$ and $f\lambda = c = 1/\sqrt{\mu\epsilon}$. Thus

$$\lambda < 2b. \tag{8.40}$$

Therefore *unless the mode is transverse electric, the longest wavelength carried by the waveguide is determined by the shortest dimension b.*

Assume then that the wave is *transverse electric*, where the electric field is transverse to the direction of propagation:

$$E = \sin(\omega t - \gamma z)\ [u(x,y)\mathbf{i} + v(x,y)\mathbf{j}\ ]. \tag{8.41}$$

Then as in (8.38) above, imposing the wave equation (8.36) yields the two eigenvalue problems

$$\nabla^2 u = -(\mu\epsilon\omega^2 - \gamma^2)u \tag{8.42}$$

and

$$\nabla^2 v = -(\mu\epsilon\omega^2 - \gamma^2)v \tag{8.43}$$

on the rectangle $0 < x < a$, $0 < y < b$. Because E has zero tangential conditions, the Laplacian of (8.42) is subject to zero boundary conditions along the two edges $y = 0$ and $y = b$ while the Laplacian of (8.43) is subject to zero conditions along $x = 0$ and $x = a$. Because $\nabla \cdot E = 0$, $u_x + v_y = 0$. It then follows from the conditions on (8.43) that the Laplacian of (8.42) is subject to zero normal conditions along $x = 0$ and $x = a$. But then the eigenmodes of the Laplacian of (8.42) are

$$\phi_{mn} = \cos m\pi x/a \ \sin n\pi y/b$$

with dominant mode $m = 0$, $n = 1$, again giving (8.40). So *the longest wavelength with nonzero* **i** *component carried by the waveguide is again determined by the shortest dimension b.*

This leaves the simple transverse mode form

$$E = \sin(\omega t - \gamma z)\ v(x, y)\mathbf{j}. \tag{8.44}$$

Because $\nabla \cdot E = 0$, $v = v(x)$. Similarly to (8.42), the Laplacian on v is subject to zero conditions on the edges $x = 0$ and $x = a$. Thus the modes of (8.43) are

$$\psi_m = \sin m\pi x/a,$$

with dominant mode $m = 1$ giving that

$$\mu\epsilon\omega^2 - \gamma^2 \geq \frac{\pi^2}{a^2},$$

i.e.,

$$\lambda < 2a. \tag{8.45}$$

And in fact, choosing this dominant mode

$$E = \sin(\omega t - \gamma z)\ \sin \pi x/a\,\mathbf{j} \tag{8.46}$$

of radio frequency

$$f = \frac{\omega}{2\pi} = \frac{\nu}{2a\sqrt{\nu^2\mu\epsilon - 1}}, \tag{8.47}$$

where

$$\gamma = \frac{\omega}{\nu}$$

yields a solution for any wavelength $\lambda < 2a$. There are two ways to see this. One is to deduce from (8.35) and (8.46) that H

$$= \frac{-\gamma}{\omega\mu} \sin(\omega t - \gamma z)\ \sin \pi x/a\,\mathbf{i} + \frac{\pi}{a\omega\mu} \cos(\omega t - \gamma z)\ \cos \pi x/a\,\mathbf{k} \tag{8.48}$$

and then laboriously establish (Exercise 8.24) that E and H do indeed satisfy their respective boundary conditions as well as Maxwell's Equations (8.35). Alternatively, armed with Exercise 8.32, we are able to see by inspection that (8.46) is a solution.

This simple mode (8.46) is called the *dominant transverse electric mode* $\mathbf{TE_{10}}$ and is in fact employed in nearly all microwave applications. Note that the waveguide can support a continuum of frequencies f, each propagating at a velocity ν peculiar to that frequency.

Exercises.

The first 33 exercises elaborate the material of Chapter 8. More routine exercises begin with 8.34.

8.1 Again solve BVP 7 when all but one face of the block is insulated. Afterwards carefully justify that you have indeed obtained the unique strong solution.

8.2 Argue intuitively that sinusoidal variations of temperature at ground surface will propagate downward as sinusoidal variations of diminishing amplitude and increasing phase delay.

8.3 Solve $u_t = u_{xx}$ on $0 < x < 1$ subject to $u(0,t) = f(t)$ and $u_x(1,t) = 0$ where initially $u(x,0) = 0$.

Answer:

$$u = f(t) - \sum_{k=1}^{\infty} a_k(f(0)e^{\lambda_k t} + e^{\lambda_k t} * f(t))\,\phi_k$$

where

$$1 = \sum_{k=1}^{\infty} a_k\phi_k,$$

and where

$$\lambda_k = -(2k-1)^2\pi^2/4 \text{ and } \phi_k = \sin\frac{(2k-1)\pi x}{2}.$$

Hint: Set $v(x,t) = u(x,t) - f(t)$.

Recall that the *convolution* of two functions $f * g$ is defined as

$$(f * g)t = \int_0^t f(\alpha)g(t-\alpha)\,d\alpha.$$

8.4 Show that the eigenfunctions ϕ_n obtained in the previous exercise form a complete orthogonal basis of $L^2(0,1)$.

Hint: Show the resolvent is compact and Hermitian at $\lambda = 0$. See §14.

8.5 Deduce from the Exercise 8.3 that after transients die away, subsurface ground temperature $u(x,t)$ at depth x when pumped by a sinusoidal surface temperature $u(0,t) = \sin \omega t$ is of the form

$$u(x,t) = X(x)\ \sin(\omega t - \theta(x)).$$

8.6 Substantiate that one estimate of the error in the skin effect formula (8.9) is indeed given by (8.11).

8.7 Typical moist soil may have conductivity $\kappa = 0.58$ Btu/h-ft-F, specific heat $c = 0.2$ Btu/lb-F, and density $\rho = 100$ lb/cuft. What percentage of annual temperature variations are felt at 8 feet of depth? What is the phase delay at this depth? What is the influence of daily variations at this depth?

Answer: 41% , 52 days, and 4×10^{-6}% respectively.

8.8 The four fundamental **Equations of Maxwell** describing all electromagnetic phenomena (in homogeneous isotropic linear stationary media) are

$$\begin{aligned} \nabla \cdot D &= \rho \\ \nabla \times H &= J + \frac{\partial D}{\partial t} \\ \nabla \times E &= -\frac{\partial B}{\partial t} \\ \nabla \cdot B &= 0, \end{aligned}$$

together with the ancillary $D = \epsilon E,\ B = \mu H,\ J = \sigma E$. By simple manipulation, show

$$\nabla^2 E - \nabla\nabla \cdot E = \mu\sigma\frac{\partial E}{\partial t} + \mu\epsilon\frac{\partial^2 E}{\partial t^2}$$

and

$$\nabla^2 H = \mu\sigma\frac{\partial H}{\partial t} + \mu\epsilon\frac{\partial^2 H}{\partial t^2}.$$

Hint: $\nabla \times (\nabla \times E) = \nabla\nabla \cdot E - \nabla^2 E$.

8.9 Determine that 1 MHz current penetrates only $\delta = 0.066$ mm into flat copper conductors.

Facts: At 20C, copper has resistivity $1/\sigma = 1.724 \times 10^{-8}$ Ohms/m and permeability $\mu = \mu_0 = 4\pi \times 10^{-7}$ Henrys/m (Newton/Ampere).

8.10 Find the nondimensional general form of the wave function of a quantum particle trapped within the unit cube $0 < x, y, z < 1$ wherein the potential $a(x)$ is identically 1.

Answer:

$$\psi = \sum_{k,l,m=1}^{\infty} c_{klm} e^{-it(1+\pi^2(k^2+l^2+m^2))} \sin k\pi x \, \sin l\pi y \, \sin m\pi z.$$

8.11 Show that the Schrödinger operator

$$H = -\nabla^2 + V$$

of a particle trapped in the bounded domain Ω with bounded non-negative potential $a(x)$ is Hermitian positive definite with compact resolvent. The eigenfunctions of H form a complete orthogonal basis of $L^2(\Omega)$.

Outline: An additive bounded perturbation by B of an operator A possesses the resolvent

$$(\lambda - A - B)^{-1} = R(\lambda)(I - BR(\lambda))^{-1}.$$

Apply Exercise 13.12.

8.12 Verify the eigenvalue equation (13.25) and its ancillary (13.26).

8.13 Estimate the number N of tunneling stationary states as a function of barrier height a_0 for the step potential barrier (8.22).

Answer: $N \sim [a_0]$.

8.14 Find the tunneling stationary states of a particle trapped in $[-\pi, \pi]$ when the potential is the thin fence

$$a(x) = \begin{cases} \infty & \text{if } x \le -\pi \\ 0 & \text{if } -\pi < x < -\epsilon \\ 1 & \text{if } -\epsilon \le x \le \epsilon \\ 0 & \text{if } \epsilon < x < \pi \\ \infty & \text{if } \pi \le x \end{cases}$$

Shortcut: Think about even and odd solutions.

8.15 The spatial operator $\nabla^4 = \partial^4/\partial x^4$ of the Bernoulli-Euler beam is self-adjoint and positive semidefinite with compact resolvent for each of the 9 combinations of clamped, pinned, or free boundary conditions. Establish one of the 9 cases. Which of the 9 are positive definite?

8.16 Verify that the cantilevered beam eigenmodes are of the form (8.31).

8.17 Verify the clamped-free eigenvalue equations (8.32) and (8.33).

8.18 Solve for the coefficients of the clamped-free beam (8.28) when the initial deflection is $f(x) = x(1-x)$.

8.19 It is widely believed that when a 'structural' damping term

$$c\frac{\partial^3 u}{\partial t \partial x^2}$$

is added to the undamped Euler-Bernoulli beam model, that damping is modal. That is, the eigenmodes ϕ_n of the undamped beam (8.27) are eigenmodes of the damping operator $c\frac{\partial^2}{\partial x^2}$. Show this widely held belief is incorrect.

8.20 Consider an ideal transmission line of series distributed inductance L and shunt distributed capacitance C extending from the ideal voltage source at $x = 0$ to the termination Z at $x = l$. See Figure 1.4 and Exercise 1.31. Suppose the termination is a capacitor of capacitance C_0. Suppose further that this system is now running free — the generator is quiescent, i.e., $V(0,t) = 0$ for $t > 0$. Show that the voltage $V(x,t)$ and the current $I(x,t)$ on the line are given by the orthogonal series

$$V(x,t) = \sum_{n=1}^{\infty} a_n(t) \sin \omega_n x$$

and

$$I(x,t) = \sum_{n=1}^{\infty} b_n(t) \cos \omega_n x,$$

where

$$C\,\dot{a}_n = \omega_n b_n, \quad L\,\dot{b}_n = -\omega_n a_n,$$

and where the positive increasing ω_n are determined by the rule

$$\Omega_n \tan l\omega_n = C/C_0.$$

Hint: $C_0\, V_t(l,t) = I(l,t)$.

8.21 (O. Heaviside) A lossy transmission line consists of cascaded infinitesimal sections as shown in Figure 4.5 — series distributed resistance R and inductance L with distributed shunt capacitance C and resistance $1/G$ per unit length. Show that the voltage $V(x,t)$ and current $I(x,t)$ on this line satisfy the *telegrapher's equations*

$$C\frac{\partial V}{\partial t} = -\frac{\partial I}{\partial x} - GV$$

and

$$L\frac{\partial I}{\partial t} = -\frac{\partial V}{\partial x} - RI.$$

Compare with Exercise 1.31. Will such a lossy line enjoy an orthogonal series decomposition as in Exercise 8.20?

8.22 Establish (8.38).

8.23 Again establish (8.45) by starting with H in the form (8.37) and imposing the zero normal boundary conditions on w.

8.24 Show directly that E and H of (8.46)–(8.48) satisfy Maxwell's Equations (8.35) as well as the required boundary conditions.

8.25 Show that no frequency lower than 1.5 GHz can be supported in rectangular waveguide of width 10 cm and height 5 cm.

8.26 In (8.36), the wave appears to travel down the waveguide at a velocity $\nu > 1/\sqrt{\mu\epsilon} = c$ exceeding the speed of light. How can this be?

8.27 Find the electric and magnetic field intensities E and H for the simplest *transverse magnetic mode* ($\mathbf{TM}_{11}$) when H propagates perpendicular to the walls of the waveguide.

Hint: Let $H = \sin(\omega t - \gamma z)\,[u\mathbf{i} + v\mathbf{j}\,]$. Impose the normal and tangential zero boundary conditions on H and E.

8.28 Find the expected position of the electron in ground state ψ_1 of the step barrier $a_0 = 3$ in Figure 8.5a.

Answer: $q = \langle x\psi_1, \psi_1 \rangle \approx -1.2410290$.

8.29 Find all possible free undamped transverse vibrations of a floating (homogeneous isotropic) beam.

8.30 The universally accepted model for vibrations of an isotropic homogeneous elastic solid is

$$\rho \frac{\partial^2 D}{\partial t^2} = (\lambda + \mu)\nabla\ \nabla \cdot D + \mu \nabla^2 D$$

where $D = u\mathbf{i} + v\mathbf{j} + w\mathbf{k}$ is particle displacement, and where λ and μ are *Lamé's* constants.

Show by simple manipulation that

$$\rho \frac{\partial^2 (\nabla \cdot D)}{\partial t^2} = (\lambda + 2\mu)\nabla^2(\nabla \cdot D)$$

and

$$\rho \frac{\partial^2 (\nabla \times D)}{\partial t^2} = \mu \nabla^2 (\nabla \times D).$$

Thus in the language of seismology, the *Pressure* wave $p = \nabla \cdot D$ from a distant shock — traveling with velocity $\sqrt{(\lambda + 2\mu)/\rho}$ — arrives first, followed later by the *Shear* wave $S = \nabla \times D$ traveling at the lower velocity $\sqrt{\mu/\rho}$. Bringing up the rear in this impulse train is the destructive surface-traveling *Longwave* allowed by the *very* complex boundary conditions at the shear-free surface. Continuum Mechanics is not easy. See [Kolsky], [Rayleigh], or [Courant and Hilbert].

8.31 If there is no component of the electric field E tangent to a surface S, then the flux of the magnetic field H leaving through S is constant.

Outline: The line integral of E about any closed curve C in S is zero. Thus by Stokes, Maxwell, and Leibniz,

$$0 = \oint_C E \circ d\mathbf{r} = \int_S \nabla \times E \cdot \mathbf{n} d\sigma = -\mu \frac{d}{dt} \int_S H \circ \mathbf{n} d\sigma.$$

8.32 Show that for waveguides, it is sufficient to find an electric field intensity E with zero tangential boundary values that solves

$$\mu\epsilon \frac{\partial^2 E}{\partial t^2} = \nabla^2 E$$

and

$$\nabla \cdot E = 0$$

to guarantee the existence of an accompanying magnetic field intensity H with zero normal boundary values that together solve Maxwell's equations.

8.33 What is the resonant frequency of a rectangular cavity resonator of dimensions $a = b > c$?

Answer:

$$f = \frac{1}{a\sqrt{2\mu\epsilon}}.$$

8.34 Carefully rework BVP 1 of Chapter 5. Justify *every* step.

8.35 Solve the Fourier Ring Problem (BVP 3 Chapter 5) for an initial temperature distribution $f(x) = x$.

8.36 A rod is initially at temperature $u(x, 0) = x$ with boundary conditions $u_x(0, t) = 0$ and $u(1, t) = 0$. Find the future interior temperatures of this rod.

8.37 Find the temperature $u = u(x, t)$ of a Trombe wall $0 < x < 1$, initially at uniform temperature 0, that is subjected to a flux pulse of insolation: $u_x(0, t) = 1$ when $0 \leq t < 1$, but 0 otherwise. Assume the inner wall surface is delivering heat to the interior according the rule $u_x(1, t) = -u(1, t)$.

Hint: Superimpose two solutions — to a step change, then to a delayed negative step change.

8.38 Find the temperatures of a relaxing rectangle $-a < x < a$, $-b < y < b$, initially at temperature $u(x, y, 0) = xy$, subject to zero boundary conditions.

Answer: u

$$= \sum_{m,n=1}^{\infty} c_{mn} e^{-\alpha((m\pi/2a)^2 + (n\pi/2b)^2)t} \sin \frac{m\pi(x + a)}{2a} \sin \frac{n\pi(y + b)}{2b}$$

where

$$c_{mn} = \frac{4ab(1 + (-1)^m)(1 + (-1)^n)}{mn\pi^2}.$$

8.39 Rework the Exercise 8.38 under the conditions that all sides but the side $x = a$ are insulated, while $x = a$ is held at 0 temperature.

8.40 Find the natural frequencies of the rectangular drum $-a < x < a$, $-b < y < b$.

8.41 Find the fundamental frequencies of a quivering cube of Jello^{TM}.

Hint: Ignore shear forces. Each component of the displacement must then satisfy the wave equation. Think carefully about the boundary conditions at the face touching the plate versus the conditions at the remaining 5 faces. Compare with Exercise 8.30.

8.42 Find the natural frequencies of vibration of the system

$$\frac{\partial^2 u}{\partial t^2} = c^2 \nabla^2 u$$

on the square $-1 < x, y < 1$ subject to zero normal boundary conditions, i.e.,

$$\frac{du}{d\mathbf{n}} = 0.$$

8.43 Find how the interior temperatures of the square $-2 < x, y < 2$ with zero boundary temperatures relax to 0 given that the initial temperature distribution is $u(x, y, 0) = 1$ on the smaller square $-1 < x, y < 1$, 0 outside.

8.44 (A. J. Hull) A viscous damper (shock absorber, dash pot) resists proportionally to the velocity applied. What is a practical method for measuring its *coefficient of friction* (its constant of proportionality)?

Outline: Rap a pinned-free beam to find its natural longitudinal frequencies of vibration. Attach the damper at the free end. Rap the damped beam. Deduce the coefficient of friction.

8.45 (D. T. Johnson) A gel chamber is initially free of glucose. At each end is an identical membrane permeable to glucose. At $t = 0$ the left end at $x = 0$ is placed in contact with a source of glucose at concentration C_0 while the right end at $x = L$ is put in contact with a sink of 0 concentration. Find the future concentration $C(x, t)$ of glucose within the gel chamber. See [Bird, Stewart, and Lightfoot].

8.46 Show that in spite of popular belief, the electric field E need not be perpendicular to the magnetic field H.

Hint: Superimpose the two simplest modes of a square waveguide.

8.47 (An *inverse problem*) Given a solid isotropic homogeneous cube of some material, design an experiment and data manipulation to obtain the diffusivity α of this material.

Chapter 9

Bessel Functions

Let us study the enormously useful Bessel functions — their power series and integral representations, identities, asymptotic behavior, and completeness. These Bessel functions arise in physical problems with cylindrically shaped spatial domains. Throughout this Chapter, n will generically denote a nonnegative integer.

§9.1 Power Series Representation

The Bessel functions are solutions of the deceptively simple *Bessel's Equation*

$$x^2y'' + xy' + (x^2 - n^2)y = 0. \tag{9.1}$$

Picard's Theorem — the fundamental existence and uniqueness theorem for ODEs — guarantees two local independent solutions of Bessel's Equation (9.1) about each $x = x_0$, except possibly at $x = 0$ [Brauer-Nohel]. Nevertheless insist upon an analytic solution at $x = 0$:

$$y = \sum_{k=0}^{\infty} a_k x^k. \tag{9.2}$$

Placing this series into Bessel's Equation and gathering like powers of x yields the *recursion*

$$(k^2 - n^2)a_k = -a_{k-2}, \tag{9.3}$$

once we agree that all negatively subscripted a_k equal 0 (Exercise 9.1).

What conclusions can we draw from the recursion (9.3) about the form of a power series solution? Certainly all $a_k = 0$ when $k < n$. Even more, $a_{n+1} = a_{n+3} = \ldots = 0$. Finally, all subsequent a_k are determined by the value of a_n where it is traditional to take

$$a_n = 1/2^n n!\,. \tag{9.4}$$

We now have a solution.

Result A. For each non-negative integer n,

$$J_n(x) = \sum_{k=0}^{\infty} \frac{(-1)^k}{k!(k+n)!} (\frac{x}{2})^{2k+n} \tag{9.5}$$

is a solution of Bessel's Equation (9.1). The series possesses an infinite radius of convergence.

Proof. Exercise 9.2.

J_n is called *the n-th Bessel function of the first kind.*

§9.2 Standard Formulae

We may differentiate or integrate term-by-term, rearrange, etc to easily obtain the following standard facts:

$$J_n(-x) = (-1)^n J_n(x) \tag{9.6}$$

$$(x^n J_n(x))' = x^n J_{n-1}(x) \tag{9.7}$$

$$(x^{-n} J_n(x))' = -x^{-n} J_{n+1}(x) \tag{9.8}$$

$$x J_n'(x) = n J_n(x) - x J_{n+1}(x) \tag{9.9}$$

$$x J_n'(x) = -n J_n(x) + x J_{n-1}(x) \tag{9.10}$$

$$\int_0^x t^n J_{n-1}(t)\, dt = x^n J_n(x). \tag{9.11}$$

In particular(Exercises 9.3 and 9.4),

$$J_0'(x) = -J_1(x) \tag{9.12}$$

and

$$\int_0^x t J_0(t)\, dt = x J_1(x). \tag{9.13}$$

§9.3 Integral Representation

Let us compute the integral

$$I = \int_{-\pi}^{\pi} e^{ix \sin\theta}\, d\theta \tag{9.14}$$

in two different ways. On the one hand, because sine is odd we see

$$I = \int_{-\pi}^{\pi} \cos(x \sin\theta)\, d\theta. \tag{9.15}$$

On the other hand, using the Taylor expansion

$$e^{ix\sin\theta} = \sum_{k=1}^{\infty} \frac{i^k x^k}{k!} \sin^k\theta \tag{9.16}$$

and integrating term-by-term, we obtain

$$I = \sum_{k=0}^{\infty} \frac{(-1)^k}{(2k)!} x^{2k} \int_{-\pi}^{\pi} \sin^{2k}\theta\, d\theta, \tag{9.17}$$

since odd-power terms do not survive the integration. But by Exercise 9.5,

$$\int_{-\pi}^{\pi} \sin^{2k}\theta\, d\theta = \frac{2\pi(2k)!}{(k!2^k)^2}\,, \tag{9.18}$$

giving the famous integral representation

$$J_0(x) = \frac{1}{\pi}\int_0^{\pi} \cos(x\sin\theta)\, d\theta. \tag{9.19}$$

By induction on n (Exercise 9.6),

$$J_n(x) = \frac{1}{\pi}\int_0^{\pi} \cos(n\theta - x\sin\theta)\, d\theta. \tag{9.20}$$

§9.4 Asymptotics

Note that J_{2m} is an even function while J_{2m+1} is odd. Also note that $J_0(0) = 1$ while for $n > 1$, $J_n(0) = 0$. In fact, J_n possesses a zero at $x = 0$ of multiplicity n. Moreover from the integral representation (9.20) it is clear that $|J_n(x)| \leq 1$ for all real x.

A graphing routine (Exercise 9.8) will reveal that each J_n decays away for large x; see Figure 9.1. Moreover zeros appear more or less evenly spaced and interlaced. The interlacing is immediate.

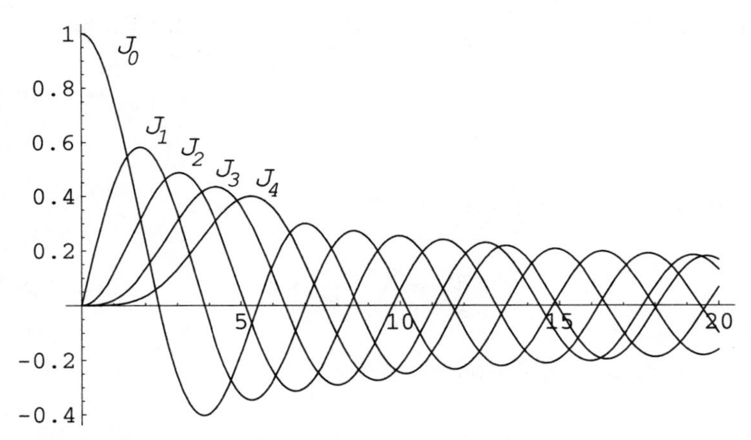

Figure 9.1 The graphs of $y = J_n(x)$, $n = 0, 1, 2, 3, 4$ superimposed. Note the decreasing amplitudes and the interlacing of zeros.

Result B. Let $\alpha_1^{(n)} < \alpha_2^{(n)} < \alpha_3^{(n)} < \ldots$ denote the positive roots of $J_n(x) = 0$. Then these zeros interlace according to the rule

$$\alpha_k^{(n)} < \alpha_k^{(n+1)} < \alpha_{k+1}^{(n)} < a_{k+1}^{(n+1)}. \tag{9.21}$$

Proof. By Rolle's Theorem and by the standard formula (9.8), it is certain that J_{n+1} must vanish between each pair of positive zeros of J_n. Likewise because of (9.7), J_n must vanish between each pair of zeros of J_{n+1}. Because of (9.9), J_n and J_{n+1} cannot share positive zeros since *positive zeros are simple zeros* (Exercise 9.13).

This interlacing is an instance of a deep principle that appears in many disguises under many names, e.g., the *Rayleigh inclusion principle, Cauchy's interlacing theorem, Sturm's Theorem, Weyl's maximum-minimum principle*, and others.

We will see in the next subsection that J_0 (hence each J_n) does in fact possess an increasing unbounded sequence of positive zeros — yet another consequence of Rellich's Theorem.

Result C.

$$\lim_{x \to \infty} J_n(x) = 0. \tag{9.22}$$

Proof. $J_0(x)$

$$= \frac{1}{\pi} \int_0^{\pi} \cos(x \sin\theta)\, d\theta = \frac{2}{\pi} \int_0^{\pi/2} \cos(x \cos\theta)\, d\theta = \frac{2}{\pi} \int_0^1 \frac{\cos xy}{\sqrt{1-y^2}}\, dy,$$

which according to the Riemann-Lebesgue Lemma (7.34) must have limit 0 as $x \to \infty$.

Apply this same approach to

$$J_n(x) = \frac{1}{\pi} \int_0^{\pi} \cos(n\theta - x \sin\theta)\, d\theta$$

$$= \frac{1}{\pi} \int_0^{\pi} \cos n\theta \ \cos(x \sin\theta) + \ \sin n\theta \ \sin(x \sin\theta)\, d\theta. \tag{9.23}$$

There are many interesting, far deeper asymptotic formulae for the values of Bessel functions at large x. Perhaps the most well known is

$$J_0(x) = \sqrt{\frac{2}{\pi x}} \cos(x - \pi/4) + o(\sqrt{x}), \tag{9.24}$$

which immediately establishes the existence of an unbounded increasing sequence of zeros [Bowman]. See the definitive *Handbook of Mathematical Functions* by Abramowitz and Stegun, a must-own.

§9.5* Orthogonality and Completeness

Consider the Hilbert space $X = L^2(\mathbf{D})$ of all square-integrable real-valued functions f on the open unit disk $\mathbf{D} = \{r < 1\}$ in the plane under the usual inner product.

By Rellich's Theorem, the Laplacian $A = \nabla^2$ on X with zero boundary conditions is negative definite, Hermitian with a compact resolvent, and its (continuous) eigenfunctions form a complete orthogonal basis of X.

Consider the closed subspace $X_0 \subset X$ of all radially symmetric $f = \phi(r)$. Cut both A and its resolvent back to X_0 where A remains negative definite, Hermitian, with compact resolvent (Exercise 9.24). Hence *the radially symmetric eigenfunctions form a complete orthogonal basis of the subspace of radially symmetric functions.*

Let us translate all this into polar coordinates. Consider only radially symmetric $\phi = \phi(r)$. The Laplacian with zero boundary values becomes in polar coordinates the differential operator (Exercise 1.19)

$$A\phi = \phi'' + \frac{1}{r}\phi' \tag{9.25}$$

subject to $\phi(1) = 0$ as well as a tacit condition revealed below. The inner product becomes in polar coordinates (within the multiple 2π),

$$\langle \phi, \psi \rangle = \int_0^1 \phi(r)\psi(r)\, rdr. \tag{9.26}$$

The subspace X_0, a Hilbert space in its own right, consists of all ϕ with

$$\int_0^1 \phi(r)^2\, rdr < \infty.$$

We have already seen in our informal solution of the Circular Drum in §5.5 that the eigenvalue equation $A\phi = \lambda\phi$, which in this case is $\phi'' + (1/r)\phi' = \lambda\phi = -\alpha^2\phi$, becomes, after the substitution $x = \alpha r$, Bessel's Equation $x^2y'' + xy' + x^2y = 0$ (Exercise 5.6). As discussed in the next subsection, any bounded solution of Bessel's Equation for $n = 0$ must be a constant multiple of $J_0(x)$. But since the eigenfunctions are continuous, they are bounded. Thus $y = J_0(x)$ and $J_0(\alpha) = 0$. We have obtained an important result.

Theorem A. The radially symmetric eigenfunctions of the Laplacian with zero boundary conditions on the unit disk **D** are

$$\phi_k = J_0(\alpha_k r), \qquad k = 1, 2, 3, \ldots$$

where $\alpha_1 < \alpha_2 < \alpha_3 < \ldots$ are the positive zeros of J_0. These eigenfunctions form a complete orthogonal basis of X_0, the Hilbert space of all radially symmetric, square-integrable functions on the unit disk. That is, for $k \neq l$,

$$\langle \phi_k, \phi_l \rangle = \int_0^1 J_0(\alpha_k r) J_0(\alpha_l r)\, rdr = 0, \tag{9.27}$$

and for any f with

$$\langle f, f \rangle = \int_0^1 f(r)^2\, rdr < \infty,$$

we have the *Bessel-Fourier expansion*

$$f(r) = \sum_{k=1}^{\infty} c_k J_0(\alpha_k r), \tag{9.28}$$

where

$$c_k = \frac{\langle f, \phi_k \rangle}{\langle \phi_k, \phi_k \rangle}. \tag{9.29}$$

It is a traditional student task, here Exercise 9.14 and 9.15, to show the normalization factor

$$\langle \phi_k, \phi_k \rangle = \frac{J_1(\alpha_k)^2}{2}. \tag{9.30}$$

Let us repeat the above development, not on the radially symmetric functions of X_0, but instead on the the closed subspace X_n of X of all functions of the special form

$$h(r,\theta) = f(r)\cos n\theta + g(r)\sin n\theta$$

for a fixed positive integer n. Again when the Laplacian is cut back to X_n, it is Hermitian with compact resolvent, and hence its eigenfunctions within X_n, namely (Exercise 9.16)

$$J_n(\alpha_k^{(n)} r)\cos n\theta \ \text{ and } \ J_n(\alpha_k^{(n)} r)\sin n\theta, \quad k = 1, 2, 3, \ldots,$$

comprise a complete orthogonal basis of X_n. This yields a companion to Theorem A.

Theorem B. The functions

$$\phi_k = J_n(\alpha_k^{(n)} r), \quad k = 1, 2, 3, \ldots$$

also form a complete orthogonal basis of the radially symmetric square-integrable functions $f(r)$ of X_0; there is the Bessel-Fourier expansion

$$f = \sum_{k=1}^{\infty} c_k J_n(\alpha_k^{(n)} \cdot) \tag{9.31}$$

where

$$c_k = \frac{\langle f, J_n(\alpha_k^{(n)} \cdot)\rangle}{\langle J_n(\alpha_k^{(n)} \cdot), J_n(\alpha_k^{(n)} \cdot)\rangle}, \tag{9.32}$$

where the $\alpha_k^{(n)}$, $k = 1, 2, 3, \ldots$ are the positive zeros of J_n.

By Exercise 9.15, the normalization factor in the denominator of (9.32) is

$$\int_0^1 J_n(\alpha_k r)^2 \, r dr = \frac{J_n'(\alpha_k)^2}{2} = \frac{J_{n+1}(\alpha_k)^2}{2} = \frac{J_{n-1}(\alpha_k)^2}{2}. \tag{9.33}$$

Theorem C. The eigenfunctions of the Laplacian with zero boundary conditions on the unit disk **D** are

$$J_0(\alpha_k^{(0)} r),\ J_n(\alpha_k^{(n)} r)\cos n\theta,\ J_n(\alpha_k^{(n)} r)\sin n\theta, \tag{9.34}$$

$$k = 1, 2, 3, \ldots, \quad n = 1, 2, \ldots,$$

belonging to the eigenvalues

$$\lambda_{k,n} = -(\alpha_k^{(n)})^2, \tag{9.35}$$

where $\alpha_1^{(n)} < \alpha_2^{(n)} < \alpha_3^{(n)} < \ldots$ are the positive zeros of J_n. These eigenfunctions form a complete orthogonal basis for $X = L^2(\mathbf{D})$, the Hilbert space of all square-integrable functions on **D**.

Proof. By Rellich, we know the eigenfunctions form a complete orthogonal basis. Each function of (9.34) is an eigenfunction of ∇^2 with eigenvalue $\lambda_{k,n}$ (Exercise 9.17). Because the disk in polar coordinates is a cartesian product $\mathbf{D} = (0, 1) \times (0, 2\pi)$, as in Exercise 6.22 every element $h(r, \theta)$ of X has a Fourier expansion of the form

$$h = \sum_{n=0}^{\infty} f_n(r) \cos n\theta + g_n(x) \sin n\theta.$$

But by Theorems A and B, each f_n and g_n has an expansion in terms of $J_n(\alpha_k^{(n)} r)$, $k = 1, 2, 3, \ldots$. Thus the eigenfunctions of (9.34) comprise a complete basis and hence make up a complete list of all the eigenmodes.

The underlying inner product in Theorem C is of course the usual one for the disk, which in polar coordinates becomes

$$\langle f, g \rangle = \int_0^{2\pi} \int_0^1 f(r, \theta) g(r, \theta) \, r dr \, d\theta. \tag{9.36}$$

§9.6 Other Bessel Functions

The *Bessel functions of the first kind* J_ν are defined for any real ν as follows: for ν not a negative integer,

$$J_\nu(x) = \sum_{k=0}^{\infty} \frac{(-1)^k}{k!\Gamma(k+\nu+1)} (\frac{x}{2})^{2k+\nu}, \tag{9.37}$$

while for positive integers n,

$$J_{-n}(x) = (-1)^n J_n(x). \tag{9.38}$$

$J_\nu(x)$ is a bounded solution of (Exercise 9.20)

$$x^2 y'' + xy' + (x^2 - \nu^2)y = 0, \quad \nu \geq 0. \tag{9.39}$$

The n-th *modified Bessel function of the first kind* is

$$I_n(x) = i^{-n} J_n(ix) \tag{9.40}$$

which is a solution bounded near 0 of

$$x^2 y'' + xy' - (x^2 + n^2)y = 0. \tag{9.41}$$

The power series for I_n is the nonalternating version of the power series (9.5) for J_n.

Picard of course guarantees a second solution of Bessel's Equation (9.1) independent of J_n. Let us briefly touch on this second solution Y_n, variously called *Bessel's function of the second kind*, or *Weber's function*, or *Neumann's Bessel function*. By the classic method of *Reduction of Order* found in any ODE text, we can bootstrap from the solution J_n to the solution

$$y = aJ_n(x) + bJ_n(x) \int^x \frac{dx}{xJ_n(x)^2}. \tag{9.42}$$

If we expand the integrand in powers of x,

$$\frac{1}{xJ_n(x)^2} = \sum_{k=-(2n+1)}^{\infty} c_k x^k,$$

integrate, then multiply by $J_n(x)$ with its zero of order n, we obtain solutions of the form

$$Y_n(x) = cJ_n(x) \log x + \sum_{k=-n}^{\infty} a_k x^k. \tag{9.44}$$

Once traditional choices for the first few coefficients are made, Y_n becomes an unbounded near 0 solution of Bessel's Equation (9.1). Every solution of (9.1) is a linear combination of these two fundamental solutions J_n and Y_n.

See [Abramowitz and Stegun] for detailed expansions and tabulations of J_n and Y_n. The classic reference on Bessel functions is the 1922 book by Watson. An easy read is the short book by Bowman. Some have spent their life investigating these rich functions [Bateman].

Exercises

In most of the identities that follow, equality holds pointwise as well as in any norm sense whatsoever. But several infinite series expansions may hold only pointwise or almost everywhere, others uniformly on compact subsets, while still others hold in the L^2 sense. As part of each solution you must specify in what sense equality holds.

9.1 Derive the recursion (9.3) of a power series solution of Bessel's Equation. Write out the first 6 terms of the series for $n = 0, 1, 2$.

9.2 Show via the ratio test that the power series (9.5) has an infinite radius of convergence. Establish that $J_n(x)$ does indeed solve Bessel's Equation (9.1).

9.3 Using standard manipulations of power series, verify the standard formulae (9.6)–(9.8).

9.4 Deduce the remaining formulae (9.9)–(9.11) from (9.7) and (9.8).

9.5 Prove that

$$\int_0^{2\pi} \sin^{2k}\theta \, d\theta = \frac{2\pi(2k)!}{(k!2^k)^2}.$$

Hint: Go by induction on k. Alternatively employ the useful

$$\int_0^{\pi/2} \cos^m\theta \, \sin^n\theta \, d\theta = \frac{\Gamma(\frac{m+1}{2})\Gamma(\frac{n+1}{2})}{2\Gamma(\frac{m+n}{2}+1)}.$$

9.6 Deduce the integral representation (9.20) for J_n from the integral representation (9.19) for J_0.

Hint: Go by induction on n using the standard formula (9.9), Leibniz's Rule, and Integration by Parts.

9.7 Display your skill with Integration by Parts and Leibniz's Rule by showing directly that the integral of (9.19) satisfies Bessel's Equation with $n = 0$. Attempt the corresponding calculation for $n > 0$.

9.8 Write two routines for computing $J_n(x)$ — first by truncating the series representation, then by (say) an adaptive Simpson's Rule applied to the integral representation. Which is more costly for a given accuracy? Graph $y = J_n(x)$.

9.9 Write a routine for approximating the zeros of J_n. Check your estimates against Table 9.1. Note the interlacing predicted by Result B.

Table 9.1. The first several zeros of J_n

n	$\alpha_1^{(n)}$	$\alpha_2^{(n)}$	$\alpha_3^{(n)}$	$\alpha_4^{(n)}$	$\alpha_5^{(n)}$
0	2.40483	5.52009	8.65373	11.79153	14.93092
1	3.83171	7.01559	10.17347	13.32369	16.47063
2	5.13562	8.41724	11.61984	14.79595	17.95982
3	6.38016	9.76102	13.01520	16.22347	19.40942
4	7.58834	11.06741	14.37254	17.61597	20.82693
5	8.77148	12.33860	15.70017	18.98013	22.21780

9.10 Prove that for $0 < |t| < \infty$,

$$e^{x(t-1/t)/2} = J_0(x) + \sum_{k=1}^{\infty} J_k(x)(t^k + (-1)^k t^{-k}).$$

Hint: Multiply out the series representations of both factors of the product $e^{xt/2}e^{-x/2t}$.

Deduce the fundamental *heterodyne formula* of a diode mixer [Perlow]:

$$e^{x\cos\theta} = I_0(x) + 2\sum_{k=1}^{\infty} I_k(x)\cos k\theta.$$

9.11 Set $t = i$ in Exercise 9.10 to obtain

$$\cos x = J_0(x) + 2\sum_{k=1}^{\infty}(-1)^k J_{2k}(x)$$

and

$$\sin x = 2\sum_{k=0}^{\infty}(-1)^k J_{2k+1}(x).$$

9.12 Put $t = e^{i\theta}$ in Exercise 9.10 to see that

$$\cos(x\sin\theta) = J_0(x) + 2\sum_{k=1}^{\infty} J_{2k}(x)\cos 2k\theta$$

and

$$\sin(x\sin\theta) = 2\sum_{k=0}^{\infty} J_{2k+1}(x)\sin(2k+1)\theta.$$

Deduce

$$J_{2n}(x) = \frac{1}{\pi}\int_0^{\pi} \cos(x\sin\theta)\ \cos 2n\theta\ d\theta$$

and

$$J_{2n+1}(x) = \frac{1}{\pi}\int_0^{\pi} \sin(x\sin\theta)\ \sin(2n+1)\theta\ d\theta.$$

9.13 Show that no positive zero of J_n is repeated.

Hint: Suppose $J_n(\alpha) = J_n'(\alpha) = 0$. Now bootstrap to $J_n^{(k)}(\alpha) = 0$ for all k using (9.1).

9.14 Show

$$\int_0^{\alpha} J_n(x)^2\ x\ dx = \frac{(\alpha^2 - n^2)J_n(\alpha)^2 + \alpha^2 J_n'(\alpha)^2}{2}.$$

Outline: Integrate the identity

$$xJ_n'(x)(xJ_n'(x))' = J_n'(x)(x^2J_n''(x)+xJ_n'(x)) = (n^2-x^2)J_n'(x)J_n(x).$$

9.15 Prove that if α is a positive root of $J_n(x) = 0$,

$$\int_0^1 J_n(\alpha r)^2\ rdr = \frac{J_n'(\alpha)^2}{2} = \frac{J_{n+1}(\alpha)^2}{2} = \frac{J_{n-1}(\alpha)^2}{2}.$$

Outline: Make the substitution $x = \alpha r$ and apply the Exercise 9.14, then (9.9) and (9.10).

9.16 Let ϕ be any eigenfunction of ∇^2 with zero boundary conditions on $\mathbf{D}$ of the decoupled form $\phi = R(r)\Theta(\theta)$. Show

$$\phi = aJ_n(\alpha_k^{(n)} r)\cos n\theta + bJ_n(\alpha_k^{(n)} r)\sin n\theta.$$

Hint: $\Theta(\theta)$ is periodic.

9.17 Show that each function of (9.34) is an eigenfunction of the Laplacian and conversely.

9.18 Show that if α_k are the positive zeros of J_n, for $r < 1$, then

$$r^n = 2\sum_{k=1}^{\infty} \frac{J_n(\alpha_k r)}{\alpha_k J_{n+1}(\alpha_k)}.$$

9.19 In contrast with Exercise 9.18, deduce from Exercise 9.12 that

$$1 = J_0(x) + 2\sum_{k=1}^{\infty} J_{2k}(x).$$

In what sense of equality does this hold?

9.20 Obtain the Bessel function of the first kind $J_\nu(x)$ given in (9.37) via the **Method of Frobenius**: assume a series solution of Bessel's equation (9.1) with n replaced by ν (ν not a negative integer) of the form

$$y = x^r \sum_{k=0}^{\infty} a_k x^k$$

and proceed to a recursion. Check that your steps reverse giving that J_ν is indeed a solution.

9.21 Show

$$J_{1/2}(x) = \sqrt{\frac{2}{\pi x}}\sin x, \qquad I_{1/2}(x) = \sqrt{\frac{2}{\pi x}}\sinh x,$$

$$J_{-1/2}(x) = \sqrt{\frac{2}{\pi x}}\cos x, \qquad I_{-1/2}(x) = \sqrt{\frac{2}{\pi x}}\cosh x.$$

9.22 Show $J_n(x) = 0$ has only real roots.

Outline: Even on a complex function space, the operator

$$A\phi = \phi'' + \frac{1}{r}\phi' - \frac{n^2}{r^2}\phi$$

subject to $\phi(1) = 0$ is negative definite and hence by Exercise 5.22 has only negative eigenvalues. Take $\phi = J_n(\alpha r)$ where $J_n(\alpha) = 0$, $\alpha \neq 0$, and deduce that $-\alpha^2 < 0$.

Summary: J_n has $x = 0$ as a zero of multiplicity n. All other zeros are simple and real, and are in fact, $\pm\alpha_1^{(n)}, \pm\alpha_2^{(n)}, \ldots$.

9.23 Find the general solution of

$$y'' - \frac{3}{x}y' + 9x^4 y = 0.$$

Answer:

$$y = c_1 x^2 J_{2/3}(x^3) + c_2 x^2 Y_{2/3}(x^3).$$

Approach: Try

$$y = c_1 x^a J_n(bx^c) + c_2 x^a Y_n(bx^c).$$

9.24* Show that ∇^2 with zero boundary conditions on the unit disk $\mathbf{D} = \{r < 1\}$ and its resolvent $R(\lambda)$ map radially symmetric functions to radially symmetric functions.

Outline: An informal proof proceeds by commuting partials:

$$\frac{\partial}{\partial\theta}\nabla^2\phi = \nabla^2\frac{\partial\phi}{\partial\theta} = \frac{\partial f(r)}{\partial\theta} = 0.$$

A formal proof requires the Sobolev ideas of Chapter 19 since $\mathrm{dom}\nabla^2$ is defined by certain linear functionals on $W_0^{1,2}(\mathbf{D})$.

Chapter 10
Cylindrical Problems

As a rule, problems with cylindrical spatial domains Ω give rise to Bessel eigenfunctions. Let us now work through a list of illustrative cylindrical problems.

§10.1 Quenching a Solid Cylinder (BVP 14)

Suppose a homogeneous isotropic solid cylinder, initially at a uniform temperature throughout, is dropped into a bath of lower temperature. Find the interior temperatures as the cylinder cools.

Intuitively, interior temperatures must be radially symmetric. Thus in non-dimensional variables the problem becomes in cylindrical coordinates

$$\frac{\partial u}{\partial t} = \nabla^2 u, \quad r < 1, \quad 0 < z < 1, \tag{10.1}$$

subject to

i) $u(r, z, 0) = 1,$

ii) $u(1, z, t) = 0,$

iii) $u(r, 0, t) = 0 = u(r, 1, t).$

In cylindrical coordinates the Laplacian becomes

$$\nabla^2 u = \frac{\partial^2 u}{\partial r^2} + \frac{1}{r}\frac{\partial u}{\partial r} + \frac{\partial^2 u}{\partial z^2} \tag{10.2}$$

(Exercise 1.19). Assuming variables separate

$$u = q(t)R(r)Z(z),$$

we obtain

$$\frac{\dot{q}(t)}{q(t)} = \frac{R'' + (1/r)R'}{R} + \frac{Z''}{Z} = \lambda,$$

and so

$$\frac{R'' + (1/r)R'}{R} = -\alpha^2$$

and

$$\frac{Z''}{Z} = -\beta^2.$$

That is,

$$r^2R'' + rR' + \alpha^2 r^2 R = 0 \tag{10.3}$$

and

$$Z'' + \beta^2 Z = 0. \tag{10.4}$$

(Here as usual the prime $'$ will denote differentiation with respect to the single appropriate variable).

As we saw in our informal solution in §5.5, a change of variables $x = \alpha r$ will bring (10.3) to Bessel's equation

$$x^2y'' + xy' + x^2y = 0,$$

and thus

$$R(r) = J_0(\alpha r). \tag{10.5}$$

Imposing the zero boundary condition ii) on the curved sides of the cylinder yields

$$J_0(\alpha) = 0,$$

and so

$$R_m = J_0(\alpha_m r), \tag{10.6}$$

where $a_1 < \alpha_2 < \cdots$ are the positive zeros of J_0. Imposing the zero boundary condition iii) at the flat faces yields as in BVP 1,

$$Z_n = \sin n\pi z. \tag{10.7}$$

Therefore the solution to our problem must be of the form

$$u = \sum_{m,n=1}^{\infty} c_{mn} e^{-(\alpha_m^2 + n^2\pi^2)t} J_0(\alpha_{mr}) \sin n\pi z. \tag{10.8}$$

Imposing the initial condition i), we have

$$1 = \sum_{m,n=1}^{\infty} c_{mn} J_0(\alpha_m r) \sin n\pi z. \tag{10.9}$$

Using the usual inner product but in cylindrical coordinates, we obtain as usual

$$\begin{aligned}&\langle 1, J_0(\alpha_m r)\sin n\pi z\rangle \\ &= c_{mn}\langle J_0(\alpha_m r)\sin n\pi z,\ \ J_0(\alpha_m r)\sin n\pi z\rangle,\end{aligned} \tag{10.10}$$

i.e.,

$$\begin{aligned}&\int_0^1 J_0(\alpha_m r) r\, dr \int_0^1 \sin n\pi z\, dz \\ &= c_{mn}\int_0^1 J_0(\alpha_m r)^2 r\, dr \int_0^1 \sin^2 n\pi z\, dz.\end{aligned} \tag{10.11}$$

But recall (9.30):

$$\int_0^1 J_0(\alpha_m r)^2 r\, dr = \frac{J_1(\alpha_m)^2}{2}.$$

Moreover by (9.13),

$$\int_0^1 J_0(\alpha_m r) r\, dr = \frac{1}{\alpha_m^2}\int_0^{\alpha_m} J_0(x) x\, dx = \frac{J_1(\alpha_m)}{\alpha_m}.$$

Thus

$$c_{mn} = \frac{4}{\alpha_m J_1(\alpha_m)} \cdot \frac{1-(-1)^n}{n\pi}. \tag{10.12}$$

There are many interesting and useful variations of this problem posed in the Exercises.

Let us examine the underpinnings of the above computations. As noted in §6.7, H. Weyl's result on the growth of the eigenvalues of the Laplacian together with Rellich's theorem guarantees a strong solution of the form

$$u = \sum_k c_k e^{\lambda_k t}\phi_k, \tag{10.13}$$

where the ϕ_k are the eigenfunction of the Laplacian belonging to the values λ_k. But $\phi_{mn} = J_0(\alpha_m r)\sin n\pi z$ are indeed eigenfunctions. Moreover, because the cylinder is the cartesian product of the unit disk with the unit interval, since the $J_0(\alpha_m r)$ are complete for the disc (§9.5, Theorem A), and since the $\sin n\pi z$ are complete for the unit interval (§7.1, Result A), it follows from the Lemma on Product Domains (Exercise 6.22) that the separated eigenfunctions

$$R_m Z_n = J_0(\alpha_m r)\sin n\pi z$$

must form a complete list of the eigenfunctions guaranteed by Rellich's Theorem. Thus the guaranteed unique strong solution (10.13) is (10.8) with coefficients given by (10.12).

§10.2 A Circular Drum Revisited (BVP 5)

Again sound a circular drum, but not at its center. Let $u = u(r, \theta, t)$ be the vertical deflection of the stretched membrane. In nondimensional variables,

$$\frac{\partial^2 u}{\partial t^2} = \nabla^2 u, \quad r < 1, \quad 0 \le \theta \le 2\pi \tag{10.14}$$

subject to

i) $u(r, \theta, 0) = f(r, \theta)$,

ii) $u_t(r, \theta, 0) = 0$,

iii) $u(1, \theta, t) = 0$.

Recall that in polar coordinates

$$\nabla^2 u = \frac{\partial^2 u}{\partial r^2} + \frac{1}{r}\frac{\partial u}{\partial r} + \frac{1}{r^2}\frac{\partial^2 u}{\partial \theta^2}. \tag{10.15}$$

Assuming variable separate

$$u = q(t)R(r)\Theta(\theta), \tag{10.16}$$

we obtain

$$\frac{\ddot{q}}{q} = \frac{R'' + (1/r)R'}{R} + \frac{1}{r^2}\frac{\Theta''}{\Theta} \tag{10.17}$$

where as usual, the prime $'$ denotes differentiation with respect to the single appropriate variable.

Because $\Theta(\theta)$ is periodic modulo 2π,

$$\Theta = a \cos n\theta + b \sin n\theta, \quad n = 0, 1, 2, \dots ,$$

and hence

$$r^2 R'' + rR' + (\alpha^2 r^2 - n^2)R = 0,$$

with bounded solution

$$R = J_n(\alpha r).$$

Imposing the zero boundary conditions at the rim of the drum means

$$J_n(\alpha) = 0.$$

Superimposing our separated solutions we obtain

$$u = \sum_{m=1,n=0}^{\infty} (c_{mn} \cos \alpha_m^{(n)} t + d_{mn} \sin \alpha_m^{(n)} t) \cdot$$

$$(a_{mn} J_n(\alpha_m^{(n)} r) \cos n\theta + b_{mn} J_n(\alpha_m^{(n)} r) \sin n\theta). \tag{10.18}$$

Since the initial deformation starts from rest (condition ii), $d_{mn} = 0$. Therefore our solution beginning from the initial deformation $f(r, z)$ is

$$u = \sum_{m=1,n=0}^{\infty} \cos \alpha_m^{(n)} t \; J_n(\alpha_m^{(n)} r)(a_{mn} \cos n\theta + b_{mn} \sin n\theta) \tag{10.19}$$

where

$$a_{mn} = \frac{(f(r, \theta), J_n(\alpha_m^{(n)} r) \cos n\theta)}{(J_n(\alpha_m^{(n)} r) \cos n\theta, J_n(\alpha_m^{(n)} r) \cos n\theta)} \tag{10.20}$$

and

$$b_{mn} = \frac{(f(r, \theta), J_n(\alpha_m^{(n)} \sigma) \sin n\theta)}{(J_n(\alpha_m^{(n)} \sigma) \sin n\theta, J_n(\alpha_m^{(n)} \sigma) \sin n\theta)}. \tag{10.21}$$

As long as the initial shape f is smooth enough so that

$$\sum_{m=1,n=0}^{\infty} \alpha_m^{(n)2} (a_{mn}^2 + b_{mn}^2) < \infty,$$

all the arguments of §6.3 hold giving that (10.18)–(10.21) is the (unique) strong solution of (10.14).

It is fascinating to watch animations of these vibrations (see Chapter 15). Bereft of damping, the sum (10.18) of non-consonant overtones $\cos \alpha_m^{(n)} t$ produce never-repeating shape trajectories of infinite variety.

§10.3 The Closed-Loop Heat Pump Revisited (SSP 7)

Recall our scheme proposed in Chapter 2 for heating buildings by extracting ground heat via buried exchangers — see Figure 2.7. Let us make a first attempt at modeling the transient ground temperatures during the heating season.

Again let us assume earth temperatures at the far-field distance $r = b$ are unaffected and remain at the yearly average temperature B. Let the radius of the exchanger be $r = a < b$. Lastly assume a constant load — the heat pump extracts energy at the constant rate of $\dot{Q}$ per unit exchanger length.

How will the surrounding earth cool over time?

Our model of earth temperatures $u = u(r,t)$ about one unit length of the exchanger is

$$\frac{\partial u}{\partial t} = \frac{\kappa}{c\rho}\left(\frac{\partial^2 u}{\partial r^2} + \frac{1}{r}\frac{\partial u}{\partial r}\right) \tag{10.22}$$

subject to

i) $u(r,0) = B$,

ii) $u(b,t) = B$,

iii) $2\pi a\kappa\, u_r(a,t) = \dot{Q}$,

where κ, c, ρ is the conductivity, specific heat, and density of the earth respectively.

Our first task is to transform the problem to one with homogeneous boundary conditions by *subtracting off the steady state.* To wit, set

$$v = u - (B - \frac{\dot{Q}}{2\pi\kappa}\log b/r).$$

See (2.15). Then our problem (10.22) becomes (Exercise 10.13)

$$\frac{\partial v}{\partial t} = \frac{\kappa}{c\rho}\left(\frac{\partial^2 v}{\partial r^2} + \frac{1}{r}\frac{\partial v}{\partial r}\right) \tag{10.23}$$

subject to

i)$'$ $v(r,0) = (\dot{Q}/2\pi\kappa)\log b/r$,

ii)$'$ $v(b,t) = 0$,

iii)′ $v_r(a,t) = 0$.

By assuming a separated solution, $v = q(t)R(r)$,

$$\frac{c\rho}{\kappa}\frac{\dot{q}}{q} = \frac{R'' + (1/r)R'}{R} = \lambda = -\alpha^2,$$

and thus

$$R = cJ_0(\alpha r) + dY_0(\alpha r) \tag{10.24}$$

and

$$q = e^{-\kappa\alpha^2 t/c\rho}.$$

Imposing the zero boundary conditions ii)′ and iii)′ yields

$$\begin{aligned} 0 &= cJ_0'(\alpha a) + dY_0'(\alpha a) \\ 0 &= cJ_0(\alpha b) + dY_0(\alpha b) \end{aligned} \tag{10.25}$$

with non-trivial solution exactly when the determinant

$$\begin{vmatrix} J_0'(\alpha a) & Y_0'(\alpha a) \\ J_0(\alpha b) & Y_0(\alpha b) \end{vmatrix} = 0. \tag{10.26}$$

By the spectral theorem we are guaranteed (Exercise 10.28) an increasing sequence

$$\alpha_1 < \alpha_2 < \alpha_3 < \dots$$

of positive roots $\alpha = \alpha_n$ of the determinant (10.26) yielding a complete set of orthogonal eigenfunctions

$$R_n = Y_0(\alpha_n b)J_0(\alpha_n r) - J_0(\alpha_n b)Y_0(\alpha_n r),$$

orthogonal with respect to the inner product

$$\langle f, g \rangle = \int_a^b f(r)g(r)r\, dr.$$

Seasonal ground temperatures are therefore given by the rule

$$u = B - \frac{\dot{Q}}{2\pi\kappa}\log b/r$$

$$+ \sum_{n=1}^{\infty} c_n e^{-\kappa\alpha_n^2 t/c\rho}(Y_0(\alpha_n a)J_0(\alpha_n r) - J_0(\alpha_n a)Y_0(\alpha_n r)), \tag{10.27}$$

where

$$c_n = \frac{\dot{Q}}{2\pi\kappa} \frac{\langle \log b/r, Y_0(\alpha_n b)J_0(\alpha_n r) - J_0(\alpha_n b)Y_0(\alpha_n r)\rangle}{\|Y_0(\alpha_n b)J_0(\alpha_n r) - J_0(\alpha_n b)Y_0(\alpha_n r)\|^2}. \tag{10.28}$$

Let us not proceed further with this rather complicated solution. We will derive in Chapter 16 a simpler but accurate transient model by letting the far field distance $b \to \infty$ and the exchanger radius $a \to 0$ to obtain the *Kelvin line source* solution.

§10.4 Current within a Round Conductor (BVP 15)

As we saw in BVP 9 of §8.3, the electric field intensity E within any good conductor satisfies to great accuracy

$$\mu\sigma \frac{\partial E}{\partial t} = \nabla^2 E. \tag{10.29}$$

Suppose the conductor is round of radius $r = a$. Let z be the distance down the conductor from the sinusoidal voltage source of frequency $f = \omega/2\pi$ Hz. Assume start-up transients have long since died away. Because of the high conductivity σ, E more or less points down the conductor. Let then

$$u = u(r, z, t) \tag{10.30}$$

be the component of E in this direction. Then from (10.29) within the conductor,

$$\mu\sigma \frac{\partial u}{\partial t} = \frac{\partial^2 u}{\partial r^2} + \frac{1}{r}\frac{\partial u}{\partial r} + \frac{\partial^2 u}{\partial z^2}. \tag{10.31}$$

When the wavelength is long compared to the radius $r = a$, we may neglect the z-term in the Laplacian (10.31). So for other than millimeter wavelengths it is accurate to model using simply

$$\mu\sigma \frac{\partial u}{\partial t} = \frac{\partial^2 u}{\partial r^2} + \frac{1}{r}\frac{\partial u}{\partial r}. \tag{10.32}$$

Moreover currents and voltages further down the conductor are only delayed versions of those closer to the source. Thus we may as well assume $u = u(r, t)$.

Intuitively, and as verified in Exercise 10.15,

$$u = X(r)\sin(\omega t - \theta(r)),$$

i.e., intensity within a conductor is merely delayed and of smaller amplitude than surface intensity.

Since we are solving a periodic steady state problem, again turn to the Helmholtz phasor method of §8.2: set

$$\rho = X(r)e^{i(\omega t - \theta(r))} = Y(r)e^{i\omega t},$$

and impose the equation (10.32) of diffusion to obtain

$$i\omega\sigma\mu Y = Y'' + \frac{1}{r}Y', \tag{10.33}$$

i.e,

$$r^2Y'' + rY' - i\omega\mu\sigma Y = 0. \tag{10.34}$$

But then (Exercise 10.16), relative amplitude and phase at radius r within the conductor is given by

$$Y(r) = \frac{I_0(\beta r)}{I_0(\beta a)} \tag{10.35}$$

where

$$\beta = \sqrt{\pi f \mu\sigma}(1 + i). \tag{10.36}$$

We know from examples (Exercise 8.7) that the depth of penetration into a flat conductor is extremely shallow, so shallow that our round conductor would appear relatively flat. So we would expect from (8.16) that

$$|Y(r)| \approx e^{-\sqrt{\pi f \mu\sigma}}, \tag{10.37}$$

which is indeed the case (Exercise 10.17).

Exercises

All solids are assumed homogeneous and isotropic.

10.1 Quench a solid cylinder as in BVP 14 but where one flat face (say at $z = 1$) is insulated.

Answer:

$$u = 4 \sum_{m=1,n=0}^{\infty} e^{-(\alpha_m^2 + (n+1/2)^2\pi^2)t} \cdot \frac{J_0(\alpha_m r)}{\alpha_m J_1(\alpha_m)} \frac{\sin(n+1/2)\pi z}{(n+1/2)\pi}$$

where $\alpha_1 < \alpha_2 < \cdots$ are the positive zeros of J_0.

10.2 Quench a solid cylinder initially at temperature $u = f(r)$ whose curved side is insulated.

Answer:

$$u = \sum_{m=0,n=1} c_{mn} e^{-(\alpha_m^2 + n^2\pi^2)t} J_0(\alpha_m r) \sin n\pi z$$

with

$$c_{mn} = \frac{4(1-(-1)^n)}{n\pi} \cdot \frac{\int_0^1 f(r) J_0(\alpha_m r) r\omega, dr}{J_0(\alpha_m)^2},$$

where

$$\alpha_0 = 0 < \alpha_1 < \alpha_2 < \cdots$$

are the non-negative zeros of J_1.

10.3 Apply Exercise 10.2 when the cylinder is initially at uniform temperature $f(r) = 1$. Find c_{mn}.

Check: This is BVP 1 in disguise.

10.4 Quench a solid cylinder with insulated end faces, of initial temperature $u(r, \theta, 0) = f(r, \theta)$ in a bath of temperature 0.

Answer:

$$u = \sum_{m=1,n=0}^{\infty} \mathrm{e}^{-\alpha_m^{(n)2} t} (a_{mn} \cos n\theta + b_{mn} \sin n\theta) J_n(\alpha_m^{(n)} r),$$

where $\alpha_0 = 0 < \alpha_1^{(n)} < \alpha_2^{(n)} < \ldots$ are the non-negative zeros of J_n.

10.5 Apply the result of the previous exercise for the initial temperature profile $f(r, \theta) = r\theta, \ -\pi \leq \theta < \pi$.

10.6 Find the steady state temperature within a solid cylinder of radius and height 1 when the flat faces are held at temperature 0 while the curved side is held at temperature 1.

Answer:

$$u = 4 \sum_{\substack{n=1 \\ n \text{ odd}}}^{\infty} \frac{I_0(n\pi r)}{I_0(n\pi)} \frac{\sin n\pi z}{n\pi}.$$

10.7 Find the steady state temperature within a half cylinder $0 \le r \le 1$, $0 \le \theta \le \pi$, $0 \le z \le 1$ where the flat faces are held at temperature 0 and the curved side at temperature 1.

Answer:

$$u = (16/\pi) \sum_{\substack{m,n=1 \\ m,n \text{ odd}}}^{\infty} \frac{I_m(n\pi r)}{I_m(n\pi)} \cdot \frac{\sin m\theta}{m} \cdot \frac{\sin n\pi z}{n\pi}.$$

10.8 Quench the half cylinder of the previous exercise, given an initial uniform temperature of 1 and a bath at temperature 0.

Answer:

$$u = \sum_{k,m,n=1}^{\infty} c_{kmn} e^{-(\alpha_k^{(m)2} + n^2\pi^2)t} J_m(\alpha_k^{(m)} r) \sin m\theta \ \sin n\pi z,$$

where $\alpha_k^{(m)}$ is the k-th positive zero of J_m.

10.9 A solid cylinder of radius $r = a$ with insulated flat ends, initially at the temperature of $u(r, 0) = f(r)$, is losing heat to its surroundings from its curved side by *Newton's law of cooling* — at a rate proportional to the difference between its surface temperature $u(a, t)$ and ambient temperature T_0. Find the interior temperature profiles $u = u(r, t)$ as they relax to T_0.

Answer: If κ, c, ρ, h is conductivity, specific heat, density, and surface heat transfer coefficient of the cylinder, then

$$u = \sum_{k=1}^{\infty} c_k \mathrm{e}^{-\kappa\alpha_k^2 t/c\rho} J_0(\alpha_k r) + T_0,$$

where $\alpha_1 < \alpha_2 < \ldots$ are the positive roots of

$$x\kappa J_1(xa) = hJ_0(xa),$$

and where

$$c_k = \frac{2h\kappa}{(\kappa\alpha_k)^2 + h^2} \frac{1}{J_0(\alpha_k a))}.$$

10.10 Find the coefficients c_n in the previous exercise when the initial temperature is a uniform $u(r, 0) = T_1$.

10.11 Highly selective IF bandpass filters are achieved with *electromechanical filters.* Many small ferroceramic thin disks are mounted coaxially along a thin rod. The disk at one end is excited by a transducer, the resulting vibrations propagate to each disk in turn mechanically along the coaxial rod. The output is taken from the last disk again by an electromagnetic transducer. The sizes and spacings of the disks are chosen to achieve superior bandpass characteristics.

What are the natural frequencies of a thin disk when rung at its center?

Answer: $f_k = \alpha_k^2/2\pi$ where the α_k are the positive roots of

$$I_m(x)J_m'(x) - J_m(x)I_m'(x) = 0.$$

Discussion: One might at first conjecture that the proper model is the vibrating membrane fixed at the center: $u_{tt} = \nabla^2 u$, giving the answer $f_{kn} = \beta_k^{(n)}/2\pi$ where $J_n'(\beta_k^{(n)}) = 0$, $n > 0$.

However a more accurate model would take into account the bending stiffness of the disk:

$$u_{tt} = -\nabla^2\nabla^2 u,$$

the plate equation. The boundary conditions are then $u(0,\theta,t) = 0 = u_r(0,\theta,t)$ and $\nabla^2 u(1,\theta,t) = 0 = \partial\nabla^2 u(1,\theta,t)/\partial r$. Eigenfunctions are then linear combinations of

$$J_n(\alpha r)\cos n\theta,\ J_n(\alpha r)\sin n\theta,\ I_n(\alpha r)\cos n\theta,\ I_n(\alpha r)\sin n\theta.$$

See [Southwell; Rayleigh, pp. 353–371; Radcliffe and Mote].

10.12 Solve Exercise 2.12 by Separation of Variables. Why is the solution free of Bessel functions?

Answer:

$$u = 1/2 + \frac{2}{\pi}\sum_{n=0}^{\infty} r^{2n+1}\frac{\sin(2n+1)\theta}{2n+1}.$$

10.13 Verify that subtracting away the steady state brings the problem (10.22) to the problem (10.23) with homogeneous boundary conditions.

10.14 Show that the squared norm in the denominator of (10.28) is given by

$$\|Y_0(\alpha_n b)J_0(\alpha_n r) - J_0(\alpha_n b)Y_0(\alpha_n r)\|^2$$

$$= \int_a^b (Y_0(\alpha_n b)J_0(\alpha_n r) - J_0(\alpha_n b)Y_0(\alpha_n r))^2 \, rdr$$

$$= \frac{b^2\phi'(\alpha_n b)^2 - a^2\phi(\alpha_n a)^2}{2},$$

where

$$\phi(x) = Y_0(\alpha_n b)J_0(x) - J_0(\alpha_n b)Y_0(x).$$

Hint: Redo Exercise 9.14 for any solution $y = \phi(x)$ to Bessel's equation when $n = 0$.

10.15 Suppose the temperature at the perimeter of a disk is varying sinusoidally with frequency $f = \omega/2\pi$. Show that after transients die away, the periodic steady state temperatures within the disk are of the form

$$u = X(r)\sin(\omega t - \theta(r)).$$

10.16 Verify (10.35) and (10.36).

10.17 Numerically verify the asymptotic estimate (10.37) of (10.35) for depth of penetration of current within a round conductor.

10.18 How must (10.35) be modified for millimeter wavelengths?

Answer:

$$Y(r) = \frac{I_0(\beta r)}{I_0(\beta a)}$$

where

$$\beta = \sqrt{i\omega\mu\sigma - \omega^2\mu\epsilon}.$$

Hint: To good approximation,

$$\rho(r, z, t) = \rho(r, z_0, t - (z - z_0)\sqrt{\mu\epsilon}).$$

10.19 A circular waveguide of radius a can carry radio frequencies of free space wavelength λ only if

$$\lambda < 2a\frac{\pi}{\alpha_1^{(0)}} \approx 2.612a.$$

Outline: As in §8.7, write both E and H in the form

$$\sin(\omega t - \gamma z)[\, T(x,y) + w(x,y)\mathbf{k}\,]$$

where T is transverse. Then either $\mu\epsilon\omega^2 > (\alpha_1^{(0)}/a)^2$ or $w \equiv 0$. Apply Exercise 10.32.

10.20 Find a bounded solution of the problem

$$\frac{\partial u}{\partial t} = \frac{\partial}{\partial x}(x\frac{\partial u}{\partial x}), \quad 0 < x < 1, \; 0 < t,$$

subject to $u(x,0) = f(x)$ and $u(1,t) = 0$.

Answer:

$$u = \sum_{n=1}^{\infty} c_n \mathrm{e}^{-\alpha_n^2 t/4} J_0(\alpha_n \sqrt{x})$$

where the $\alpha_1 < \alpha_2 < \ldots$ are the positive roots of J_0.

Hint: Make the substitution $4x = y^2$.

10.21* The Rellich-Weyl Principle (§6.7) guarantees a unique strong solution of the form

$$u = \sum_{n=1}^{\infty} c_n \mathrm{e}^{\lambda_n t}\phi_n$$

when we quench an elliptical cylinder $x^2/4 + y^2 < 1, \; 0 < z < 1$. What are the spatial eigenfunctions $\phi_n(x)$?

10.22 Steady flow of a viscous incompressible fluid is modeled by the **Navier-Stokes equation** [Batchelor]

$$\nabla p = -\rho(V \cdot \nabla)V + \mu\nabla^2, V$$

where p is pressure, V is fluid velocity, ρ is density, and μ is viscosity. Apply this equation to steady flow in a straight circular tube. The velocity profile across the tube increases from the boundary of the tube towards a maximum at the center. At first thought, a candidate for this profile would be $J_0(\alpha r)$. But in fact the profile is a simple quadratic function of r. Why?

Hint: By symmetry, the convective term $(V \cdot \nabla)V = 0$. Moreover pressure decreases steadily down the tube.

10.23 A solid rod at temperature 1 of radius $r = a$ is suddenly inserted into a sleeve $a < r < b$ of the same material at temperature 0. The outer surface of this sleeve is held thereafter at temperature 0. Find how the interior temperatures of this construct relax to 0. Ignore heat loss from the flat end faces.

Answer:

$$u = \frac{2a}{b} \sum_{k=1}^{\infty} \mathrm{e}^{-\kappa \alpha_k^2 t / c\rho} \frac{J_1(\alpha_k a/b)}{\alpha_k J_1(\alpha_k)^2} J_0(\alpha_k r/b)$$

where the α_k are the positive roots of $J_0(x) = 0$.

10.24 How must the solution of Exercise 10.23 be modified if the rod is made from a different material than the sleeve?

10.25 Find the potential within the cylindrical capacitor $r < 1$, $0 < z < 1$ when the end at $z = 0$ and the curved side is at voltage 0, while the other end at $z = 1$ is at voltage 1.

Answer:

$$u = 2 \sum_{n=1}^{\infty} \frac{J_0(\alpha_n r)}{\alpha_n J_1(\alpha_n)} \frac{\sinh \alpha_n z}{\sinh \alpha_n}$$

where the α_n are the positive zeros of J_0.

10.26 Find the steady state temperatures within the cylinder $r < 1$, $0 < z < 1$ when the bottom at $z = 0$ is held at temperature 0, the curved side is insulated, and the top is held at temperature $f(r)$.

Answer:

$$u = \sum_{n=1}^{\infty} c_n J_0(\alpha_n r) \sinh \alpha_n z,$$

where the α_n are the positive zeros of J_1 and where

$$c_n = 2 \frac{\langle f(r), J_0(\alpha_n r) \rangle}{\sinh \alpha_n \; J_0(\alpha_n)^2}.$$

What rules out solutions of the form

$$u = \sum_{n=1}^{\infty} c_n I_0(\alpha_n r) \sin \alpha_n z$$

where $\alpha_n > 0$?

10.27 What is the (lowest) resonant radio frequency of a cylindrical cavity resonator of radius a and height b when $a > b$?

Answer:

$$f = \omega/2\pi = \frac{\alpha_1^{(0)}}{2\pi a\sqrt{\mu\epsilon}}$$

Hint: Try $E = \sin \omega t \, R(r)\mathbf{k}$.

10.28* Show that the implied spatial differential operator of the problem (10.23) is indeed negative definite Hermitian with compact resolvent. Deduce from the Spectral Theorem that the determinant (10.26) has positive roots increasing to infinity yielding an orthogonal basis of eigenfunctions displayed in (10.27).

10.29 What is the acoustic resonant frequency of a rigid hollow closed cylinder of radius a and length b?

Answer: $f = c\sqrt{(\alpha/a)^2 + (n\pi/b)^2}/2\pi$ where α is a positive root of $J_1(r) = 0$.

Outline: According to the Lagrange acoustic model [Rayleigh], *particle displacement* D satisfies the wave equation

$$\frac{\partial^2 D}{\partial t^2} = c^2 \nabla^2 D$$

whereupon $p = \rho c^2 \nabla \cdot D$ is pressure and $V = \partial D/\partial t$ is particle velocity. Displacement D is assumed *irrotational,* i.e., $\nabla \times D = 0$. Argue that pressure p satisfies zero flux boundary conditions. (In contrast, if one end is open like a flute, pressure is 0 at that end).

10.30 In view of previous problems it may come as a surprise that Bessel functions play no part in the description of wave propagation down a properly terminated coaxial cable (for frequencies lower than the cutoff of the cable when viewed as an annular waveguide). Show in fact that both E and H propagate transversely with E radial and H tangential and in fact

$$E = \sin(\omega t - \omega z/\sqrt{\mu\epsilon})\nabla \log r, \ a < r < b.$$

Contrast this result to Exercise 10.32.

10.31 Find the dominant transverse electric mode of propagation down a circular waveguide of radius a.

Answer: $\mathbf{TE_{11}}$:

$$E = \sin(\omega t - \gamma z)J_1'(\alpha r)\sin\theta\ \nabla \log r + \alpha r J_1(\alpha r)\cos\theta\ \nabla\theta,$$

where α is the least positive root of $J_1'(\alpha a) = 0$.

10.32 Show that it is impossible for both E and H to propagate transversely within a waveguide that has a simply connected cross-section Ω.

Outline: Suppose both E and H are transverse. Write E in the form (8.41) and note that $\nabla\times E\cdot\mathbf{k} = \sin(\omega t - \gamma z)\ (v_x - u_y) = 0$. Thus since Ω is without holes, $u\mathbf{i} + v\mathbf{j} = \nabla f$ where, because $\nabla\cdot E = 0$, f is harmonic. But f has zero tangential boundary conditions and is therefore constant. Thus $E = 0$.

10.33 (The quantum bouncing ball) A very small bouncing ball of mass m obeys

$$i\hbar\frac{\partial\psi}{\partial t} = -\frac{\hbar^2}{2m}\frac{\partial^2\psi}{\partial x^2} + V\psi$$

where the (gravitational) potential is observed by the multiplication operator $V\psi = a\psi$ with multiplier

$$a(x) = \begin{cases} mgx & \text{if } x \geq 0 \\ \infty & \text{if } x < 0. \end{cases}$$

Find the nondimensional stationary states ψ, the solutions of

$$-\psi'' + x\psi = E\psi.$$

Answer: $\psi(x) = \mathrm{Ai}(x - E)$ with E determined by the characteristic equation $\mathrm{Ai}(-E) = 0$ where the *Airy function*

$$\mathrm{Ai}(z) = \frac{1}{3}\sqrt{z}(I_{-1/3}(\frac{2}{3}z^{3/2}) - I_{1/3}(\frac{2}{3}z^{3/2})).$$

Outline: Set $u(z) = \psi(z + E)$ to bring the problem to *Airy's equation*

$$\frac{d^2u}{dz^2} - zu = 0.$$

To solve Airy's equation set $w = 3u/\sqrt{z}$ and $y = 3z^{3/2}/2$ to transform the problem to the modified Bessel equation (9.41):

$$y^2\frac{d^2w}{dy^2} + y\frac{dw}{dy} - (y^2 + \frac{1}{9})w = 0$$

with independent solutions $I_{1/3}$ and $I_{-1/3}$. Impose the boundary conditions $\psi(0) = 0 = \psi(\infty)$. See [Davies, p. 28] and [Abramowitz and Stegun].

10.34 Show that the volume flow of a fluid down a circular pipe is half the flow obtained by using the velocity along the central axis.

Hint: Exercise 10.22.

Chapter 11
Orthogonal Polynomials

In this Chapter we study the most often used orthogonal polynomials — the Legendre, Chebyshev, Hermite, and Laguerre polynomials. We see how they arise from the Gram-Schmidt process and how they arise naturally as eigenfunctions of physically important spatial operators. We will learn their recurrence formulae, norm formulae, and differential relations.

§11.1 How They Arise

Perform the Gram-Schmidt orthogonalization process of §6.4 on

$$1,\ x,\ x^2,\ x^3,\ \ldots$$

to obtain a sequence

$$\phi_0(x),\ \phi_1(x),\ \phi_2(x),\ \phi_3(x),\ \ldots$$

of *orthonormal polynomials.* The polynomials ϕ_n depend strongly upon the inner product used.

Orthogonal polynomials are employed in Numerical Analysis for least square approximations. Several classes play a central role in Filter Design, others in Number Theory, still others in Quantum Mechanics. Our interest is that orthogonal polynomials arise as the spatial eigenfunctions of important boundary value problems.

§11.2 Weighted Inner Products

Many physically meaningful energy functions are *weighted inner products*

$$\langle f, g \rangle_w = \int_a^b f(x)g(x)\ w(x)\ dx \tag{11.1}$$

where the absolutely integrable *weight* $w(x) > 0$ for $a < x < b$.

Certain weights on certain intervals give rise to very useful classes of orthogonal polynomials.

Example 1. Take $a = -1$, $b = 1$, and $w(x) \equiv 1$. We then obtain (Exercise 11.1) the *Legendre polynomials*

$$P_0(x) = 1,\ P_1(x) = x,\ P_2(x) = (3x^2 - 1)/2,$$
$$P_3(x) = (5x^3 - 3x)/2,\ P_4(x) = (35x^4 - 30x^2 + 3)/8,\ \ldots. \qquad (11.2)$$

Alert: Rather than normalizing each P_n to norm 1 it is instead traditional to choose some constant multiple to achieve a standard value at one of the interval endpoints. For the Legendre polynomials the *standardization* is

$$P_n(1) = 1. \qquad (11.3)$$

Example 2. Again take $a = -1$, and $b = 1$ but set $w(x) = 1/\sqrt{1 - x^2}$ to obtain (Exercise 11.2) the *Chebyshev polynomials of the first kind*

$$T_0(x) = 1,\ T_1(x) = x,\ T_2(x) = 2x^2 - 1,$$

$$T_3(x) = 4x^3 - 3x,\ T_4(x) = 8x^4 - 8x^2 + 1,\ \ldots \qquad (11.4)$$

with the standardization

$$T_n(1) = 1. \qquad (11.5)$$

Chebyshev is often transliterated as *Tchebyscheff.*

Example 3. When $a = -\infty$, $b = \infty$, and $w(x) = \mathrm{e}^{-x^2}$, the resulting orthogonal polynomials (Exercise 11.4)

$$H_0(x) = 1,\ H_1(x) = 2x,\ H_2(x) = 4x^2 - 2,$$
$$H_3(x) = 8x^3 - 12x,\ H_4(x) = 16x^4 - 48x^2 + 12,\ \ldots \qquad (11.6)$$

are the *Hermite polynomials.* They are standardized via Rodrigues's formula below.

Example 4. Take $a = 0$, $b = \infty$ and $w(x) = \mathrm{e}^{-x}$ to obtain the *Laguerre polynomials*

$$L_0(x) = 1,\ L_1(x) = -x + 1,\ L_2(x) = (x^2 - 4x + 2)/2,$$

$$L_3(x) = (-x^3 + 9x^2 - 18x + 6)/6,$$

$$L_4(x) = (x^4 - 16x^3 + 72x^2 - 96x + 24)/24, \ \ldots \tag{11.7}$$

(Exercise 11.6) with standardization

$$L_n(0) = 1. \tag{11.8}$$

§11.3 Completeness

Notation. $L_w^2(a,b)$ will denote the Hilbert space of all w-square-integrable functions f:

$$\int_a^b f(x)^2 \, w(x) \, dx < \infty.$$

See Exercise 11.39.

Lemma. Suppose $1, \ x, \ x^2, \ x^3, \ldots$ are all w-square integrable, i.e., functions of $L_w^2(a,b)$. Moreover suppose that for some $r > 0$,

$$\int_a^b e^{r|x|} \, w(x) dx < \infty. \tag{11.9}$$

Then the orthonormal polynomials $\phi_0, \ \phi_1, \ \phi_2, \ \phi_3, \ \ldots$ arising from the weight w form a complete orthonormal basis of $L_w^2(a,b)$

Proof sketch. As an artifact of the Gram-Schmidt process, the span of $\{1, \ x, \ x^2, \ \ldots, x^n\}$ is identical with the span of $\{\phi_0, \ \phi_1, \phi_2, \ \ldots, \phi_n\}$. Thus it is enough to show the polynomials are dense. If they are not, some nonzero w-square integrable function f is orthogonal to all x^n. But then

$$\int_a^b e^{ixt} f(x) w(x) dx = \sum_{n=0}^{\infty} \frac{(it)^n}{n!} \int_a^b x^n f(x) w(x) dx = 0$$

assuming integrability and that the interchange of integration with summation is justified. By extending $w(x)$ to be 0 off (a,b), we have obtained that the Fourier transform of $g(x) = f(x)w(x)$ is zero. But the only absolutely integrable function with zero Fourier transform is the zero function [Rudin].

But when is the interchange of integration and summation justified? Certainly when the interval (a, b) is bounded. If the interval is not bounded, a more delicate interchange argument using an additional assumption such as (11.9) is required — see the classic *Introduction to Real Functions and Orthogonal Expansions* by Bèla Sz.-Nagy. Put as an epigram, under mild assumptions on the weight w,

the w-moments of a nonzero absolutely w-integrable function f cannot all vanish.

§11.4 Polynomial Eigenfunctions

Consider spatial differential operators of the special form

$$A\phi = p(x)\phi'' + q(x)\phi' \tag{11.10}$$

where both $p(x)$ and $q(x)$ are polynomials. Note that the constant function $\phi_0 = 1$ is always an eigenfunction belonging to the value $\lambda = 0$.

Suppose A has a polynomial eigenfunction ϕ_n of every degree n with distinct eigenvalues. Then if A is Hermitian with respect to some weighted inner product with weight w, these eigenfunctions ϕ_n are mutually w-orthogonal and therefore by induction *are the orthogonal polynomials arising from this weight w.*

But which operators $A\phi = p(x)\phi''+q(x)\phi'$ have polynomial eigenfunctions of every degree?

Consider the very special case

$$A\phi = (p_0 + p_2x^2)\phi'' + q_1x\phi', \quad p_0 \neq 0. \tag{11.11}$$

If we search for a polynomial eigenfunction of degree n

$$\phi_n = \sum_{k=0}^{n} c_k x^k$$

solving

$$A\phi_n = \lambda_n\phi_n, \tag{11.12}$$

we are led (Exercise 11.7) to the recursion

$$c_{k+2} = \frac{\lambda_n - q_1k - p_2k(k-1)}{p_0(k+1)(k+2)}c_k. \tag{11.13}$$

When n is even, set $c_0 = 1$, $c_1 = 0$, and take $\lambda_n = q_1 n + p_2 n(n-1)$ to break off the recursion at degree n, thus obtaining a polynomial solution of (11.12). If n is odd, take $c_0 = 0$, $c_1 = 1$, and λ_n as in the even case. To summarize,

Result A. The operator

$$A\phi = (p_0 + p_2 x^2)\phi'' + q_1 x\phi', \quad p_0 \neq 0$$

has a polynomial eigenfunction ϕ_n of each degree n belonging to the eigenvalue

$$\lambda_n = q_1 n + p_2 n(n-1).$$

Corollary A. The *Legendre operator*

$$A\phi = (1 - x^2)\phi'' - 2x\phi' \tag{11.14}$$

has a polynomial eigenfunction ϕ_n of each degree n with eigenvalue

$$\lambda_n = -n(n+1). \tag{11.15}$$

Corollary B. The *Chebyshev operator*

$$A\phi = (1 - x^2)\phi'' - x\phi' \tag{11.16}$$

has a polynomial eigenfunction ϕ_n of each degree n with eigenvalue

$$\lambda = -n^2. \tag{11.17}$$

Corollary C. The *Hermite operator*

$$A\phi = \phi'' - 2x\phi' \tag{11.18}$$

has a polynomial eigenfunction ϕ_n of each degree n with eigenvalue

$$\lambda = -2n. \tag{11.19}$$

And so forth. See [Abramowitz-Stegun] or [Lebedev].

Theorem A. The Legendre operator

$$A\phi = (1 - x^2)\phi'' - 2x\phi'$$

can be well conditioned to be densely defined, Hermitian, negative semi-definite on $X = L^2(-1, 1)$. The Legendre polynomials

$$P_0,\ P_1,\ P_2,\ P_3,\ \ldots$$

comprise a complete orthogonal basis of eigenfunctions of A with eigenvalues

$$\lambda_n = -n(n+1)$$

under the usual inner product, i.e., weight $w(x) = 1$.

Proof. Clearly A commutes across the inner product since A can be put in the *self-adjoint form*

$$A\phi = ((1 - x^2)\phi')', \tag{11.20}$$

giving by Parts that

$$\langle A\phi, \psi\rangle = \int_{-1}^{1} ((1 - x^2)\phi')'\psi \, dx$$

$$= (1 - x^2)\phi'\psi|_{-1}^{1} - \int_{-1}^{1} \phi' \, (1 - x^2)\psi' \, dx$$

$$= - \int_{-1}^{1} \phi' \, (1 - x^2)\psi' \, dx \tag{11.21}$$

$$= -\phi(1 - x^2)\psi'|_{-1}^{1} + \int_{-1}^{1} \phi \, ((1 - x^2)\psi')' \, dx = \langle \phi, A\psi\rangle.$$

It is clear from (11.21) that A is negative semi-definite. Since 0 is an eigenvalue, A cannot be definite.

In detail, since A commutes across the inner product, eigenfunctions belonging to distinct eigenvalues are orthogonal. By Corollary A above, A possesses polynomial eigenfunctions ϕ_n of each degree n with distinct eigenvalues λ_n. In particular ϕ_0 is a multiple of $P_0 = 1$. Also ϕ_1 is of degree 1 yet orthogonal to 1 and hence must be a multiple of $P_1 = x$. Likewise ϕ_2 is of degree 2 yet orthogonal to both P_0 and P_1, thus ϕ_2 must be a multiple of P_2, and so on. Within a constant multiple, ϕ_n is P_n.

There are many ways to see that the Legendre polynomials P_n form a complete basis. For one, there is the above Lemma. For another, as we shall see in Chapter 12, the Legendre operator arises from the Laplacian on the unit ball with zero boundary conditions; the theorem can be deduced from Rellich's Theorem.

Note that the operator

$$R(\lambda)f = \sum_{n=0}^{\infty} \frac{\langle f, P_n \rangle}{\lambda + n(n+1)} \frac{P_n}{\langle P_n, P_n \rangle} \tag{11.22}$$

is for (say) $\lambda > 0$ everywhere defined, injective, bounded on X, and inverts $\lambda - A$ at each eigenfunction P_n (Exercise 11.8). Thus A is well conditioned once we declare $R(\lambda)$ to be its resolvent — see §6.6.

Theorem B. The Chebyshev operator

$$A\phi = (1 - x^2)\phi'' - x\phi'$$

can be well conditioned to be densely defined, Hermitian, negative semi-definite on $X = L^2_w(-1, 1)$. The Chebyshev polynomials

$$T_0,\ T_1,\ T_2,\ T_3,\ \ldots$$

comprise a complete orthogonal basis of eigenfunctions of A with eigenvalues

$$\lambda_n = -n^2$$

under the weighted inner product

$$\langle f, g \rangle_w = \int_{-1}^{1} f(x)g(x)\,(1 - x^2)^{-1/2} dx.$$

Proof. It is easy to see directly that A commutes across the inner product (Exercise 11.9). Thus the polynomial eigenfunctions ϕ_n guaranteed by Corollary B are in fact the Chebyshev polynomials T_n.

It is more revealing to make the substitution $x = \cos\theta$ whereupon the operator A becomes (Exercise 11.10)

$$A\psi = \frac{d^2}{d\theta^2}\psi \tag{11.23}$$

on the space $L^2(0,\pi)$ under the usual (unweighted) inner product, subject to

$$\psi'(0) = 0 = \psi'(\pi).$$

We now recognize that the polynomial eigenfunctions T_n have become simply $\cos n\theta$ and are certainly dense (§7.1, Result B).

This explains the vague familiarity of the first few Chebyshev polynomials displayed in Example 2 — they are the common trigonometric identities

$$\cos n\theta = T_n(\cos\theta). \tag{11.24}$$

Theorem C. The Hermite operator

$$A\phi = \phi'' - 2x\phi'$$

can be well conditioned to be densely defined, Hermitian, negative semi-definite on $X = L^2_w(-\infty,\ \infty)$. The Hermite polynomials

$$H_0,\ H_1,\ H_2,\ H_3,\ \ldots$$

comprise a complete orthogonal basis of eigenfunctions of A with eigenvalues

$$\lambda_n = -2n$$

under the weighted inner product

$$\langle f, g\rangle_w = \int_{-\infty}^{\infty} f(x)g(x)\, \mathrm{e}^{-x^2}\, dx.$$

Proof. It is easy to see that A commutes across the inner product giving as before that the Hermite polynomials are the polynomial eigenfunctions guaranteed by Corollary C. Completeness follows by the Lemma. Well conditioning is provided by the resolvent

$$R(\lambda)f = \sum_{n=0}^{\infty} \frac{\langle f, H_n\rangle_w}{\lambda + 2n}\, \frac{H_n}{\langle H_n, H_n\rangle_w}.$$

Theorem D. The Laguerre operator

$$A\phi = x\phi'' + (1 - x)\phi'$$

can be well conditioned to be densely defined, Hermitian, negative semi-definite on $X = L^2_w(0,\ \infty)$. The Laguerre polynomials

$$L_0,\ L_1,\ L_2,\ L_3,\ \ldots$$

comprise a complete orthogonal basis of eigenfunctions of A with eigenvalues

$$\lambda_n = -n$$

under the weighted inner product

$$\langle f, g\rangle_w = \int_0^\infty f(x)g(x)\,\mathrm{e}^{-x}dx.$$

Proof. Exercise 11.12.

§11.5 Choosing Weights

There is actually no choice. The weight $w(x)$ required to bring the operator A of (11.10) to self-adjoint form — or in more modern language, the weight required to obtain an inner product with respect to which A is Hermitian — is the integrating factor

$$w(x) = p(x)^{-1}\exp\left(\int q(x)p(x)^{-1}\,dx\right). \tag{11.25}$$

For if

$$w(x)A\phi = w(x)p(x)\phi'' + w(x)q(x)\phi' = (w(x)p(x)\phi')', \tag{11.26}$$

then $w(x)$ is given by (11.25) (Exercise 11.13).

§11.6 Recurrence Formulae

Let $\psi_n(x)$ be the standardized orthogonal polynomials arising from the weight w on the interval (a, b).

Result B. There are coefficients $\alpha_n, \beta_n, \gamma_n, \delta_n$ (determined only within common multiples) so that for $n \geq 2$,

$$\delta_n\psi_{n+1}(x) = (\alpha_n + \beta_n x)\psi_n(x) - \gamma_n\psi_{n-1}(x). \tag{11.27}$$

Proof. We may assume $\delta_n = 1$. For an appropriate choice of α_n, β_n, and γ_n,

$$\psi_{n+1}(x) - \beta_n x\psi_n(x) - \alpha_n\psi_n(x) + \gamma_n\psi_{n-1}(x) \tag{11.28}$$

is of degree $n-2$ and hence of the form

$$\sum_{k=0}^{n-2} c_k\psi_k(x). \tag{11.29}$$

But since each ψ_n is orthogonal to all polynomials of lower degree,

$$c_k = \langle \psi_{n+1}(x) - (\alpha_n + \beta_n x)\psi_n(x) + \gamma_n\psi_{n-1}(x), \psi_k(x)\rangle_w \tag{11.30}$$

$$= -\beta_n\langle x\psi_n(x), \psi_k(x)\rangle_w = -\beta_n\langle \psi_n(x), x\psi_k(x)\rangle_w = 0. \tag{11.31}$$

Table 11.1 lists the coefficients of the recurrence formulae for the standardized Legendre, Chebyshev, Hermite, and Laguerre polynomials. These values follow from Rodrigues's formula below.

Table 11.1 The recurrence relations

ψ_n	δ_n	α_n	β_n	γ_n
P_n	$n+1$	0	$2n+1$	n
T_n	1	0	2	1
H_n	1	0	2	2n
L_n	$n+1$	$2n+1$	-1	n

§11.7 Norm Formulae

Again suppose the $\psi_n(x)$ are the standardized orthogonal polynomials arising from the weight $w(x)$ on the interval (a, b).

By the recurrence Result B, there are coefficients α_n, β_n, γ_n, δ_n, determined only within common multiples, so that

$$\delta_n\psi_{n+1}(x) = (\alpha_n + \beta_n x)\psi_n(x) - \gamma_n\psi_{n-1}(x). \tag{11.32}$$

Result C. Norms of the standardized orthogonal polynomials can be determined recursively by the rule

$$\|\psi_n\|_w^2 = \frac{\beta_{n-1}}{\beta_n}\frac{\gamma_n}{\delta_{n-1}}\|\psi_{n-1}\|_w^2, \quad n \geq 2. \tag{11.33}$$

Proof. Applying the functional $\langle . \, , \psi_{n-1}\rangle_w$ to both sides of (11.32) yields

$$\langle x\psi_n, \psi_{n-1}\rangle_w = \frac{\gamma_n}{\beta_n}\|\psi_{n-1}\|_w^2.$$

Likewise, applying the functional $\langle . \, , \psi_n\rangle_w$ to both sides of (11.32), after n has been replaced by $n-1$, yields

$$\|\psi_n\|_w^2 = \frac{\beta_{n-1}}{\delta_{n-1}}\langle x\psi_{n-1}, \psi_n\rangle_w.$$

Corollary.

$$\|\psi_n\|_w^2 = \frac{\beta_1}{\beta_n}\prod_{k=2}^{n}(\gamma_k/\delta_{k-1})\ \|\psi_1\|_w^2. \tag{11.34}$$

Assume the recurrence coefficients listed in Table 11.1 are correct. Then immediately we obtain (Exercise 11.15) the

Norm formulae:

$$\|P_n\| = \sqrt{\frac{2}{2n+1}}, \tag{11.35}$$

$$\|T_0\|_w = \sqrt{\pi}, \ \ \|T_n\|_w = \sqrt{\pi/2}, \ n > 0 \tag{11.36}$$

$$\|H_n\|_w = \sqrt{\sqrt{\pi}2^n n!}, \tag{11.37}$$

$$\|L_n\|_w = 1. \tag{11.38}$$

§11.8 Rodrigues's Formula

Again let $\psi_0(x), \ \psi_1(x), \ \psi_2(x), \ \ldots$ be the standardized orthogonal polynomials arising from the weight $w(x)$ on the interval (a, b).

A select few classes are given by *Rodrigues's formula*

$$\psi_n(x) = \frac{1}{e_n w(x)}\frac{d^n}{dx^n}(w(x)g(x)^n) \tag{11.39}$$

where e_n, $w(x)$, and $g(x)$ for the Legendre, Chebyshev, Hermite, and Laguerre polynomials are given in Table 11.2. More extensive tables appear in [Abramowitz and Stegun].

The function $g(x)$ in (11.39) is determined by the case $n = 1$, to wit

$$g(x) = c\, w(x)^{-1} \int_a^x \psi_1(x) w(x) dx. \tag{11.40}$$

Table 11.2 Rodrigues's formula

ψ_n	$w(x)$	e_n	$g(x)$
P_n	1	$(-1)^n 2^n n!$	$1 - x^2$
T_n	$(1 - x^2)^{-1/2}$	$(-1)^n 2^n \Gamma(n + 1/2)/\sqrt{\pi}$	$1 - x^2$
H_n	e^{-x^2}	$(-1)^n$	1
L_n	e^{-x}	$n!$	x

But why should one expect such a formula as Rodrigues's? It helps to change over to 'big D' operator notation

$$D^n = \frac{d^n}{dx^n}.$$

Then functions determined by Rodrigues's formula

$$f_n(x) = w(x)^{-1} D^n w(x) g(x)^n$$

are w-orthogonal to each x^m for $m < n$ provided $D^i w(x) g(x)^n$ vanishes at both endpoints a and b for $i < n$. This can be seen by repeated Integration by Parts. Thus if each f_n is a polynomial of degree n, the f_n are the orthogonal polynomials. From this point on the subject becomes *ad hoc.*

Fact. The Legendre polynomials satisfy Rodrigues's formula

$$P_n(x) = \frac{1}{(-1)^n 2^n n!} D^n (1 - x^2)^n. \tag{11.41}$$

Proof. The function

$$f_n(x) = D^n (1 - x^2)^n \tag{11.42}$$

is a polynomial of degree n that is orthogonal to all of $1,\ x,\ x^2,\ \ldots, x^{n-1}$ and hence must be a multiple of the n-th Legendre polynomial $P_n(x)$ (Exercise 11.16).

By Leibniz's binomial formulae (Exercise 11.17),

$$f_n(x) = D^n(1-x^2)^n = D^n(1-x)^n(1+x)^n$$

$$= \sum_{k=0}^{n} \frac{n!}{k!(n-k)!}(-1)^k D^k(1-x)^n D^{n-k}(1+x)^n. \qquad (11.43)$$

But the only term in the sum of (11.43) that survives the specialization $x = 1$ is the term with $k = n$. Hence

$$f_n(1) = (-1)^n 2^n n!.$$

Fact. The Legendre polynomials obey the recurrence

$$(n+1)P_{n+1}(x) = (2n+1)xP_n(x) - nP_{n-1}(x). \qquad (11.44)$$

Proof. Let $y = 1 - x^2$. According to Rodrigues's formula we must show

$$D^{n+1}y^{n+1} = -2(2n+1)xD^n y^n - 4n^2 D^{n-1}y^{n-1}.$$

Our strategy is to derive the quantity $D^n xy^n$ by three methods:

i) by integrating

$$D^n xy^n = \frac{D^{n+1}y^{n+1}}{-2(n+1)}, \qquad (11.45)$$

ii) by Leibniz's binomial rule

$$D^n xy^n = xD^n y^n + nD^{n-1}y^n, \qquad (11.46)$$

iii) and by the product rule

$$D^n xy^n = D^{n-1}Dxy^n = D^{n-1}[-2nx^2y^{n-1} + y^n]$$

$$= D^{n-1}[2ny^{n-1} - 2x^2ny^{n-1} + y^n - 2ny^{n-1}]$$

$$= (2n+1)D^{n-1}y^n - 2nD^{n-1}y^{n-1}. \qquad (11.47)$$

Combining ii) with iii) yields

$$xD^n y^n = (n+1)D^{n-1}y^n - 2nD^{n-1}y^{n-1}. \qquad (11.48)$$

Combining i) with ii) and employing (11.48) within the brackets [.] yields

$$D^{n+1}y^{n+1} = -2(n+1)xD^n y^n - 2n(n+1)D^{n-1}y^n$$

$$= -2(2n+1)xD^n y^n + 2n[xD^n y^n - (n+1)D^{n-1}y^n]$$

$$= -2(2n+1)xD^n y^n - 4n^2 D^{n-1}y^{n-1}.$$

In stark contrast, verifying the recurrence for the Chebyshev polynomials T_n is quite easy (Exercise 11.18). However verifying Rodrigues's formula is not easy (Exercise 11.19).

It is easy to show (Exercise 11.20) that a multiple of the Hermite polynomial H_n satisfies Rodrigues's formula. It is then traditional to standardize the Hermite polynomials with this formula. Showing the recurrence relation is easier yet (Exercise 11.21).

Neither verification is easy for the Laguerre polynomial L_n (Exercise 11.22).

§11.9 NonPolynomial Eigenfunctions

Re-examine the recursion (11.13):

$$c_{k+2} = \frac{\lambda_n - q_1 k - p_2 k(k-1)}{p_0(k+1)(k+2)} c_k,$$

obtained when we sought polynomial eigenfunctions

$$\phi_n = \sum_{k=0}^{n} c_k x^k$$

of the operator

$$A\phi = (p_0 + p_2 x^2)\phi'' + q_1 x\phi', \quad p_0 \neq 0.$$

Recall that we set $\lambda_n = q_1 n + p_2 n(n-1)$ and put $c_1 = 0$ if n is even, $c_0 = 0$ when n is odd. Thus the recursion will break off at $k = n$, i.e., all subsequent coefficients $c_{n+k} = 0$.

But if we reverse parity by setting $c_0 = 0$ and $c_1 = 1$ when n is even, $c_0 = 1$ and $c_1 = 0$ when n is odd, *the recursion will not break off* — we will have obtained an eigenfunction *of a second kind* represented by an infinite series. The two kinds of eigenfunctions are of opposite parity: where the polynomial ψ_n is an even function for n even, the infinite series is odd, and *visa versa*.

Example 5. Recall that the Legendre operator $A\phi = (1 - x^2)\phi'' - 2x\phi'$ gives rise to the recursion

$$c_{k+2} = \frac{\lambda_n + k(k+1)}{(k+1)(k+2)} c_k. \tag{11.49}$$

Reverse parity and set the first nonzero coefficient equal to the (nonstandard) value 1. Then the *Legendre functions of the second kind* are (Exercise 11.25)

$$Q_0(x) = x + \frac{x^3}{3} + \frac{x^5}{5} + \frac{x^7}{7} + \cdots, \tag{11.50}$$

$$Q_1(x) = 1 - x^2 - \frac{x^4}{3} - \frac{x^6}{5} - \frac{x^8}{7} - \cdots, \tag{11.51}$$

$$Q_2(x) = x - \frac{2x^3}{3} - \frac{2x^5}{5} - \frac{8x^5}{35} - \cdots, \tag{11.52}$$

and so forth. Note that by the Ratio Test, the series for the Q_n have radius of convergence 1.

Alert. The more traditional choice for the first nonzero coefficient of Q_n is

$$c_1 = \frac{(-1)^k 2^{2k} (k)!^2}{(2k)!} \quad \text{for } n = 2k$$

and

$$c_0 = \frac{(-1)^k 2^{2k} k!^2}{(2k+1)!} \quad \text{for } n = 2k + 1.$$

Remark. Taking λ not of the form $-n(n+1)$ in the Legendre recursion yields convergent series that are eigenfunctions with value λ. Thus without careful conditioning, the Legendre operator has every λ as an eigenvalue.

§11.10 The Differential Relations

The orthogonal polynomials satisfy often useful differential relations

$$g_2(x)\psi_n'(x) = g_1(x)\psi_n(x) + g_0(x)\psi_{n-1}(x), \tag{11.53}$$

with coefficient functions shown in Table 11.3.

Table 11.3 The differential relations

ψ_n	g_2	g_1	g_0
P_n	$1-x^2$	$-nx$	n
T_n	$1-x^2$	$-nx$	n
H_n	1	0	$2n$
$L_n^{(m)}$	x	n	$-(n+m)$

Table 11.4 The differential equations

class	equation	soln.	λ
Legendre	$(1-x^2)y'' - 2xy' = \lambda y$	$P_n(x)$	$-n(n+1)$
Chebyshev	$(1-x^2)y'' - xy' = \lambda y$	$T_n(x)$	$-n^2$
Hermite	$y'' - 2xy' = \lambda y$	$H_n(x)$	$-2n$
Laguerre	$xy'' + (1-x)y' = \lambda y$	$L_n(x)$	$-n$

Exercises

11.1 Obtain the first 5 Legendre polynomials (11.2) by the Gram-Schmidt procedure of §6.4 but standardize so that $P_n(1) = 1$. Establish that at least for these first five, $\|P_n\|^2 = 2/(2n+1)$.

11.2 Verify that the first five Chebyshev polynomials are displayed correctly in (11.4).

Hint: Merely show they are mutually orthogonal and deduce the result with no further computation.

11.3 Compute the norms of the first five Chebyshev polynomials in (11.4).

Answer: $\|T_0\|^2 = \pi$, $\|T_n\|^2 = \pi/2$ for $n > 0$.

Hint: Make the substitution $x = \cos\theta$.

11.4 Show that the first five Hermite polynomials are correctly displayed within constant multiples in (11.6).

11.5 Show directly for the first five Hermite polynomials of (11.6) that $\|H_n\|^2 = \sqrt{\pi}2^n n!$.

Hint: The substitution $y = x^2$ brings terms to multiples of the Gamma function

$$\Gamma(s) = \int_0^\infty y^{s-1} \mathrm{e}^{-y}\, dy.$$

Recall that $\Gamma(s+1) = s\Gamma(s)$, $\Gamma(n+1) = n!$, and $\Gamma(1/2) = \sqrt{\pi}$.

11.6 Show that the first five Laguerre polynomials are correctly displayed in (11.7).

11.7 Verify the recursion (11.13) and deduce that polynomial eigenfunctions of every degree exist.

11.8 Show that (11.22) defines an everywhere defined injective bounded operator $R(\lambda)$ that is the inverse of $\lambda - A$ on at least the span of the Legendre polynomials.

11.9 Show directly that the Chebyshev operator (11.16) commutes across the inner product weighted by $w(x) = (1 - x^2)^{-1/2}$.

11.10 Show that the substitution $x = \cos\theta$ carries the Chebyshev operator to (11.23). Deduce that the Chebyshev polynomial T_n is carried to $\cos n\theta$.

11.11 Compute the resolvent of the transformed Chebyshev operator (11.23). Deduce the resolvent $R(\lambda)$ that will well condition the original Chebyshev operator (11.16). Show $R(\lambda)$ also enjoys the series expansion

$$R(\lambda)f = \sum_{n=0}^{\infty} \frac{\langle f, T_n\rangle_w}{\lambda + n^2} \frac{T_n}{\langle T_n, T_n\rangle_w}.$$

11.12 Prove Theorem D on the Laguerre operator.

Outline: The coefficients of polynomial eigenfunctions of the Laguerre operator satisfy the recursion

$$c_{k+1} = (\lambda + k)c_k/(k+1)^2.$$

11.13 Prove that the weight required to bring A to self-adjoint form (11.26) is given by (11.25).

11.14 Verify directly that the formula (11.25) yields the correct weights $w(x)$ for the Legendre, Chebyshev, Hermite, and Laguerre operators displayed in Table 11.2.

11.15 Deduce the norm formulae (11.35)–(11.38) from (11.34) and Table 11.1.

11.16 Prove that the function given by (11.42) is a polynomial of degree n that is orthogonal to each x^k, $0 \le k \le n-1$.

11.17 Prove **Leibniz's binomial formula** for repeated differentiation of a product

$$D^n(fg) = \sum_{k=0}^{n} \frac{n!}{k!(n-k)!} D^k f \cdot D^{n-k} g.$$

11.18 Show the recurrence $T_{n+1}(x) = 2xT_n(x) - T_{n-1}(x)$.

Hint: $\cos n\theta = T_n(\cos\theta)$.

11.19 Show that T_n is given by Rodrigues's formula in Table 11.2.

Hint: When $x = \cos\theta$,

$$\sqrt{1-x^2}\,\frac{d^n}{dx^n} = (-1)^n \frac{d^n}{d\theta^n}.$$

11.20 Show that functions defined by Rodrigues's formula for H_n in Table 11.2 are polynomials of degree n that are orthogonal to $1,\ x,\ x^2,\ \ldots, x^{n-1}$ with respect to the weight $w(x) = \mathrm{e}^{-x^2}$ on the interval $(-\infty, \infty)$.

11.21 Show the recurrence $H_{n+1}(x) = 2xH_n(x) - 2nH_{n-1}(x)$.

Hint: Apply Rodrigues, then Leibniz's binomial formula.

11.22 Show that the Laguerre polynomials L_n satisfy both the recurrence of Table 11.1 as well as Rodrigues's formula of Table 11.2.

11.23 Show $P'_{n+1}(x) - P'_{n-1}(x) = (2n+1)P_n(x)$.

Hint: Write

$$P'_{n+1}(x) - P'_{n-1}(x) = \sum_{j=0}^{n} c_j P_j(x).$$

11.24 Show $(1-x^2)P'_n(x) = -nxP_n(x) + nP_{n-1}(x)$.

11.25 Verify that the first three Legendre functions of the second kind are displayed correctly in (11.50)–(11.52). Find the first 5 nonzero terms of $Q_3(x)$ assuming that the first non-zero coefficient is 1.

For Exercises 11.26–11.32, let $\phi_0(x)$, $\phi_1(x)$, $\phi_2(x)$, $\phi_3(x)$, ... denote the orthonormal polynomials arising from the weight $w(x)$ on the interval (a, b).

11.26 Let the leading coefficient of $\phi_n(x)$ be denoted by $k_n > 0$. Show that in the recurrence formula (11.27),

$$\beta_n = \frac{k_n}{k_{n-1}} \quad \text{and} \quad \gamma_n = \frac{k_n k_{n-2}}{k_{n-1}^2}.$$

11.27 **(Christoffel-Darboux)** Show

$$\sum_{k=0}^{n} \phi_k(x)\phi_k(y) = \frac{k_n}{k_{n+1}} \frac{\phi_{n+1}(x)\phi_n(y) - \phi_n(x)\phi_{n+1}(y)}{x-y}.$$

Hint: Go by induction on n using the recurrence formula and the preceding exercise.

11.28 Prove

$$\sum_{k=0}^{n} |\phi_k(x)|^2 = \frac{k_n}{k_{n+1}} [\phi'_{n+1}(x)\phi_n(x) - \phi'_n(x)\phi_{n+1}(x)] > 0.$$

11.29 Prove that each $\phi_n(x)$ has n distinct (simple) real zeros, all within the interval (a, b). Consequently $\phi_n(x)$ exhibits exactly n sign changes within the interval (a, b).

Outline: Assume $\phi_n(x) = 0$ has exactly k distinct roots x_i of odd multiplicity in (a, b). Set $p(x) = (x - x_1) \cdots (x - x_k)$. Then $\phi_n(x)p(x)$ is of constant sign yet $(\phi_n, p)_w = 0$ unless $k = n$.

11.30 Expanding on Exercise 11.29, show moreover that these zeros *interlace*. That is, if $a < x_1^{(n)} < x_2^{(n)} < \ldots < x_n^{(n)} < b$ are the roots of $\phi_n(x) = 0$, then

$$x_i^{(n+1)} < x_i^{(n)} < x_{i+1}^{(n+1)}, \quad i = 1, 2, \ldots n,$$

and

$$x_i^{(n)} < x_{i+1}^{(n+1)} < x_{i+1}^{(n)}, \quad i = 1, 2, \ldots, n-1.$$

Hint: Apply Exercise 11.28 and count sign changes.

11.31 **(Quadrature)** Moreover, show there are constants $c_1, c_2, \ldots, c_n$ so that for any polynomial $p(x)$ of degree $2n-1$ or less,

$$\int_a^b p(x)\, w(x) dx = c_1 p(x_1^{(n)}) + c_2 p(x_2^{(n)}) + \ldots + c_n p(x_n^{(n)}).$$

Outline: Note that the remainder $r(x)$ left when $p(x)$ is divided by $\phi_n(x)$ is the Lagrange interpolating polynomial that agrees with the values of $p(x)$ at the zeros $x_i^{(n)}$. Integrating the interpolating polynomial yields the result [Szegö].

More abstractly, point evaluations at the n zeros $x_i^{(n)}$ form a basis for the linear functionals on the subspace spanned by $\{1,\ x,\ x^2,\ \ldots, x^{n-1}\}$. In particular, the functional $\zeta(f) = \langle f, 1\rangle_w = c_1 f(x_1^{(n)}) + \ldots + c_n f(x_n^{(n)})$. But $\langle p(x), 1\rangle_w = \langle r(x), 1\rangle_w$.

11.32 Show the monic polynomial $p(x) = x^n + \ldots$ of degree n that minimizes the integral

$$\int_a^b p(x)^2\, w(x) dx$$

is a multiple of the n-th orthonormal polynomial $\phi_n(x)$ arising from the weight w.

Hint: Apply Bessel's inequality.

11.33 Show for $|t| < 1$,

$$(1 - 2xt + t^2)^{1/2} = \sum_{n=0}^{\infty} P_n(x) t^n.$$

11.34 Show for $|x| < 1$, $|t| < 1$,

$$\frac{1 - xt}{1 - 2xt + t^2} = \sum_{n=0}^{\infty} T_n(x) t^n.$$

11.35 Show

$$e^{2xt - t^2} = \sum_{n=0}^{\infty} \frac{H_n(x)}{n!} t^n.$$

11.36 Show for $|t| < 1$,

$$(1 - t)^{-1} e^{-xt/(1-t)} = \sum_{n=0}^{\infty} L_n(x) t^n.$$

11.37 Show that no Legendre function of the second kind Q_n can converge at either endpoint $x = 1$ or $x = -1$.

Outline: Because the coefficients are all eventually negative, Q_n simultaneously converges or diverges at $x = 1, -1$. Suppose convergent. Then by completeness and Exercise 5.24,

$$Q_n = \sum_{k=0}^{\infty} a_k P_k = a_n P_n,$$

an impossibility.

11.38 Establish the *differential relations* of Table 11.3.

11.39 Suppose $w(x) > 0$ on (a, b). Show the inner product space $L^2_w(a, b)$ of all f with

$$\langle f, f \rangle_w = \int_a^b f(x)^2 w(x) dx < \infty$$

is complete and in fact consists of all (measurable) functions of the form $f = g/\sqrt{w}$, where g is in $L^2(a, b)$.

11.40 (The quantum harmonic oscillator). Consider a tiny frictionless mass-spring system as in Figure 11.1. Show that the quantum theoretic model becomes in nondimensional variables

$$i\frac{\partial \psi}{\partial t} = -\psi'' + x^2\psi.$$

See (8.18). Show also that the n-th stationary state ψ_n is of the form

$$\psi_n(x) = cH_n(x)\mathrm{e}^{-x^2/2}$$

of energy $E_n = 2n + 1$. The general wave function is therefore of the form

$$\psi = \sum_{n=0}^{\infty} c_n \mathrm{e}^{-x^2/2-i(2n+1)t} H_n(x).$$

Hint: Set $\psi = \mathrm{e}^{-x^2/2}\phi$ to bring the eigenequation $-\psi'' + x^2\psi = E\psi$ to the Hermite form $\phi'' - 2x\phi' = -(E-1)\phi$.

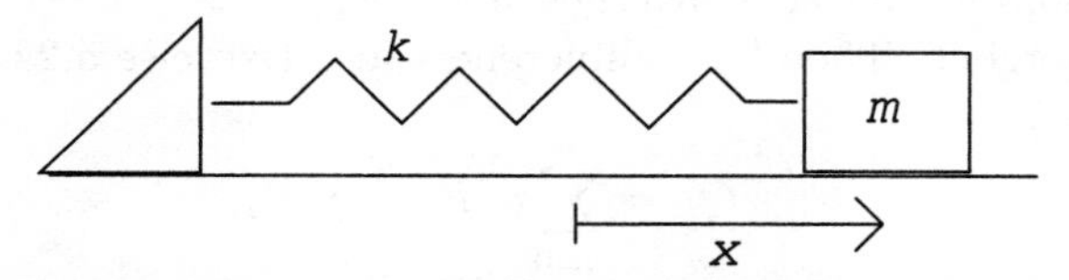

Figure 11.1 The single unforced quantum harmonic oscillator. Imagine a tiny frictionless mass-spring system with displacement from equilibrium *x*.

Chapter 12
Spherical Problems

Let us pose and solve several problems where the spatial domain Ω is a ball. Recall that in spherical coordinates the Laplacian becomes (Exercise 1.20)

$$\nabla^2 u = \frac{\partial^2 u}{\partial \rho^2} + \frac{2}{\rho}\frac{\partial u}{\partial \rho} + \frac{1}{\rho^2 \sin^2 \varphi}\frac{\partial^2 u}{\partial \theta^2} + \frac{1}{\rho^2}\frac{\partial^2 u}{\partial \varphi^2} + \frac{\cot \varphi}{\rho^2}\frac{\partial u}{\partial \varphi}. \quad (12.1)$$

§12.1 A Spherical Capacitor (SSP 12)

The upper hemisphere of a charge-free hollow spherical capacitor of radius a is at constant voltage V, the lower hemisphere is at voltage 0, while the equator is a thin band of insulation. Find the potential u within the capacitor.

We have solved this problem informally via Brownian motion (Exercise 2.13). Let us now solve this steady problem $\nabla^2 u = 0$ via separation of variables.

By symmetry, the potential u within our capacitor is independent of the azimuth (longitude) angle θ, i.e., $u = u(\rho, \varphi)$.

Assume variables separate

$$u = R(\rho)\Phi(\varphi), \quad (12.2)$$

bringing Laplace's equation for the potential u to

$$-\frac{\rho^2 R'' + 2\rho R'}{R} = \lambda = \frac{d^2\Phi/d\varphi^2 + \cot\varphi \, d\Phi/d\varphi}{\Phi}. \quad (12.3)$$

Under the change of variables $x = \cos\varphi$,

$$\frac{d\Phi}{d\varphi} = -\frac{d\Phi}{dx}\sin\varphi \quad (12.4)$$

and

$$\frac{d^2\Phi}{d\varphi^2} = \frac{d^2\Phi}{dx^2}\sin^2\varphi - \frac{d\Phi}{dx}\cos\varphi \quad (12.5)$$

(Exercise 12.1). Thus (12.3) becomes Legendre's

$$(1-x^2)\frac{d^2\Phi}{dx^2} - 2x\frac{d\Phi}{dx} = \lambda\Phi \tag{12.6}$$

and Euler's

$$\rho^2 R'' + 2\rho R' + \lambda R = 0. \tag{12.7}$$

But since the potential is bounded (Exercise 12.2),

$$\lambda = -n(n+1) \tag{12.8}$$

and

$$\Phi = P_n(x) \tag{12.9}$$

where P_n is the n-th Legendre polynomial. Moreover (Exercise 12.3)

$$R = \rho^n. \tag{12.10}$$

Therefore the potential within the charge-free sphere is given by

$$u = \sum_{n=0}^{\infty} c_n \rho^n P_n(x) = \sum_{n=0}^{\infty} c_n \rho^n P_n(\cos\varphi). \tag{12.11}$$

Imposing the boundary voltages existing on the upper and lower hemispheres yields the requirement

$$\sum_{n=0}^{\infty} c_n a^n P(x) = \begin{cases} V & \text{for} \quad 0 < x < 1 \\ 0 & \text{for} \quad -1 < x < 0 \end{cases} \tag{12.12}$$

and hence (Exercise 12.4)

$$c_0 = \frac{V}{2},$$

and for $n > 0$,

$$c_n = \frac{V}{a^n \|P_n\|^2} \int_0^1 P_n(x)\, dx = \frac{P_{n-1}(0) - P_{n+1}(0)}{2} \frac{V}{a^n}. \tag{12.13}$$

Since $P_{2k+1}(0) = 0$, no other even term appears except the first — not unexpected since on the sphere $\rho = a$, $u - V/2$ is an odd function of x. In summary

$$u = \frac{V}{2} + \frac{V}{2} \sum_{k=0}^{\infty} \frac{\rho^{2k+1}}{a^{2k+1}} (P_{2k}(0) - P_{2k+2}(0)) P_{2k+1}(\cos\varphi). \tag{12.14}$$

§12.2 Quenching a Ball Revisited (BVP 6)

Again drop a warm ball into a cold bath. Unlike before in §5.6, suppose initial interior temperatures are not uniform, but (say) depend upon ρ and φ yet not on θ. It is then clear that thereafter, interior temperatures will be independent of θ and of the form $u = u(\rho, \varphi, t)$.

Consider then the problem

$$\frac{\partial u}{\partial t} = \nabla^2 u, \quad 0 \le \rho < 1, \quad 0 \le \varphi \le \pi, \tag{12.15}$$

subject to

i) $u(\rho, \varphi, 0) = f(\rho, \varphi)$,

ii) $u(1, \varphi, t) = 0$.

Rellich and Weyl guarantee a unique strong solution of the form

$$u = \sum_{n=1}^{\infty} c_n e^{\lambda_n t} \psi_n, \tag{12.16}$$

where $\psi_n = \psi_n(\rho, \varphi)$ are the spatial eigenfunctions of the Laplacian with zero boundary conditions.

To search for this guaranteed solution suppose variables separate

$$u = T(t)R(\rho)\Phi(\varphi), \tag{12.17}$$

giving that

$$\frac{\dot{T}}{T} = \lambda = \frac{R'' + (2/\rho)R'}{R} + \frac{\Phi'' + \cot\varphi\,\Phi'}{\rho^2\Phi} = -\alpha^2, \tag{12.18}$$

where the prime $'$ denotes differentiation with respect to the single appropriate variable. As we saw in the previous problem, because (12.18) implies that

$$\frac{\Phi'' + \cot\varphi\,\Phi'}{\Phi}$$

is constant, for some non-negative integer n

$$\Phi = P_n(\cos\varphi). \tag{12.19}$$

But then from (12.18),

$$\rho^2 R'' + 2\rho R' = (n(n+1) - \alpha^2\rho^2)R. \tag{12.20}$$

As before in cylindrical problems, set $x = \alpha\rho$, $y = R(x/\alpha)$, to bring (12.20) to

$$x^2 y'' + 2xy' + (x^2 - n(n+1))y = 0. \tag{12.21}$$

An attempt at a power series solution (Exercise 12.7 and 12.8) will convince you to turn to the substitution

$$w = \sqrt{x}y \tag{17.22}$$

transforming (12.21) to Bessel's

$$x^2 w'' + xw' + (x^2 - (n + \frac{1}{2})^2)y = 0 \tag{12.23}$$

(Exercise 12.9) with bounded solution

$$w = J_{n+\frac{1}{2}}(x). \tag{12.24}$$

Therefore

$$R(\rho) = \frac{J_{n+\frac{1}{2}}(\alpha\rho)}{\sqrt{\alpha\rho}}. \tag{12.25}$$

Rellich guarantees (details omitted) an increasing sequence of positive zeros $\alpha_1 < \alpha_2 < \ldots$ of $J_{n+\frac{1}{2}}$. Moreover the eigenfunctions

$$\psi_{mn} = \frac{J_{n+\frac{1}{2}}(\alpha_m\rho)}{\sqrt{\alpha\rho}} P_n(\cos\varphi) \tag{12.26}$$

form an orthogonal basis of the subspace of all square-integrable functions on the unit ball independent of the variable θ.

Our solution then to the cooling ball (12.15) is

$$u = \sum_{\substack{m=1 \\ n=0}}^{\infty} c_n \frac{e^{-\alpha_m^2 t} J_{n+\frac{1}{2}}(\alpha_m\rho)}{\sqrt{\alpha\rho}} P_n(\cos\varphi). \tag{12.27}$$

The coefficients c_{mn} are determined in the usual way via the usual inner product (in spherical coordinates)

$$\langle f, g \rangle = \int_0^\pi \int_0^1 f(\rho,\varphi)g(\rho,\varphi)\rho^2 d\rho \, \sin\varphi \, d\varphi. \tag{12.28}$$

§12.3 A Spherical Bell (BVP 16)

Think of a hollow bronze spherical bell. Rap this bell with a mallet. What frequencies will we hear?

To keep complexities manageable, assume the bell is evacuated so that internal acoustic resonances are absent. Ignore shear forces within the brass so that vibrations in the first approximation are given by the wave equation. It is reasonable (and most likely accurate) to assume that the acoustic frequencies heard are a result of the radial deflections alone. So if

$$u = u(\varphi, \theta, t) \tag{12.29}$$

is radial deflection from rest, our problem becomes (Exercise 12.11) the dimensionless problem

$$\frac{\partial^2 u}{\partial t^2} = \nabla^2 u, \quad 0 \leq \varphi \leq \pi, \ 0 \leq \theta \leq 2\pi, \tag{12.30}$$

subject to

i) $u(\varphi, \theta, 0) = f(\varphi, \theta)$,

ii) $u_t(\varphi, \theta, 0) = 0$,

iii) u periodic in φ and θ.

In analogy with striking the circular drum at its center, if the initial deformation $f(\varphi, \theta)$ is symmetric about a point of the sphere, we may rotate axes so that this point of impact is the north pole, and thus vibrations thereafter are independent of θ. Thus by separation of variables (Exercise 12.12),

$$u = \sum_{n=0}^{\infty} c_n \cos t\sqrt{n(n+1)}\, P_n(\cos\varphi). \tag{12.31}$$

Therefore the frequencies heard are

$$f_n = \frac{\sqrt{n(n+1)}}{2\pi}, \quad n = 0,\ 1,\ 2,\ 3,\ \ldots\ . \tag{12.32}$$

We will see amusing animations of these vibrations in Chapter 15.

But as we saw in the reprise of the circular drum in §10.2, if the initial deformation is not symmetric, the vibrations are far more complex. The spherical bell is no different. For when we attack (12.30) with separation of variables

$$u = T(t)\Phi(\varphi)\Theta(\theta), \tag{12.33}$$

we obtain the more complicated

$$\frac{\ddot{T}}{T} = \lambda = \frac{1}{\sin^2\varphi}\frac{\Theta''}{\Theta} + \frac{\Phi'' + \cot\varphi\,\Phi'}{\Phi}. \tag{12.34}$$

One thing is clear from (12.34): Θ''/Θ is constant. So since Θ is periodic,

$$\Theta = a\cos m\theta + b\sin m\theta, \quad m = 0,\ 1,\ 2,\ \ldots. \tag{12.35}$$

Because of the initial condition ii),

$$q(t) = \cos\sqrt{-\lambda}t. \tag{17.36}$$

From (12.34 and (12.35) comes

$$\frac{d^2\Phi}{d\varphi^2} + \cot\varphi\,\frac{d\Phi}{d\varphi} = (\lambda + \frac{m^2}{\sin^2\varphi})\Phi, \tag{12.37}$$

which is transformed by the standard change of variables $x = \cos\varphi$, $y = \Phi(\varphi)$, to *Legendre's associated equation*

$$(1 - x^2)y'' - 2xy' = (\lambda + \frac{m^2}{1 - x^2})y. \tag{12.38}$$

Make the famous substitution

$$y = (1 - x^2)^{m/2}z \tag{12.39}$$

to bring Legendre's associated equation (17.38) to (Exercise 12.13)

$$(1 - x^2)z'' - 2x(m + 1)z' = (\lambda + m(m + 1))z. \tag{12.40}$$

But as seen in Exercise 12.14, this transformed equation is solved by the m-th derivative of any solution to Legendre's equation

$$(1 - x^2)y'' - 2xy' = \lambda y.$$

Since we search for a bounded solution, Exercise 12.2 implies

$$\lambda = -n(n + 1) \tag{12.41}$$

and

$$z = P_n^{(m)}(x) \tag{12.42}$$

for $n \geq m$.

Therefore the radial vibrations of our spherical bell are

$$u = \sum_{\substack{n=0 \\ 0 \leq m \leq n}}^{\infty} \cos t\sqrt{n(n+1)}\, P_n^{(m)}(\cos\varphi)\cdot$$

$$\sin^m \varphi\, (a_{mn} \cos m\theta + b_{mn} \sin m\theta). \tag{12.43}$$

Surprisingly no new frequencies are sounded other than those arising from a symmetric rap at the north pole (12.32).

The values for the coefficients c_{mn} are obtained in the usual way via the usual inner product in spherical coordinates

$$\langle f, g \rangle = \int_0^{2\pi} \int_0^{\pi} f(\varphi,\theta) g(\varphi,\theta) \sin\varphi \, d\varphi \, d\theta. \tag{12.44}$$

And now, a triumph of this subject.

§12.4 The Hydrogen Atom (BVP 17)

In his doctoral thesis, Erwin Schrödinger proposed in 1924 that the wave function ψ of the electron of the hydrogen atom must obey

$$i\hbar\frac{\partial\psi}{\partial t} = -\frac{\hbar^2}{2m}\nabla^2\psi + V\psi \tag{12.45}$$

[van der Waerden]. The Hermitian instrument V observing potential energy (see BVPs 10 and 11) is multiplication by a Coulomb electrostatic potential term:

$$V\psi = -\frac{e^2}{4\pi\epsilon_0}\frac{1}{\rho}\psi. \tag{12.46}$$

Then (12.45) becomes the nondimensional problem (Exercise 12.15)

$$i\frac{\partial\psi}{\partial t} = -\nabla^2\psi - \frac{1}{\rho}\psi, \tag{12.47}$$

where $\psi = \psi(\rho, \varphi, \theta, t)$. Let us predict the allowable energy states of this electron.

We search for the stationary states ψ, i.e., the eigenfunctions of the Schrödinger operator:

$$\nabla^2\psi + \frac{1}{\rho}\psi = -E\psi \tag{12.48}$$

of energy $E < 0$. When we search for separated stationary states

$$\psi = R(\rho)\Phi(\varphi)\Theta(\theta), \tag{12.49}$$

we discover that (Exercise 12.16)

$$\frac{\rho^2 R'' + 2\rho R'}{R} + \frac{1}{\sin^2\varphi}\frac{\Theta''}{\Theta} + \frac{\Phi'' + \cot\varphi\,\Phi'}{\Phi} = -\rho^2 E - \rho, \tag{12.50}$$

where in the common abuse of notation, the prime $'$ denotes differentiation with respect to the single appropriate variable. As in the previous problem,

$$\frac{\Theta''}{\Theta} = -m^2, \quad m = 0,\ 1,\ 2,\ \ldots$$

and

$$\Theta = a\cos m\theta + b\sin m\theta. \tag{12.51}$$

Thus also from (12.50), for some constant λ comes

$$\Phi'' + \cot\varphi\,\Phi' = (\lambda + \frac{m^2}{\sin^2\varphi})\Phi \tag{12.52}$$

and

$$\rho^2\frac{d^2R}{d\rho^2} + 2\rho\frac{dR}{d\rho} = (-\rho^2 E - \rho - \lambda)R. \tag{12.53}$$

As we saw in the previous problem,

$$\lambda = -n(n+1)$$

and

$$\Phi = \sin^m\varphi\, P_n^{(m)}(\cos\varphi). \tag{12.54}$$

Only the radial equation (12.53) remains to be solved.

A sequence of three hard-won substitutions reveal that the radial equation (12.53) is an associate of Laguerre's equation.

Step 1. Set $k = 1/2\sqrt{-E}$ and $kx = \rho$. This substitution brings (12.53) to (Exercise 12.17)

$$\frac{d^2R}{dx^2} + \frac{2}{x}\frac{dR}{dx} = (\frac{1}{4} - \frac{k}{x} + \frac{n(n+1)}{x^2})R. \tag{12.55}$$

Step 2. Set $y = x^{-n}R$. This brings (12.55) to (Exercise 12.18)

$$y'' + \frac{2n+2}{x}y' = \frac{y}{4} - \frac{k}{x}y. \tag{12.56}$$

Step 3. Set $z = e^{x/2}y$ to bring (12.56) to (Exercise 12.19)

$$xz'' + (2n+2-x)z' = (n+1-k)z, \tag{12.57}$$

which is an instance of *Laguerre's associated equation.*

For if y is any solution of Laguerre's

$$xy'' + (1-x)y' = \mu y,$$

then its j-th derivative $w = y^{(j)}$ solves (Exercise 12.20)

$$xw'' + (j+1-x)w' = (\mu + j)w.$$

Apply this observation to our transformed radial equation (12.57). The solution z must be the $j = (2n+1)$-th derivative of a Laguerre function belonging to the eigenvalue $\mu = -n-k$. Because of a physically dictated growth restriction on z, I know that the eigenvalue μ is a negative integer (Exercise 12.22), hence that k is integral, and hence that the allowable energy levels E come in the discrete packets

$$E = \frac{-1}{4k^2}. \tag{12.58}$$

These values were observed experimentally long before the model (12.45) was proposed. Note k must satisfy

$$k \geq n+1 \tag{12.59}$$

since we search for nonzero solutions

$$z = L_{k+n}^{(2n+1)}(x) \tag{12.60}$$

of (12.57).

Inverting the three substitutions bringing (12.53) to (12.57), together with the solution (12.60), yields (Exercise 12.21) the radial function (within a constant multiple)

$$R = \rho^n \, e^{-\rho/2k} \, L_{k+n}^{(2n+1)}(\rho/k). \tag{12.61}$$

The stationary states ψ belonging to the energy E are the normalized products $R\Phi\Theta = \psi$ where R, Φ, Θ are given in (12.61), (12.54), and (12.51).

Summary. Each allowable energy level E arises from a unique *principal quantum number* k. For each k there are stationary states at energy E for each combination of the *azimuthal quantum number* $n = 0, 1, , \ldots, k-1$ and the *magnetic quantum number* $m = 0, 1, \ldots, n-1$, a total of $k(k+1)/2$ states. As the principle quantum number k increases, energy E increases towards 0, and the expected location of the electron is further and further from the nucleus (Exercise 12.25).

Critique. A long list of mathematical difficulties have been ignored during the above calculation. For example, what guarantees $-E > 0$ in the eigenequation (12.48)? This positiveness was used strongly during the three substitutions leading to (12.61). After all, the Laplacian is usually negative definite (although multiplication by $1/\rho$ is clearly positive definite). What guarantees all stationary states are of the separated form (12.49)? Rellich's theorem is of no use since the electron can, with nonzero probability, be arbitrarily distant from the nucleus — we are not dealing with a bounded spatial domain Ω. However it may be a reasonable approximation to assume the electron is trapped within some large ball, which is consistent with the exponential die-off of the radial function R of (12.61). But there must be further difficulties: if Rellich's theorem would apply, the allowable energies E of (12.58) would grow in magnitude, not cluster near 0. The trouble arises as the electron approaches the nucleus — the observer V of potential energy is not a bounded operator, at least not on the obvious Hilbert space. The resolvent of the Schrödinger operator cannot be compact. Does the resolvent in fact exist? Is this a well conditioned or even well posed problem? The resolution of these difficulties can be found in the series of books by Reed and Simon.

Quantum Mechanics helped launch and continues to drive Functional Analysis. For light reading pick up the delightful book on Quantum Mechanics by P. Davies, [Gillespie], or the classic Pauling and Wilson]. Attempt [Mackey] or go to the source of it all — *Mathematical Foundations of Quantum Mechanics* by John von Neumann, *Methods of Mathematical Physics* by Courant and Hilbert, and *The Theory of Groups and Quantum Mechanics* by Hermann Weyl.

Exercises

12.1 Show that under the substitution $x = \cos\varphi$,

$$\begin{aligned} \frac{d\Phi}{d\varphi} &= -\sin\varphi\, \frac{d\Phi}{dx}, \\ \frac{d^2\Phi}{d\varphi^2} &= \sin^2\varphi\, \frac{d^2\Phi}{dx^2} - \cos\varphi\, \frac{d\Phi}{dx}, \end{aligned}$$

and so

$$\frac{d^2\Phi}{d\varphi^2} + \cot\varphi\, \frac{d\Phi}{d\varphi} = (1-x^2)\frac{d^2\Phi}{dx^2} - 2x\frac{d\Phi}{dx}.$$

12.2 Again as in Exercise 11.37 show any nontrivial solution $y = Q(x)$ of Legendre's

$$(1-x^2)y'' - 2xy' = \lambda y,$$

where $\lambda \neq -n(n+1)$, is unbounded as $x \to \pm 1$.

12.3 Verify that any bounded solution of the Euler equation

$$\rho^2 R'' + 2\rho R' - n(n+1)R = 0$$

is indeed a multiple of $R = \rho^n$.

12.4 Verify (12.13).

Hint: Deduce from Exercise 11.23 that

$$\int_0^1 P_n(x)dx = \frac{P_{n-1}(0) - P_{n+1}(0)}{2n+1}.$$

12.5 Find the steady state temperatures within a solid hemisphere $0 \le \rho \le 1,\ 0 \le \varphi \le \pi/2$ when the base $\varphi = \pi/2$ is at temperature 0 while the curved remaining surface $\rho = 1,\ 0 \le \varphi < \pi/2$ is at temperature 1.

Answer:

$$u = \frac{1}{2}\sum_{k=0}^{\infty} \rho^{2k+1}(P_{2k}(0) - P_{2k+2}(0))P_{2k+1}(x)$$

12.6 The base $\varphi = \pi/2$ of the solid hemisphere $0 \le \rho \le 1,\ 0 \le \varphi \le \pi/2$ is insulated while elsewhere on the boundary the temperature is $f(\varphi)$. Find the steady state temperatures within the hemisphere.

Answer:

$$u = \sum_{k=0}^{\infty} c_k \rho^{2k} P_{2k}(\cos\varphi)$$

where

$$c_k = (4k+1)\int_0^{\pi/2} f(\varphi)P_{2k}(\cos\varphi)\sin\varphi\, d\varphi.$$

Hint: On the base,

$$\frac{du}{d\mathbf{n}} = \frac{\partial u}{\partial z} = \frac{\partial u}{\partial \rho \cos\varphi} = 0.$$

12.7 Show that the change of variables $z = x^{-n}y$ brings

$$x^2y'' + 2xy' + x^2y = n(n+1)y$$

to

$$x^2z'' + (2n+2)xz' + x^2z = 0.$$

12.8 Find a power series solution

$$z = \sum_{k=0}^{\infty} c_k x^k$$

with $c_0 = 1,\ c_1 = 0$, to the Bessel-like

$$xz'' + bz' + xz = 0, \quad b \ne 0.$$

Answer: The recursion is

$$c_{k+1} = -\frac{c_{k-1}}{(k+1)(k+b)}$$

giving

$$\begin{aligned} z &= \sum_{k=0}^{\infty} \frac{(-1)^k}{2^k} \frac{x^{2k}}{k!(1+b)(3+b)\cdots(2k-1+b)} \\ &= \Gamma(\frac{1+b}{2})(\frac{2}{x})^{\frac{b-1}{2}} J_{\frac{b-1}{2}}(x). \end{aligned}$$

12.9 Verify that the substitution $w = \sqrt{x}y$ will carry (12.21) to (12.23).

12.10 Find the interior temperatures of the cooling ball of (12.15) when initially

$$f(\rho, \varphi) = \begin{cases} 1 & \text{if} \quad 0 \le \varphi < \pi/2 \\ 0 & \text{if} \quad \pi/2 \le \varphi < \pi \end{cases}$$

Answer:

$$c_{mn} = \frac{2\sqrt{\alpha_m} \int_0^1 J_{n+\frac{1}{2}}(\alpha_m \rho) \rho^{\frac{3}{2}} d\rho}{\pi \int_0^1 J_{n+\frac{1}{2}}(\alpha_m \rho)^2 \rho d\rho}.$$

12.11 Argue that if the $\mathbf{i}, \mathbf{j}, \mathbf{k}$ components of the vibrating spherical bell satisfy the wave equation, then so must the radial component. (You may use physical or geometric appeals or brute manipulation).

12.12 Derive (12.31), the radial vibrations of a hollow spherical bell when carefully struck to produce a symmetrical initial indentation.

12.13 Show that the substitution $y = (1-x^2)^{\mu/2} z$ carries Legendre's associated equation

$$(1-x^2)y'' - 2xy' = (\lambda + \frac{\mu^2}{1-x^2})y$$

to

$$(1-x^2)z'' - 2x(\mu+1)z' = (\lambda + \mu(\mu+1))z.$$

12.14 Show that if $y = Q(x)$ satisfies Legendre's equation

$$(1 - x^2)y'' - 2xy' = \lambda y,$$

then its m-th derivative $z = Q^{(m)}(x)$ satisfies Legendre's associated equation

$$(1 - x^2)z'' - 2x(m + 1)z' = (\lambda + m(m + 1))z.$$

12.15 Show that Schrödinger equation (12.45) can be brought to the nondimensional form (12.47).

12.16 Verify that if variables separate as in (12.49), then the eigenequation (12.48) becomes (12.50).

12.17 Verify that the sustitution $kx = \rho$ carries (12.53) to (12.55).

12.18 Verify that the substitution $y = x^{-n}R$ carries (12.55) to (12.56).

12.19 Verify that the substitution $z = e^{x/2}$ carries (12.56) to (12.57).

12.20 Show that if y is a solution of Laguerre's equation

$$xy'' + (1 - x)y' = \lambda y,$$

then its m-th derivative $z = y^{(m)}$ satisfies Laguerre's associated equation

$$xz'' + (m + 1 - x)z' = (\lambda + m)z.$$

12.21 Resubstitute three times to obtain the radial function (12.61) from (12.60).

12.22 Unless μ is a nonpositive integer and $z = L_\mu(x)$, all other nonzero solutions z of Laguerre's

$$xz'' + (1 - x)z' = \mu z$$

possess a growth rate so large that $e^{-x/2}z \to \infty$ as $x \to \infty$. Such a growth rate would prevent R, hence the stationary state ψ of (12.49) from having a finite norm. See [Sagan].

12.23 Solve the eigenequation (12.48) for the stationary state ψ ignoring the angular variables, i.e., assume $\psi = R(\rho)$.

Answer:

$$R = e^{-\rho/2k}\, L'_k(\rho/k),$$

where

$$E = \frac{1}{4k^2}.$$

Hint: Perform the substitutions of Step 1 and Step 3, but not Step 2.

12.24 Show that in the previous exercise, as $k \to \infty$, the expected radius $\bar{\rho} = (\rho R, R)/(R, R) \to \infty$.

Outline: Substitute $x = \rho/k$ to obtain that

$$\bar{\rho} = k\frac{\langle x^2 L'_k(x), xL'_k(x)\rangle_0}{\langle xL'_k(x), xL'_k(x)\rangle_0},$$

where $\langle\ .\ ,\ .\ \rangle_0$ is the weighted inner product for the Laguerre operator. By the differential relation of Table 11.3,

$$xL'_k(x) = kL_k(x) - (k+1)L_{k-1}$$

and hence because $\|L_k\|_0 = 1$,

$$\langle xL'_k(x), xL'_k(x)\rangle_0 = \ \cdots\ = k^2 + (k+1)^2.$$

The recurrence relation of Table 11.1 rearranged becomes

$$xL_k(x) = (2k+1)L_k(x) - (k+1)L_{k+1}(x) - kL_{k-1}(x).$$

Hence

$$\langle x^2 L'_k(x), xL'_k(x)\rangle_0 = \ \cdots\ = k^2(2k+1) + 2k^2(k+1) + (k+1)^2(2k-1).$$

Therefore

$$\bar{\rho} \sim 3k^2.$$

12.25 Show from (12.61) that as the principal quantum number k increases, the mean distance $\bar{\rho} = (\rho\psi, \psi) = \langle \rho R, R\rangle/\langle R, R\rangle$ of the electron from the nucleus also increases without bound.

12.26 Lightning has been observed to excite the resonant modes of the concentric spherical cavity between the earth's surface and the Heaviside layer at an altitude of 300 km. What is the dominant resonant frequency of this cavity? (Take the radius of the earth as 6357 km).

12.27 Quench a solid hemisphere.

12.28 Find the lowest acoustic resonant frequency of a spherical chamber.

Answer: (nondimensional) $f = \omega_1/2\pi \approx 0.715$, where ω_1 is the least positive root of $\omega = \tan \omega$.

12.29 Find the lowest acoustic resonant frequency of a hemispherical chamber.

12.30 Find the stationary states of the quantum billiard ball, i.e., an electron trapped in a potential free sphere.

Chapter 13
Sturm-Liouville Problems

There are many important boundary value problems in one spatial variable whose associated spatial differential operator is second order. A *Sturm-Liouville Problem* is a search for eigenvalues and eigenfunctions of such operators subject to homogeneous boundary conditions.

§13.1 Statement of the Problem

Consider the spatial differential operator in one variable x:

$$A\phi = p(x)\phi'' + q(x)\phi' + r(x)\phi \tag{13.1}$$

operating on the at least twice-continuously differentiable functions ϕ on a bounded closed interval $[a, b]$, subject to *nontrivial* separated homogeneous boundary conditions

$$\alpha_0\phi(a) + \alpha_1\phi'(a) = 0 \tag{13.2a}$$

and

$$\beta_0\phi(b) + \beta_1\phi'(b) = 0. \tag{13.2b}$$

(*Nontrivial* means $|\alpha_0| + |\alpha_1| > 0$ and $|\beta_0| + |\beta_1| > 0$).

Question. Does A possess eigenvalues? Does it possess infinitely many? Are the eigenfunctions mutually orthogonal? Are they complete? What is the correct inner product for this operator?

These are the questions that must be answered before separation of variables can succeed in a transient problem such as

$$\dot{u} = Au.$$

The approach I have chosen to answer these questions is more consonant with the viewpoint of earlier sections and much is drawn from a manuscript of D. R. Dunninger and M. Miklavcic. For the more traditional 'influence function' approach see [Sobolev, 1964], [Courant and Hilbert], or [Naylor and Sell].

There are corresponding developments for problems with periodic boundary conditions. There are also more delicate results for problems with singular coefficients [Hajmirzaahmad and Krall]. We will consider only the problems delimited by three assumptions:

Assumption 1. $r(x)$ is bounded on $[a, b]$, i.e.,

$$|r(x)| \leq r_0 < \infty, \quad a \leq x \leq b. \tag{13.3a}$$

Assumption 2.

$$\frac{q(x)}{p(x)} \text{ is integrable over } [a, b]. \tag{13.3b}$$

Assumption 3. On $[a, b]$,

$$0 < w_0 \leq \frac{1}{p(x)} \exp\Big(\int_a^x \frac{q(y)}{p(y)} dy\Big) \leq w_1 < \infty. \tag{13.3c}$$

§13.2 The Correct Inner Product

To the operator A belongs a unique inner product.

Result A. The operator A defined by (13.1)–(13.3) is self-adjoint with respect to the weighted inner product

$$\langle f, g \rangle_w = \int_a^b f(x) g(x)\, w(x) dx \tag{13.4}$$

where the weight w is given by

$$w(x) = \frac{1}{p(x)} \exp\Big(\int_a^x \frac{q(y)}{p(y)} dy\Big). \tag{13.5}$$

All eigenvalues are therefore real.

Proof.* Assume both ϕ and ψ satisfy the separated boundary conditions (13.2). Two homogeneous linear equations have a nontrivial solution exactly when the determinant of the system is 0, thus $\phi'(a)\psi(a) - \phi(a)\psi'(a) = 0$. Likewise $\phi'(b)\psi(b) - \phi(b)\psi'(b) = 0$.

By Assumption 1, multiplication by $r(x)$ is a bounded operator that will commute across any weighted inner product, so we may as well assume $r(x) \equiv 0$.

Let

$$\eta(x) = \exp(\int_a^x \frac{q(y)}{p(y)} dy) = p(x)w(x). \tag{13.6}$$

But then, by integrating by parts twice,

$$\begin{aligned}
\langle A\phi, \psi\rangle_w &= \int_a^b (\eta(x)\phi''(x) + \frac{q(x)}{p(x)}\eta(x)\phi'(x))\psi\, dx \\
&= \int_a^b (\eta(x)\phi'(x))'\psi\, dx \\
&= \eta(x)\phi'(x)\psi(x)|_a^b - \int_a^b \phi'(x)\eta(x)\psi'(x)\, dx = \dots \\
&= \eta(b)[\phi'(b)\psi(b) - \phi(b)\psi'(b)] - \eta(a)[\phi'(a)\psi(a) - \phi(a)\psi'(a)] \\
&\quad + \langle \phi, A\psi\rangle_w = \langle \phi, A\psi\rangle_w.
\end{aligned}$$

Thus A is self-adjoint with respect to the weighted inner product given in (13.4)–(13.5).

As we shall see in the next section, A is well conditioned by a compact resolvent.

Corollary. Each Sturm-Liouville eigenvalue problem

$$p(x)\phi'' + q(x)\phi' + r(x)\phi = \lambda\phi \tag{13.7}$$

can be rewritten into *self-adjoint form*

$$(\eta(x)\phi')' = w(x)(\lambda - r(x))\phi. \tag{13.8}$$

§13.3 Compact Resolvents

Result B. The Sturm-Liouville operator

$$A\phi = p(x)\phi'' + q(x)\phi' + r(x)\phi \tag{13.9}$$

satisfying conditions (13.1)–(13.3) can be well conditioned by a compact resolvent $R(\lambda)$. The compactness of $R(\lambda)$ is with respect to the weighted inner product

$$\langle f, g\rangle_w = \int_a^b f(x)g(x)\, w(x)dx. \tag{13.10}$$

Proof Outline.* After translation and rescaling we may assume $a = 0$ and $b = 1$. We proceed by a series of reductions given as Lemmas.

Lemma A. It is enough to show Result B for $r(x) \equiv 0$.

Proof.* Recall the resolvent formula for additive perturbations (Exercise 13.11):

$$(\lambda - A)^{-1} = (\lambda - B - K)^{-1} = R_B(\lambda)(1 - KR_B(\lambda))^{-1}$$

where

$$B\phi = p(x)\phi'' + q(x)\phi',$$
$$K\phi = r(x)\phi,$$

and

$$R_B(\lambda) = (\lambda - B)^{-1}.$$

Recall that a product of two bounded operators is compact if either factor is compact (Exercise 13.12). Moreover $(1 - KR_B(\lambda))^{-1}$ will exist and is bounded for all large λ whenever B has a compact resolvent $R_B(\lambda)$ (Exercise 13.2).

Henceforth assume $r(x) \equiv 0$.

Lemma B. For any real valued function $u(x)$ continuously differentiable on $[0, 1]$,

$$\|u\|_\infty^2 \le \|u\|_2^2 + 2\|u\|_2\|u'\|_2. \tag{13.11}$$

Proof.* By the intermediate value theorem, there is some $0 < \zeta < 1$ so that

$$|u(\zeta)|^2 = \int_0^1 u^2(x)dx.$$

Thus by Cauchy's inequality,

$$|u(x)|^2 = |u(\zeta)|^2 + 2\int_\zeta^x u(y)u'(y)dy$$

$$\leq \int_0^1 u^2(x)dx + 2\int_\zeta^x |u(y)u'(y)|dy$$

$$\leq \|u\|_2^2 + 2\|u\|_2\|u'\|_2.$$

Lemma C. For some real number c_0 and every twice continuously differentiable ϕ satisfying the boundary conditions (13.2),

$$\langle A\phi, \phi\rangle_w \leq c_0\langle \phi, \phi\rangle. \tag{13.12}$$

Proof. By Parts,

$$\langle A\phi, \phi\rangle_w = \langle (\eta\phi')', \phi\rangle = \eta(x)\phi'(x)\phi(x)|_0^1 - \langle \eta\phi', \phi'\rangle, \tag{13.13}$$

where as before

$$\eta(x) = \exp(\int_0^x \frac{q(x)}{p(x)}\, dx) = p(x)w(x).$$

But because of the boundary conditions (Exercise 13.3),

$$\eta(x)\phi'(x)\phi(x)|_0^1 \leq c\|\phi\|_\infty^2 \tag{13.14}$$

Moreover

$$\langle \eta\phi', \phi'\rangle \geq \eta_m\|\phi'\|_2^2$$

where η_m is the least value taken on by $\eta(x)$ on the interval $[0, 1]$.

Thus by Lemma B and after completing the square,

$$\langle A\phi, \phi\rangle_w \leq c\|\phi\|_\infty^2 - \eta_m\|\phi'\|_2^2$$

$$\leq c\|\phi\|_2^2 + 2c\|\phi\|_2\|\phi'\|_2 - \eta_m\|\phi'\|_2^2 \leq (c/\eta_m)\|\phi\|_2^2.$$

In view of Lemma C and Lemma A, we may as well henceforth assume A is negative definite, i.e., for $\phi \neq 0$,

$$\langle A\phi, \phi \rangle_w < 0.$$

Proof Sketch* of Result B. Let us parlay Picard's Existence and Uniqueness theorem on initial value problems into this result on second order separated boundary value problems. The method is reminiscent of Variation of Parameters.

Under the notation and reductions in force, let

$$g = V(\eta^{-1}) \tag{13.15}$$

where V is the Volterra operator

$$(Vf)x = \int_0^x f(y)dy.$$

Note that

$$g(0) = 0, \quad g'(0) = 1, \tag{13.16}$$

and

$$\eta(x)g'(x) \equiv 1. \tag{13.17}$$

Our goal will be to show that the resolvent of

$$A\phi = \frac{(\eta\phi')'}{w} = p\phi'' + q\phi'$$

exists and is compact at $\lambda = 0$. We do this by explicitly displaying the resolvent — it is the operator

$$\begin{aligned} Rf &= V(gfw) - gV(fw) \\ &+ \xi(c_{12}fw - c_{11}gfw) - g\xi(c_{22}fw - c_{21}gfw), \end{aligned} \tag{13.18}$$

where ξ is the constant function valued operator

$$(\xi f)x = \int_0^1 f(x)dx,$$

and where, as we shall see, the c_{ij} are determined by the boundary conditions (13.2). By Assumption 3, R is a compact operator on $L^2_w(0,1)$ (Exercise 13.4). In fact, the range of R consists of functions

$\phi = Rf$ with $\eta\phi'$ absolutely continuous (Exercise 13.5). Assume for the moment that these functions $\phi = Rf$ satisfy the boundary conditions (13.2).

For f in $L^2_w(0,1)$ let

$$\phi = Rf.$$

Then because of (13.17),

$$A\phi = (\eta\phi')'/w$$

$$= (\eta g f w - \eta g f w - \eta g' V(fw) - \eta g' \xi(c_{22} f w - c_{21} g f w))'/w = -f,$$

i.e.,

$$ARf = -f. \tag{13.19}$$

But since A is negative definite, A is an injective map. Hence on the range of R,

$$RA\phi = -\phi. \tag{13.20}$$

It remains to show any $\phi = Rf$ satisfies the boundary conditions (13.2).

Consider the 'boundary operators'

$$B_0\phi = \alpha_0\phi(0) + \alpha_1\phi'(0) \tag{13.21a}$$

and

$$B_1\phi = \beta_0\phi(1) + \beta_1\phi(1). \tag{13.21b}$$

It is clear that the matrix

$$\begin{pmatrix} B_0 1 & B_0 g \\ B_1 1 & B_1 g \end{pmatrix}$$

is nonsingular. Otherwise some linear combination ψ of 1 and g would satisfy both boundary conditions (13.2) as well as the ODE $(\eta\psi')' = 0$ and thus would be an eigenfunction of A with eigenvalue 0. But A is negative definite. Therefore it is possible to find $c_{i,j}$ solving

$$B_0 c_{1,j} + B_0 g c_{2,j} = 0, \tag{13.22a}$$

$$B_1 c_{1,1} + B_1 g c_{2,1} = B_1 1, \tag{13.22b}$$

and

$$B_1 c_{1,2} + B_1 g c_{2,2} = B_1 g. \tag{13.22c}$$

Laborious algebra (Exercise 13.6) now establishes that each $\phi = Rf$ does indeed satisfy the boundary conditions (13.2).

Because R is compact and self-adjoint, the spectral theorem of §6.6 guarantees an orthogonal basis of eigenfunctions. Thus A is densely defined.

§13.4 The Fundamental Theorem

Let us summarize what we have obtained.

Theorem. The eigenfunctions ϕ_n of a Sturm-Liouville operator (13.1)–(13.6) form a complete orthonormal basis of the weighted Hilbert space $L^2_w(a, b)$ of all w-square integrable functions f, i.e., all functions f where

$$\langle f, f \rangle_w = \int_a^b f^2(x)\, w(x) dx < \infty,$$

where the weight w is given by

$$w(x) = \frac{1}{p(x)} \exp\left(\int_a^x \frac{q(y)}{p(y)} dy\right).$$

The eigenvalues λ_n of A, none of which are repeated, have

$$\lim_{n\to\infty} \lambda_n = -\infty.$$

See Exercise 13.14.

§13.5 Examples

Example 1. The fundamental theorem on Sturm-Liouville problems re-establishes that the Laplacian

$$A\phi = \phi''$$

on $L^2[0, 1]$ with zero boundary conditions has a complete orthogonal basis of eigenfunctions $\phi_n = \sin n\pi x$, an instance of Rellich's Theorem. But even more, this same differential operator, subject to *any* nontrivial separated boundary conditions (13.2), will have a complete orthonormal basis of eigenfunctions with respect to the usual inner product. In particular this shows that the eigenfunctions of BVP 2 obtained in §5.2 do indeed form a complete orthogonal basis.

Example 2. Consider the problem

$$A\phi = \phi'' + 2\phi' \tag{13.23}$$

on [0,1] subject to the zero boundary conditions $\phi(0) = 0 = \phi(1)$. Then the weight is

$$w(x) = \exp(\int_0^x 2dy) = e^{2x}$$

and the eigenfunctions satisfy

$$\phi'' + 2\phi' - \lambda\phi = 0,$$

giving eigenfunction candidates (Exercise 13.7)

$$\phi = c_1 e^{x(-1+\sqrt{1+\lambda})} + c_2 e^{x(-1-\sqrt{1+\lambda})}. \tag{13.24}$$

Imposing the zero boundary conditions gives (Exercise 13.7) the complete w-orthogonal basis

$$\phi_n = e^{-x} \sin n\pi x \tag{13.25}$$

belonging to the values

$$\lambda_n = -1 - n^2\pi^2, \quad n = 1,\ 2,\ \dots . \tag{13.26}$$

Example 3. Consider the problem

$$\phi'' + \frac{1}{x}\phi' = \lambda\phi \tag{13.27}$$

subject to the boundary conditions

$$\phi'(1) = -\phi(1) \text{ and } \phi(2) = 0.$$

Physically this is the spatial operator of a transient version of SSP 4, where a buried pipe carrying hot fluid is leaking heat to the surrounding earth and where far-field ground temperatures remain at the annual average temperature.

The weight is by the standard formula (13.5)

$$w(x) = \exp(\int_1^x dy/y) = x,$$

giving the self-adjoint form of the problem

$$x\phi'' + \phi' = \lambda x\phi.$$

Candidate eigenfunctions are thus

$$\phi = c_1 J_0(\alpha x) + c_2 Y_0(\alpha x) \tag{13.28}$$

where $\lambda = -\alpha^2$. The fundamental theorem of Sturm-Liouville problems now guarantees a sequence of $\alpha_n > 0$ with limit ∞, $c_1^{(n)}$ and $c_1^{(n)}$ so that functions

$$\phi_n = c_1^{(n)} J_0(\alpha_n x) + c_2^{(n)} Y_0(\alpha_n x) \tag{13.29}$$

satisfy the boundary conditions and form a complete orthonormal basis of $L_w^2(1,2)$ under the inner product of polar coordinates

$$\langle f, g\rangle_w = \int_1^2 f(x)g(x)\, x dx.$$

Exercises

13.1 Argue that the weight w given in (13.5) is (within a constant multiple) the only possible choice making A self-adjoint.

13.2* Suppose $R(\lambda)$ is the compact resolvent of a self-adjoint operator A. Suppose K is a bounded operator. Show $(1-KR(\lambda))^{-1}$ is a bounded operator for all large λ.

Outline:

$$R(\lambda)f = \sum_{n=1}^{\infty} \frac{\langle f, \phi_n\rangle}{\lambda - \lambda_n} \frac{\phi_n}{\langle \phi_n, \phi_n\rangle}$$

and so

$$\lim_{|\lambda|\to\infty} KR(\lambda) = 0.$$

But for any bounded operator C with $\|C\| < 1$,

$$(1-C)^{-1} = 1 + C + C^2 + C^3 + \ldots.$$

13.3 Deduce the inequality (13.14) from the *nontrivial* boundary conditions (13.2).

13.4* Prove that the operator R defined in (13.18) is a compact operator on $L^2_w(0,1)$.

13.5* Show that the range of R defined by (13.18) consists only of functions $\phi = Rf$ where $\eta\phi'$ is absolutely continuous.

13.6 Show that when the constants $c_{i,j}$ of R in (13.18) are determined by (13.21) and (13.22), then $\phi = Rf$ satisfies the boundary conditions (13.2).

13.7 Show that the eigenfunctions of (13.23) in Example 2 are indeed given by (13.25) with eigenvalues (13.26).

13.8 Prove directly that the eigenfunctions (13.25) of Example 2 make up a complete orthogonal basis for $L^2_w(0,1)$.

Hint: $\phi = \sin n\pi x$ form a complete orthogonal basis of $L^2(0,1)$.

13.9 Solve the Sturm-Liouville problem $\phi'' + 2\phi' = \lambda\phi$ on [0,1] with boundary conditions $\phi(0) = 0 = \phi'(0)$.

13.10 Find the eigenvalues of the problem $x^2\phi'' + x\phi' = \lambda\phi$ subject to $\phi(1) = 0 = \phi(2)$.

Hint: The problem is of Euler type.

13.11 Solve the problem $\phi'' = \lambda\phi$ subject to $\phi(0) + \phi'(0) = 0$ and $\phi(1) + \phi'(1) = 0$.

13.12* Show that the product of two bounded operators AB is compact if either factor is compact.

13.13 Find the eigenfunctions and the corresponding eigenvalues of $A\phi = \phi'' - p\phi'$, $(p > 0)$ subject to $\phi'(0) = 0 = \phi(1)$. (This is a reprise of Exercise 6.51.)

Hint: There are three distinct cases: $0 < p < 2$, $p = 2$, $p > 2$.

13.14 Prove that Sturm-Liouville problems (13.1)–(13.3) have no repeated eigenvalues.

Outline: Let ϕ and ψ be two eigenfunctions belonging to the identical eigenvalue λ. Show directly that the *Wronskian*

$$W(x) = \det \begin{bmatrix} \phi(x) & \phi'(x) \\ \psi(x) & \psi'(x) \end{bmatrix}$$

satisfies $\eta W' = -\eta' W$ and so $\eta W = ce^x$. Note that $W(a) = 0$ and so $W \equiv 0$. Find $d \neq 0$ such that $\chi = \phi - d\psi$ is also an eigenfunction belonging to λ with $\chi(a) = \chi'(a) = 0$. But then by Picard uniqueness, $\chi \equiv 0$.

13.15 Prove that zeros of eigenfunctions are simple, i.e., if $\phi(x_0) = 0$ where $a < x_0 < b$, then $\phi'(x_0) \neq 0$.

Hint: Picard uniqueness.

13.16 Solve the problem $x^2\phi'' + 2x\phi' = \lambda\phi$ subject to $\phi(1) = 0 = \phi(2)$.

13.17 Solve the problem $\phi'' - x\phi = \lambda\phi$ subject to $\phi(0) = 0 = \phi(\infty)$.

13.18 Solve the problem $\phi'' - x^2\phi = \lambda\phi$ subject to $\phi(-\infty) = 0 = \phi(\infty)$.

13.19 Solve the problem $\phi'' + x\phi' = \lambda\phi$ subject to $\phi(0) = 0 = \phi(1)$.

13.20 Numerically estimate the first two eigenvalues of the problem $\phi'' - x^2\phi' = \lambda\phi$ subject to $\phi(-1) = 0 = \phi(1)$.

13.21 Solve the problems $\phi'' = \lambda\phi$ subject to

a) $\phi(0) = 0 = \phi(1)$

b) $\phi(0) = 0,\ \phi'(1) = 0$

c) $\phi(0) = 0,\ \phi'(1) = -\phi(0)$

d) $\phi'(0) = 0,\ \phi'(1) = -\phi(0)$

e) $\phi'(0) = -\phi(0),\ \phi'(1) = -\phi(0)$

13.22 A new phenomenon appears with unbounded domains Ω: *continuous spectrum*—spectrum that is not point spectrum (eigenvalues). Let A be the operator $A\phi = \phi'$ on $L^2(0, \infty)$ subject to $\phi(0) = 0$. Show A has no point spectrum, yet A has spectrum, i.e., points where the resolvent of A fails to exist, e.g., $\lambda = 0$.

Chapter 14
Choosing Inner Products

There are important boundary value problems that possess a full set of eigenfunctions but where the eigenfunctions are not mutually orthogonal — problems with non self-adjoint spatial operators. It is folklore that this non-orthogonality is a minor matter, correctable by simply switching to the 'correct' inner product. As an epigram,

if a problem possesses a meaningful discrete spectral structure, then there exists a physically natural inner product that separates variables orthogonally.

Let us first work through several examples that bear out this folk belief. Then I will give good reasons why such a belief is well founded.

§14.1 Examples

Example 1. As we have seen for orthogonal polynomials and Sturm-Liouville problems in Chapters 12 and 13, many second order operators

$$A\phi = p(x)\phi'' + q(x)\phi' + r(x)\phi$$

can be brought to self-adjoint form by rechoosing the inner product to be the weighted inner product $\langle . \, , \, . \rangle_w$ where the weight $w(x)$ is the integrating factor

$$w(x) = p(x)^{-1}\exp\left(\int_a^x \frac{q(x)}{p(x)}dx\right).$$

Example 2. A robot arm

In one model of the transverse vibrations of a flexible robot arm attached to a rotating hub, the arm/hub coupling induced by shear leads to the stiffness operator

$$A\phi = \phi^{(4)} + x\phi^{(2)}(0) \tag{14.1}$$

on $L^2(0,1)$ with boundary conditions

$$\phi(0) = \phi'(0) = \phi''(1) = \phi'''(1) = 0. \tag{14.2}$$

See [Chait et al.].

The operator A is certainly not self-adjoint with respect to the usual inner product (Exercise 14.1)

$$\langle f, g \rangle = \int_0^1 f(x)g(x)dx. \tag{14.3}$$

However the operator does possess an eigenstructure with eigenfunctions (Exercise 14.2):

$$\phi_n = \cos\omega_n x - \cosh\omega_n x + \mu_n(\sin\omega_n x - \sinh\omega_n x)$$
$$+2\sinh\omega_n x\,/\omega_n^3 - 2x/\omega_n^2 \tag{14.4}$$

with

$$\mu_n = \sin\omega_n - \sinh\omega_n + 2(\cosh\omega_n)/\omega_n^3(\cos\omega_n + \cosh\omega_n), \tag{14.5}$$

and where ω_n is determined by the characteristic equation

$$1 + \cos\omega_n\cosh\omega_n = (\cos\omega_n\sinh\omega_n - \sin\omega_n\cosh\omega_n)/\omega_n^3. \tag{14.6}$$

But there is this folklore

Rule of Thumb. When the correct inner product $\langle .\ ,\ . \rangle_0$ is employed in vibrational problems $\ddot{u} = -Au$, potential energy will appear as $\langle Au, u \rangle_0$.

In this model $\langle \phi^{(2)}, \phi^{(2)} \rangle$ is potential energy. So by the Rule of Thumb,

$$\langle A\phi, \phi \rangle_0 = \langle \phi'', \phi'' \rangle. \tag{14.7}$$

Since (Exercise 14.3)

$$\phi''(0) = \langle \phi^{(4)}, x \rangle \text{ and } \langle \phi'', \psi'' \rangle = \langle \phi^{(4)}, \psi \rangle, \tag{14.8}$$

we see that (Exercise 14.4)

$$\langle \phi, \psi \rangle_0 + \langle \phi, x)(x, \psi \rangle_0 = \langle \phi, \psi \rangle. \tag{14.9}$$

Hence

$$\langle \phi, \psi \rangle_0 = \langle \phi, \psi \rangle - (3/4)\langle \phi, x \rangle \langle x, \psi \rangle \tag{14.10}$$

is the correct inner product for the problem. And indeed, A becomes Hermitian with respect to $\langle . , . \rangle_0$ (Exercise 14.5). As one can check, the usual and correct inner product are equivalent (Exercise 14.6).

As one can also check (Exercise 14.7), the resolvent $R(\lambda) = (\lambda I - A)^{-1}$ at $\lambda = 0$ is in fact the (compact) Hilbert-Schmidt integral operator

$$(R(0)f)x = -1/8(x^2 - x^3/2 + x^5/20) \int_0^1 xf(x)dx + x/2 \int_0^x y^2 f(y)dy$$

$$-1/6 \int_0^x y^3 f(y)dy + x^2/2 \int_x^1 yf(y)dy - x^3/6 \int_x^1 f(y)dy. \tag{14.11}$$

The Hilbert-Schmidt and spectral theorems of §6.6 guarantee that the eigenfunctions of A form a complete orthogonal system in $L^2(0,1)$ when endowed with the new inner product $\langle . , . \rangle_0$.

The modal energy distribution predicted by this new equivalent inner product was confirmed for at least the first three modes during stroboscopic observations of a small model of this robot arm.

Example 3. Wing flutter

Let us consider the classic problem of aeroelasticity (wing flutter). One traditional approach [Bisplinghoff et al.] models the wing dynamics with a system of type (in homogeneous form)

$$2\ddot{\omega} - \ddot{\theta} = -\omega'''' \tag{14.12a}$$

$$-\ddot{\omega} + \ddot{\theta} = \theta'' \tag{14.12b}$$

subject to

$$\omega(0) = \omega'(0) = 0 = \theta(0) \tag{14.13a}$$

and

$$\omega''(1) = \omega'''(1) = 0 = \theta'(1). \tag{14.13b}$$

Recasting this system into operator format we obtain the second order system

$$\begin{vmatrix} \omega \\ \theta \end{vmatrix}^{\cdot\cdot} = A \begin{vmatrix} \omega \\ \theta \end{vmatrix} \tag{14.14}$$

where the densely defined spatial operator A on $L^2(0,1) \times L^2(0,1)$ is

$$A\begin{vmatrix} \omega \\ \theta \end{vmatrix} = C\begin{vmatrix} -\omega^{(4)} \\ \theta^{(2)} \end{vmatrix} \tag{14.15}$$

and where

$$C = \begin{pmatrix} 2 & -1 \\ -1 & 1 \end{pmatrix}. \tag{14.16}$$

Rather than the traditional non self-adjoint approach consider the inner product

$$\langle \begin{vmatrix} \omega \\ \theta \end{vmatrix}, \begin{vmatrix} \eta \\ \phi \end{vmatrix} \rangle_0 = \langle C\begin{vmatrix} \omega \\ \theta \end{vmatrix}, \begin{vmatrix} \eta \\ \phi \end{vmatrix} \rangle \tag{14.17}$$

where $\langle . , . \rangle$ is the usual inner product

$$\langle \begin{vmatrix} \omega \\ \theta \end{vmatrix}, \begin{vmatrix} \eta \\ \phi \end{vmatrix} \rangle = \langle \omega, \eta \rangle + \langle \theta, \phi \rangle = \int_0^1 \omega\eta \, dx + \int_0^1 \theta\phi \, dx. \tag{14.18}$$

As one can check, A now becomes self-adjoint under this new equivalent inner product $\langle . , . \rangle_0$ (Exercise 14.8). Moreover, the resolvent of A is diagonal with products of Volterra operators as entries and is therefore compact (Exercise 14.9). Consequently the eigenfunctions

$$\phi_n = \begin{vmatrix} \omega_n \\ \theta_n \end{vmatrix}$$

of A form a complete orthogonal basis for the space $L^2(0,1) \times L^2(0,1)$ of trajectories. This trick may date back to Sobolev [Goldstein, pp. 133, 178].

Example 4. Telegrapher's equation

Suppose we wish to design an observer for active noise control of a one-dimensional duct. Or suppose we wish to study transient voltages on a mismatched ideal transmission line. In either case the telegrapher's equation (Exercise 4.4) in operator format gives rise to a spatial operator of type

$$A\begin{vmatrix} V \\ I \end{vmatrix} = \begin{vmatrix} I' \\ V' \end{vmatrix} \tag{14.19}$$

on $X = L^2(0,1) \times L^2(0,1)$ subject to

$$V(0) = 0 \text{ and } V(1) = kI(1). \tag{14.20}$$

This operator has eigenvalues (Exercise 14.10)

$$\lambda_n = -(1/2)\log|\rho| - n\pi i, \quad n = \ldots -3, -2, -1, 0, 1, 2, 3, \ldots, \tag{14.21}$$

with eigenfunctions

$$\phi_n = \left| \begin{array}{c} e^{\lambda_n x} - e^{-\lambda_n x} \\ e^{\lambda_n x} + e^{-\lambda_n x} \end{array} \right| \tag{14.22}$$

where $\rho = (k-1)/(k+1)$ is called the *reflection coefficient.*

But consider the physically natural inner product of power in watts on the space of forward and reflected voltages. Taking this inner product through a change of basis back to net voltage and current yields the equivalent inner product

$$\langle \left| \begin{array}{c} V_1 \\ I_1 \end{array} \right|, \left| \begin{array}{c} V_2 \\ I_2 \end{array} \right| \rangle_0 = \int_{-1}^{1} (V_1(x) + I_1(x))\overline{(V_2(x) + I_2(x))}|\rho|^x dx/8 \tag{14.23}$$

defined on a subspace X_0 of $L^2(-1,1) \times L^2(-1,1)$. We have extended our original function space X to the space X_0 obtained by adjoining a virtual transmission line from 0 to -1 and by extending voltage V and current I oddly and evenly respectively to all of $[-1,1]$.

A now possesses a compact normal resolvent $R(\lambda)$ on X_0 and variables separate. Truncations of the resulting series solution yield simulations of remarkable accuracy and agreement with experiment [Hull and Radcliffe].

In summary, this problem lies in the more natural function space X_0 under the more natural inner product $\langle . \, , \, . \rangle_0$.

Example 5. A flexible arm with tip mass

To illustrate a more subtle point, consider a flexible arm clamped at one end, the other end terminated by a point mass. One is led to a spatial operator of type

$$A\phi = \phi^{(4)} \tag{14.24}$$

subject to

$$\phi(0) = \phi'(0) = 0, \tag{14.25a}$$

$$\phi''(1) = 0, \quad \text{and} \quad \phi^{(3)}(1) = \phi^{(4)}(1). \tag{14.25b}$$

As in Example 2, the Rule of Thumb

$$\langle A\phi, \phi\rangle_0 = \langle \phi'', \phi''\rangle \tag{14.26}$$

yields (Exercise 14.11) for sufficiently differentiable ϕ and ψ satisfying the boundary conditions (14.25),

$$\langle \phi^{(4)}, \psi\rangle_0 = \langle \phi^{(4)}, \psi\rangle + \phi^{(4)}(1)\psi(1). \tag{14.27}$$

Thus (Exercise 14.12)

$$\langle \phi, \psi\rangle_0 = \langle \phi, \psi\rangle + \phi(1)\psi(1). \tag{14.28}$$

And indeed,

$$\langle A\phi, \psi\rangle_0 = \langle \phi, A\psi\rangle_0 \tag{14.29}$$

for all sufficiently differentiable ϕ and ψ satisfying the boundary conditions (14.25). This new correct inner product (14.28) contains the *boundary term* $\phi(1)\psi(1)$.

There now arises an interesting identification/uniqueness question often ignored. Let X_0 denote the completion of the linear span of the eigenfunctions ϕ_n of A under the new inner product $\langle . \,, .\rangle_0$. Here on the Hilbert space X_0, A becomes Hermitian with compact resolvent. But where is this space X_0?

There is of course the natural embedding of X_0 into $X_1 = L^2(0,1) \times \mathbf{C}$ given by extending the rule

$$\alpha \mapsto \langle \alpha, \alpha(1)\rangle.$$

The operator A_1 on X_1 induced by A via the embedding is of course still self-adjoint. In this and similar problems, A_1 has a compact resolvent and therefore the eigenfunctions $\langle \phi_n, \phi_n(1)\rangle$ span. Consequently (the embedded image of) $X_0 = X_1$.

This is startling news for it means that the eigenfunctions ϕ_n are complete in $X = L^2(0,1)$ under the usual inner product, but do not yield unique representations! This rare phenomenon arises because the resolvent of A is not compact on X. Instead one must employ the 'correct' inner product (14.28) on the correct space X_1. The natural space for this problem is not $X = L^2(0,1)$ but is instead $X_1 = L^2(0,1) \times \mathbf{C}$.

§14.2 Adjoints and Biorthogonal Series

Let us assume from this point on that A is a well-conditioned densely defined operator on the Hilbert space X with compact resolvent $R(\lambda)$. Assume also that the span of the eigenfunctions ϕ_n of A are dense in X. This is the common situation in physically meaningful problems with discrete spectral structure.

Were A Hermitian, the eigenvectors would form a complete orthogonal basis; the given inner product is the correct inner product. But what if not?

Without loss of generality assume that the resolvent $R = R(\lambda)$ exists at $\lambda = 0$. Here are some standard facts from Functional Analysis:

Compact Operators Principles. Suppose R is compact. Then so is its adjoint R^*. Each non-zero eigenvalue $1/\lambda$ of R has a finite dimensional eigenspace. The eigenvalues of R have no nonzero point of accumulation. Either $1/\lambda - R$ is invertible or $1/\lambda$ is an eigenvalue. If $1/\lambda$ is an eigenvalue of R, then its complex conjugate $1/\bar{\lambda}$ is an eigenvalue of R^*. The (norm) limit of compact operators is again compact. See Yoshida [1980].

Let $\lambda_1, \lambda_2, \lambda_3, \ldots$ be the (non-zero) eigenvalues of A belonging to the eigenvectors $\phi_1, \phi_2, \phi_3, \ldots$. (A repeated eigenvalue λ_k of multiplicity m will be listed m times with the corresponding ϕ forming a basis of the eigenspace of λ_k.) But then by the principles for compact operators, $\bar{\lambda}_1, \bar{\lambda}_2, \bar{\lambda}_2, \ldots$ must be the eigenvalues of A^*, (the adjoint of A well conditioned by R^*), belonging to eigenvectors $\psi_1, \psi_2, \psi_3, \ldots$ of A^*.

But (Exercise 14.13)

$$\lambda_m \neq \bar{\lambda}_n \text{ implies } \langle \phi_m, \psi_n \rangle = 0. \tag{14.30}$$

On the other hand it is impossible that $\langle \phi_m, \psi_n \rangle = 0$ for all m since the span of the ϕ_m is dense in the space X. Thus after checking a minor detail in the presence of repeated eigenvalues (Exercise 14.14),

$$\langle \phi_m, \psi_n \rangle = \begin{cases} e_n \neq 0 & \text{if } m = n \\ 0 & \text{otherwise.} \end{cases} \tag{14.31}$$

The sequence of pairs ϕ_n, ψ_n is called a *biorthogonal sequence.*

Result. Given a biorthogonal sequence ϕ_n, ψ_n, it follows that any norm convergent representation

$$f = \sum_{n=1}^{\infty} c_n \phi_n \tag{14.32}$$

is unique. In fact

$$c_n = \frac{\langle f, \psi_n \rangle}{\langle \phi_n, \psi_n \rangle}. \tag{14.33}$$

Equation (14.33) is merely a generalization of the familiar formula (6.44) for orthogonal series. This result explains why the phenomenon of nonunique expansions exhibited in Example 5 is so rare — the adjoint insures that representations are unique. Even when the span of eigenfunctions ϕ_n is not dense, the above uniqueness proof and subsequent observations go through by cutting back to the spanned subspace. Compactness is the key property.

It is not clear (and sometimes false) that every f in X has an expansion of the form (14.32) even though the span of the ϕ_n is dense.

Let us replace each ψ_n and ψ_n by constant multiples so that

$$\langle \phi_m, \psi_n \rangle = \begin{cases} 1 & \text{if } m = n \\ 0 & \text{otherwise.} \end{cases} \tag{14.34}$$

The sequence of pairs ϕ_n, ψ_n is now called *biorthonormal.*

§14.3 The Correct Inner Product

Let the notations of the previous subsection continue in force. Ideally,

the correct inner product should yield modal energy distribution.

For after all, if the modeled system has a discrete frequency of vibration with an associated spatial vibrational mode, and if this mode can be excited alone, then the energy of this mode is decoupled from all other modes — there is no 'spill up' or 'spill down' of energy onto other modes. With this in mind, the correct inner product $\langle . , . \rangle_0$ should then satisfy

$$\langle \sum_k a_k \phi_k, \sum_m b_m \phi_m \rangle_0 = \sum_n a_n b_n = \langle \sum_k a_k \phi_k, \sum_m b_m \psi_m \rangle \tag{14.35}$$

for at least all finite sums. So the required change of inner product is given by

$$\langle f, g\rangle_0 = \langle f, Pg\rangle, \tag{14.36}$$

where the operator P is defined by the rule

$$P\phi_n = \psi_n \tag{14.37}$$

and extended linearly to at least all finite sums of the ϕ_n. If this change of basis P is bounded with bounded inverse, the correct (new) inner product is equivalent to the old. If P is merely bounded, the old dominates the new inner product. If P^{-1} is bounded, the new dominates the old. In Example 5, the new inner product dominates the old but not conversely.

The natural space of the problem is the closure X_0 of the subspace of all finite sums

$$\sum_n c_n\phi_n$$

under the new inner product $\langle . , .\rangle_0$. As we saw in Example 5, the correct space X_0 of the problem may bear little resemblance to the original space X, other than each convergent series

$$\sum_{n=1}^{\infty} c_n\phi_n,$$

convergent with respect to the original norm and with square summable coefficients c_n, must still converge with respect to the new norm $\|\cdot\|_0$.

Many of these matters are elegantly discussed in detail in R. M. Young's superb *An Introduction to Nonharmonic Fourier Series.* See also [Miller] and [MacCluer and Chait].

Exercises

14.1 Show that the operator A of (14.1) and (14.2) is not self-adjoint with respect to the usual inner product (14.3).

14.2 Show however that this operator does possess the eigenfunctions (14.4) with eigenvalues $\lambda_n = \omega_n^4$ where ω_n is determined by the eigenequation (14.6).

14.3 Show that the two identities of (14.8) obtain for any ϕ and ψ with fourth derivatives satisfying the boundary conditions (14.2).

14.4 Establish (14.9).

14.5 Show that the operator A given in (14.1) and (14.2) is self-adjoint with respect to the 'correct' inner product (14.10).

14.6 Show the usual inner product (14.3) and the new inner product (14.10) are *equivalent*, i.e., each *dominates* the other, i.e., that there are non-zero constants c_i so that

$$\langle f, f\rangle < c_1\langle f, f\rangle_0 \text{ and } \langle f, f\rangle_0 < c_2\langle f, f\rangle.$$

14.7 Moreover show that the operator R given in (14.11) will serve as the resolvent of A at $\lambda = 0$. That is, for all f in $L^2(0,1)$ and all ϕ in the range of R, $ARf = -f$ and $RA\phi = -\phi$.

14.8 Show that the operator A of (14.14) and (14.13) is self-adjoint under the new inner product (14.17).

14.9 Establish that the operator A of Exercise 14.8 has a compact resolvent.

14.10 Verify the values (14.21).

14.11 Establish (14.28).

14.12 Show certain functions of $L^2(0,1)$ have two distinct representations as norm convergent series of eigenfunctions of the operator A of (14.24) and (14.25).

14.13 Show that if $A\phi = \lambda\phi$ and $A^*\psi = \mu\psi$, then either $\lambda = \bar{\mu}$ or $\langle\phi, \psi\rangle = 0$, where A^* is the operator *adjoint* to A, i.e., the operator for which $\langle A\phi, \psi\rangle = \langle\phi, A^*\psi\rangle$ for all $\phi \in \text{dom}A$ and $\psi \in \text{dom}A^*$.

14.14 Suppose A is a linear operator on the finite dimensional space E that is entirely an eigenspace of A, that is, E has a basis $\phi_1, \phi_2, \ldots, \phi_n$ of eigenvectors all belonging to the same value λ. Show that E also has a basis $\psi_1, \psi_2, \ldots, \psi_n$ of eigenvectors of A^*, (the conjugate transpose of A), all belonging to the value $\bar{\lambda}$. Show moreover that the ψ_k can be renumbered and rescaled so that $\langle\phi_k, \psi_k\rangle = 1$ while $\langle\phi_k, \psi_m\rangle = 0$ for all $k \neq m$.

14.15 In a vibrational system $M\ddot{u} = -Ku$ like the mass/spring system of Exercise 4.9 and Figure 4.2, where the mass M and stiffness K are both positive definite self-adjoint, the operator $A = -M^{-1}K$ may not be self-adjoint and its eigenvectors may fail to be mutually orthogonal. However this is a minor matter. Find an inner product $\langle . \, , \, . \rangle_0$ for which A becomes self-adjoint and the solutions to the **generalized eigenvalue problem**

$$\lambda M\phi = -K\phi$$

form an orthonormal basis.

Answer: $\langle \phi, \psi \rangle_0 = \langle M\phi, \psi \rangle$. Alternatively, $\langle \phi, \psi \rangle_0 = \langle K\phi, \psi \rangle$ will work. The eigenvectors in either case are identical within normalizations when the eigenvalues are distinct.

14.16 Find a change of variables $x = a\tilde{x} + b\tilde{y}$, $y = c\tilde{x} + d\tilde{y}$ that carries the spatial operator

$$\mathbf{A}\phi = \frac{\partial^2 \phi}{\partial x^2} + 2\frac{\partial^2 \phi}{\partial x \partial y} + 4\frac{\partial^2 \phi}{\partial y^2}$$

on Ω to the Laplacian ∇^2. What is the natural inner product for this operator $\mathbf{A}$?

Answer: $x = \tilde{x} - \tilde{y}/\sqrt{3}$, $y = \tilde{y}/\sqrt{3}$.

Approach: The matrix Q of the 'quadratic form'

$$\frac{\partial^2}{\partial x^2} + 2\frac{\partial^2}{\partial x \partial y} + 4\frac{\partial^2}{\partial y^2} = \begin{pmatrix} \frac{\partial}{\partial x} \\ \frac{\partial}{\partial y} \end{pmatrix}^T \begin{pmatrix} 1 & 1 \\ 1 & 4 \end{pmatrix} \begin{pmatrix} \frac{\partial}{\partial x} \\ \frac{\partial}{\partial y} \end{pmatrix}$$

can be brought to the identity $I = P^T Q P$ via **Sylvester's Law of Inertia** [Brown]. The matrix P is the required change of basis. The correct inner product, under which $\mathbf{A}$ becomes Hermitian, is

$$\langle \phi, \psi \rangle_0 = \int_{P^{-1}\Omega} \phi\psi \; d\tilde{x} d\tilde{y}.$$

14.17 Prove that the eigenfunctions of the operator

$$\mathbf{A}\phi = 3\frac{\partial^2 \phi}{\partial x^2} + 2\frac{\partial^2 \phi}{\partial x \partial y} + 5\frac{\partial^2 \phi}{\partial y^2}$$

with zero boundary conditions on the unit disk $x^2 + y^2 < 1$ form a complete orthogonal basis with respect to the 'correct' inner product.

14.18 (P. M. FitzSimons) Find the correct inner product for the transverse vibrations of a rod that is free at $x = 1$ but welded to a mass at $x = 0$. That is, determine the correct inner product $\langle . , . \rangle_0$ for the spatial operator

$$A\phi = \phi^{(4)} - \phi^{(3)}(0)$$

subject to

$$\phi(0) = \phi'(0) = 0 = \phi''(1).$$

Answer: $\langle \phi, \psi \rangle_0 = \langle \phi, \psi \rangle - (1/2)\langle \phi, 1 \rangle \langle 1, \psi \rangle$.

14.19 (J. V. Beck) What is the correct inner product for the Sturm-Liouville problem

$$\phi'' - p\phi' - \lambda\phi = 0$$

subject to $\phi(0) = 0 = \phi(1)$?

Answer:

$$\langle f, g \rangle = \int_0^1 f(x)g(x)e^{-px}\, dx.$$

14.20* For any mutually orthogonal $\phi_1, \phi_2, \ldots$ and scalars λ_n with λ_n / n bounded away from 0, prove that

$$R(\lambda)f = \sum_n \frac{\langle f, \phi_n \rangle}{\lambda - \lambda_n} \frac{\phi_n}{\langle \phi_n, \phi_n \rangle}$$

is a compact operator at any λ away from the λ_n.

Chapter 15

Symbolic Manipulation

The digital computer not only enables raw number crunching but nowadays provides the user with enormous power for manipulation and visualization. Several uses of the 'silicon assistant' *Mathematica* by Wolfram Research Inc. of Champaign, IL are presented in this section. The scope and power of such packages increase daily.

§15.1 Special Functions

Data on all the common special functions is on line. For example the command

```
In[1]:= LegendreP[6,x]
```

will return the 6-th Lengendre polynomial $P_6(x)$:

```
Out[1]= (-5 + 105 x^2 - 315 x^4 + 231 x^6)/16.
```

Likewise on call are the Chebyshev $T_n(x)$, Hermite $H_n(x)$, Laguerre $L_n(x)$, and other orthogonal polynomials of all degrees. The associated polynomials and functions of the second kind are also on call.

One can find all the zeros of an orthogonal polynomial for use in Gaussian quadrature (Exercise 11.31) with a command like

```
In[2]:= NSolve[ChebyshevT[5,x]==0 ,x]
```

which will return the five (real) zeros of $T_5(x)$: -0.951057, -0.587785, 0, 0.587785, and 0.951057. As guaranteed by Exercise 11.29, all zeros lie in the interval $(-1, 1)$.

Data on the Bessel functions J_n, I_n, Y_n, K_n, $H_n^{(1,2)}$ is also on line. To find the (say) 7th positive root α_7 of $J_0(x) = 0$, first obtain an estimate of α_7 by graphing $J_0(x)$ for $0 \leq x \leq 30$ with the command

```
In[3]:= Plot[BesselJ[0,x], {x,0,30}]
```

to obtain the graph of Figure 15.1.

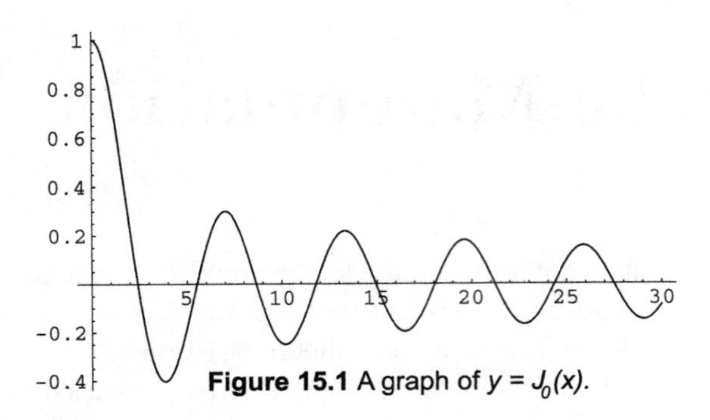

Figure 15.1 A graph of $y = J_0(x)$.

From the graph estimate that α_7 is about 21.0. Then find a more accurate approximate numerical value (with initial guess 21) by the command

```
In[4]:= FindRoot[a7 = x;BesselJ[0,x]==0, {x,21}]
```

to obtain

```
Out[4]= {x -> 21.2116}.
```

Request more accuracy be displayed with the command

```
In[5]:= a7 = N[a7,14]
```

which returns

```
Out[5]= 21.211636629879.
```

We may now obtain the coefficient c_7 of (say) the orthogonal Bessel-Fourier expansion

$$\sqrt{r} = \sum_{1}^{\infty} c_n J_0(\alpha_n r), \tag{15.1}$$

namely

$$c_7 = \frac{\langle \sqrt{r}, J_0(\alpha_7 r)\rangle}{\langle J_0\langle \alpha_7 r), J_0(\alpha_7 r)\rangle} = 2\frac{\langle \sqrt{r}, J_0(\alpha_7 r)\rangle}{J_1(\alpha_7 r)^2},$$

by numerically integrating

$$\langle \sqrt{r},\ J_0(r)\rangle = \int_0^1 \sqrt{r} J_0(\alpha_7 r) r dr$$

with the command

```
In[6]:= NIntegrate[Sqrt[r] BesselJ[0,a7 r] r,{r,0,1}]
```

which returns

```
Out[6]= 0.00791152 .
```

Complete the computation of c_7 by doubling the previous answer % and dividing by $J_1(\alpha_7)^2$:

```
In[7]:= 2 % /BesselJ[1,a7]^2
```

to obtain c_7 :

```
Out[7]= 0.527065,
```

or with more accuracy,

```
In[8]:= N[%,14]
```

which returns with

```
Out[8]=0.52706454230036.
```

§15.2 Fourier Series

Mathematica has a Fourier series package. To invoke this package enter

```
In[1]:= <<Calculus`FourierTransform`
```

To obtain say the 12-th partial sum of an easy expansion such as

$$x = 2\sum_{n=1}^{\infty}(-1)^{n+1}\frac{\sin nx}{n} \tag{15.2}$$

on $[-\pi, \pi]$ enter

```
In[2]:= f[x_] = x
In[3]:= FourierTrigSeries[f[x],{x,-Pi,Pi},12]
```

and the machine returns with the first 12 terms.

However for functions with several jump discontinuities one must force the manipulator to evaluate the coefficients numerically, otherwise it will return with their symbolic representation. Take for instance the expansion (12.15) on $[-\pi, \pi]$ of the unit step

$$\frac{\text{sgn } x + 1}{2} = 1/2 + (1/\pi)\sum_{n=1}^{\infty}(1 - (-1)^n)\frac{\sin nx}{n}. \tag{15.3}$$

The routine `FourierTrigSeries` will decline to perform this expansion. Instead a short script is necessary:

```
In[4]:= f[x_ ] = (Sign[x] + 1)/2
In[5]:= a = Table[NIntegrate[f[x] Cos[n x], {x,-Pi,Pi}]/
          Pi, {n,1,12}]
In[6]:= a0 = (1/(2 Pi)) NIntegrate[f[x],{x,-Pi,Pi}]
In[7]:= b = Table[NIntegrate[f[x] Sin[n x], {x,-Pi,Pi}]/
          Pi, {n,1,12}]
In[8]:= fou[x_] = a0 + Sum[a[[n]] Cos[n x]
```

```
                    + b[[n]] Sin[n x], {n,1,12}]
```

The numerical integration package (apparently an adaptive Simpson's rule) will show small but nonzero coefficients for some cosine terms and for some even subscripted sine terms.

One might now investigate Gibbs' phenomenon graphically near $x = 0$ by graphing the function with the partial series superimposed:

```
In[9]:= Plot[{f[x],fou[x]}, {x, -0.5, 0.5}]
```

to obtain the graph of Figure 15.2.

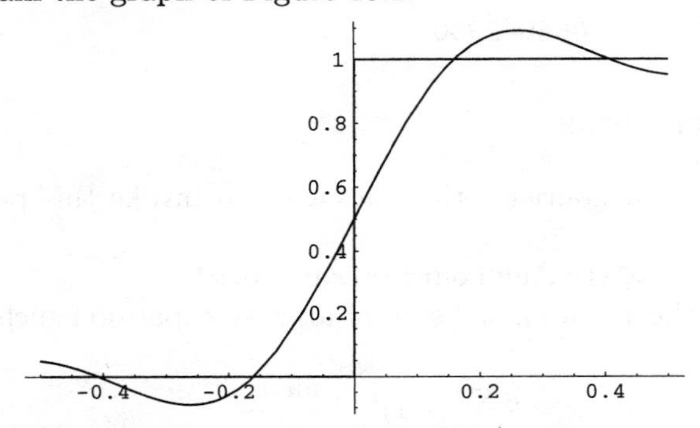

Figure 15.2 A graph of the 12th partial sum of the Fourier series of the step function near *x* = *0*.

A manipulator may be used to guess expansion coefficients. For example, to discover the coefficients of the expansion

$$|x| = \sum_{n=0}^{\infty} a_n \cos nx \tag{15.4}$$

one might experiment by employing the script

```
In[10]:= g[x_] = Abs[x]
In[11]:= a = Table[NIntegrate[g[x] Cos[n x],{x,-Pi,Pi}],
                   {n,0,12}]
```

which returns with $2\pi a_0$ and the next 11 values of πa_n:

```
Out[12]= {9.8696, -4., 0., -0.444444, 0., -0.16, 0.,
            -0.0816327, 0., -0.0493827, 0., -0.0330579, 0.}
```

From this numerical evidence and some time with a calculator leap to the conclusion

$$|x| = \frac{\pi}{2} - \frac{4}{\pi} \sum_{k=0}^{\infty} \frac{\cos(2k+1)x}{n^2}.$$

The inquiry

```
In[13]:= ?*Fourier*
```
asks for all Fourier standard functions (commands) that are included within the package at hand.

§15.3 Fourier-Bessel Series

Let us automate Fourier-Bessel developments such as

$$r = \sum_{k=1}^{\infty} c_k J_0(\alpha_k r) \quad 0 \le r < 1 \tag{15.5}$$

where α_1, α_2, α_3, ... are the positive roots of $J_0(x) = 0$.

We prepare a table of the first 12 roots α_k using the initial guess $k\pi + 3\pi/4$ provided by the asymptotic formula (14.23):
```
In[1]:= Table[FindRoot[BesselJ[0,x]==0,{x,k Pi+ 3 Pi/4}],
{k,0,11}]
In[2]:= x/.%
In[3]:= a = %
```
Thus is generated an array a whose entries are the first 12 zeros of J_0. For example, a[[1]] = 2.40483 and a[[2]] = 5.52008. (To generate a table of zeros of other Bessel functions without asymptotic formulae such as (14.23) you must first prepare a table of approximate values using `Plot`, then apply `FindRoot` with these initial guesses — see [Abell and Braselton]).

We next develop the 12-th partial sum of the expansion (15.5) by defining the function,
```
In[4]:= f[r_] = r
```
computing the inner product $(f(r), J_0(\alpha_k r))$,
```
In[5]:= Table[NIntegrate[f[r] BesselJ[0,a[[k]] r] r,
        {r,0,1}], {k,1,12}]
```
and naming the table of values as "innerprod':
```
In[6]:= innerprod = %
```
Next prepare a table of the square norms $\|J_0(\alpha_k r)\|^2$ with
```
In[7]:= Table[NIntegrate[BesselJ[0, a[[k]] r]^2 r,
        {r,0,1}], {k,1,12}]
```
or equivalently because of (9.33),
```
In[7]:= Table[BesselJ[1,a[[k]]]]^2 /2,{k,1,12}]
```
and name the table
```
In[8]:= sqnorm = %
```
Finally form the 12-th partial sum

```
In[9]:= bess[r_] = Sum[innerprod[[k]]
        *BesselJ[0,a[[k]] r]/sqnorm[[k]], {k,1,12}]
```

We may compare the pointwise accuracy of this 12-th partial sum to the actual by asking for a graph of their difference:

```
In[10]:= Plot[f[r] - bess[r], {r,0,1}]
```

to obtain Figure 15.3. Note the expected error at $r = 0$ and $r = 1$. (Why?)

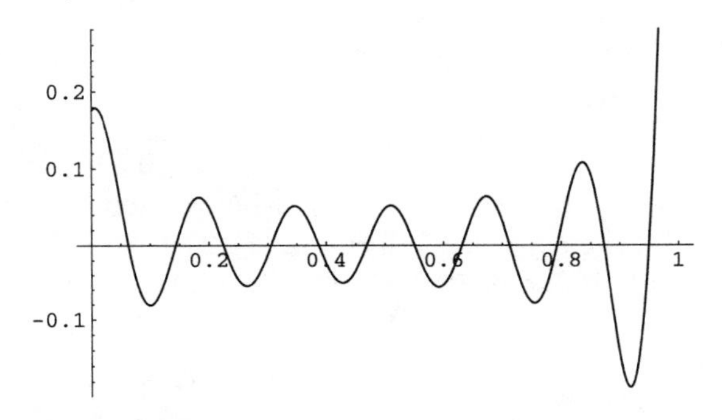

Figure 15.3 A graph of the error between *r* and the 12th partial sum of its Fourier-Bessel expansion.

There are add-on packages available for Bessel and other expansions. *Mathematica* comes off the shelf with extensive Taylor series manipulative power.

§15.4 Steady Dirichlet Problems

The manipulators can do symbolic vector analysis. For example, to check that

$$u(x,y) = \frac{1 - x^2 - y^2}{(1-x)^2 + y^2}$$

is indeed harmonic within the unit disk $x^2 + y^2 < 1$, (Exercise 1.4) call up the package with

```
In[1]:= <<Calculus`VectorAnalysis`
```

and choose coordinate systems

```
In[2]:= SetCoordinates[Cartesian[x,y,z]]
```

Define the function with

```
In[3]:= u= (1-x^2 - y^2)/((1-x)^2 + y^2)
```

and perform the Laplacian

```
In[4]:= Grad[u]
In[5]:= Div[%]
In[6]:= Simplify[%]
```

to obtain

```
Out[6]= 0
```

In the course of writing the book *Harmonic Function Theory,* the authors Axler, Bourdon, and Ramey developed *Mathematica* scripts for solving various problems such as steady Dirichlet problems

$$\nabla^2 u = 0$$

within balls in $\mathbf{R}^n$ centered at the origin, given polynomial boundary values $b(s)$. Their approach uses the Poisson (surface) integral solution

$$u(x) = \int_S b(s) \frac{1 - |x|^2}{|x - s|^n} d\sigma(s). \tag{15.6}$$

As it turns out, the *Poisson kernel*

$$P(x, s) = \frac{1 - |x|^2}{|x - s|^n}$$

has an expansion in *zonal harmonics*

$$P(x, s) = \sum_{k=0}^{\infty} Z_k(x, s)$$

where the k-th zonal harmonic $Z_k(x, s)$ is a certain spherical harmonic of degree k, i.e., one of the harmonic homogeneous polynomials of degree k. Moreover for $b(s)$ a polynomial of degree m, the Poisson integral

$$\int_S b(s) \frac{1 - |x|^2}{|x - s|^n} d\sigma(s)$$

must be a polynomial of degree at most m. Using the explicit formula for the zonal harmonics and a formula of H. Weyl, the authors wrote a script to then compute the surface integrals (15.6).

For example, one of their scripts employing their macros

```
In[1]:= SetDimension[x,5]
In[2]:= Dirichlet[ x[1]^3,x]
Out[2]= (3x[1] +4x[1]^3 - 3x[1] x[2]^2 - 3x[1] x[3]^2
        -3x[1] x[4]^2 -3x[1] x[5]^2)/7
```

reveals that the solution of $\nabla^2 u = 0$ within the unit ball of $\mathbf{R}^5$ with boundary values $b(x) = x_1^3$ is

$$u = \frac{3x_1 + 4x_1^3 - 3x_1x_2^2 - 3x_1x_3^2 - 3x_1x_4^2 - 3x_1x_5^2}{7}.$$

Their script

```
In[3]:= Dirichlet[x[1] x[2], x[3], x]
Out[3]= (14x[1] x[2] - x[3] + x[1]^2 x[3] +x[2]^2 x[3]
        + x[3]^3 +x[3] x[4]^2 + x[3] x[5]^2)/14
```

solves the Poisson problem

$$\nabla^2 u = x_3$$

within the unit ball of $\mathbf{R}^5$ with boundary values $b(x) = x_1x_2$.

Many other useful and powerful vector analytic macros are available in the software package that accompanys their book. The software package and its documentation are available free of charge at *http://math.sfsu.edu/axler.*

§15.5 Eigenmodes

Consider the simple Sturm-Liouville problem

$$y'' + y' = \lambda y \tag{15.7a}$$

subject to (say) the boundary conditions

$$y(0) = 0 \quad \text{and} \quad y(1) + y'(1) = 0. \tag{15.7b}$$

By the Fundamental theorem of §18, there are solutions $y_n = \phi_n$ only for a certain sequence of values $\lambda = \lambda_n$ tending to $-\infty$, yet these solutions y_n form a complete orthonormal basis of a certain weighted Hilbert space.

But what are the values of the eigenvalues $\lambda = \lambda_n$? What are the shapes of the corresponding eigenmodes $y = y_n$?

The common approach to answering these two questions is numerical — using shooting, finite divided differences, or the Galerkin method to transform the problem (15.7) into a numerical linear algebra problem where the eigenvalues and vectors are approximate values and solutions. See Chapter 18, [Fox], or [Ortega].

Alternatively we might employ a manipulator. With *Mathematica* one asks for solutions when $\lambda = z$ with

```
In[1]:= DSolve[y''[x]+y'[x] == z y[x], y,x]
```

which returns with the general solution as the linear combination of two independent solutions with two constants of integration `C[1]` and `C[2]`. The next two commands extract the solution from its rule/list-based answer of `Out[1]`:

```
In[2]:= y[x] /. %

In[3]:= %[[1]]
```

Choose the eigenfunction with $c_2 = 1$ with

```
In[4]:= % /. C[2]->1

In[5]:= eigenfn = %
```

Impose the zero left end condition of (15.7b) with

```
In[6]:= eigenfn /. x->0

In[7]:= bndry0 = %
```

and eliminate `C[1]` from the eigenfunction with

```
In[8]:= Solve[bndry0 == 0,C[1]]

In[9]:= eigenfn /. %

In[10]:= eigenfn = %
```

Impose the right boundary condition of (15.7b) with

```
In[11]:= eigenfn' = D[eigenfn,x]

In[12]:= eigenfn + eigenfn' /. x->1

In[13]:= bndry1 = %[[1]]
```

The characteristic equation that determines the eigenvalues $\lambda = z$ is then `bndry1==0`.

Obtain a general idea of the eigenvalue location from the plot of the modulus

```
In[14]:= ab = Abs[bndry1]

In[15]:= Plot[ab, {z,-100,10}]
```

which produces the rough sketch of Figure 15.4.

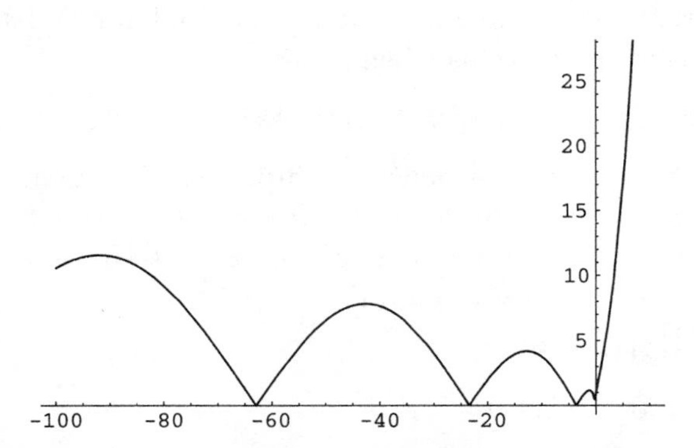

Figure 15.4 A graph of the modulus of the characteristic function.

Take a closer look at the dominant eigenvalue near 0 with

```
In[16]:= Plot[ab, {z,-5,0}]
```

to find from the resulting Figure 15.5 that apparently the dominant eigenvalue is $\lambda_1 \approx -0.25$. But from the above analytic formula for the eigenfunction of `Out[10]` or by symbolically substituting this value into the eigenfunction we see this value yields a 0 'eigenfunction' and must be discarded. Instead the dominant eigenvalue $\lambda_1 \approx -3.6$.

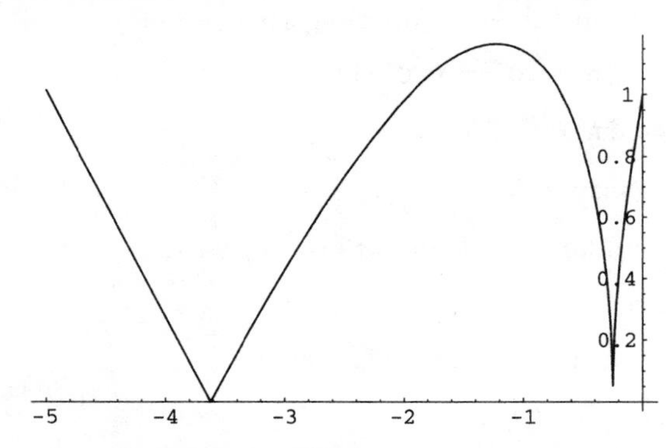

Figure 15.5 A close-up view of Figure 15.4.

Searching nearby -3.6 with

```
In[17]:= FindRoot[bndry1 == 0,{z,-3.6}]
In[18]:= N[%,14]
```

we find our dominant value $\lambda_1 = -3.6230892866262$.

Obtain the eigenmode shape $\phi_1 = y_1$ belonging to this dominant eigenvalue λ_1 with

```
In[19]:= eigenfn /.  z->-3.6230892866262
```

```
In[15]:= eigenfn1 = %
```
to obtain the dominant eigenmode

$$\phi_1 = e^{(-0.5+1.8366i)x} - e^{(-0.5-1.8366i)x} = 2ie^{-0.5x}\sin 1.8366x. \quad (15.8)$$

A rough plot of the eigenmode shape $\phi_1/2i = y_1$ belonging to this dominant mode is found with
```
In[21]:= Plot[Exp[-0.5 x] Sin[1.8366 x], {x,-0.2,1.0}]
```
to obtain the plot of Figure 15.6.

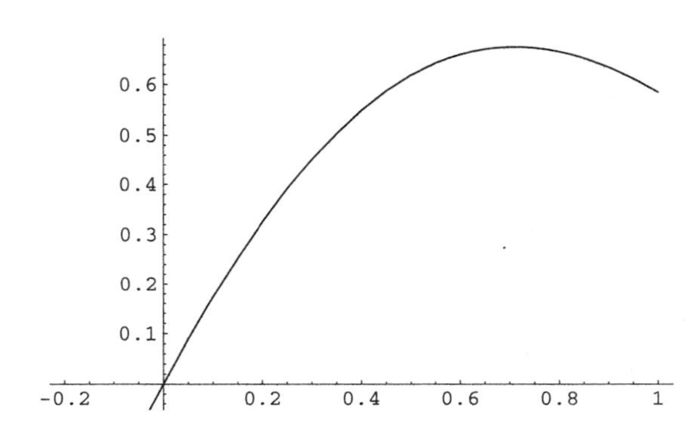

Figure 15.6 The dominant eigenmode.

Unfortunately *Mathematica* ver. 2.1 cannot solve ODEs of generic complexity much more than linear with constant coefficients. For example, the package will balk at solving

$$y'' - xy' = \lambda y \quad (15.9)$$

even though this is a slight variation of the Hermite equation. Nevertheless the assistant can often be tricked into doing work by reformulating the problem in state space:

$$y' = z \quad (15.10a)$$

$$z' = xz + \lambda y. \quad (15.10b)$$

The assistant can now solve the system when we ask
```
In[1]:= DSolve[{y'[x] == z[x], z'[x] == a y[x]
          + x z[x]}, {y[x], z[x]}, x]
```
Proceed as before with minor modifications.

§15.6 Animation of Time-Varying Solutions

Above all other capabilities, manipulators can provide visualization of time varying processes. Let us work through several examples. The scripts that follow were adapted from scripts developed by Jonathan D. Courtney.

BVP 1. The relaxation of a rod (revisited)

Recall BVP 1: the ends of a rod of length and temperature 1 are suddenly brought and held to 0. We found in §5 that interior temperatures $u = u(x,t)$ relax to 0 according to the rule (5.11):

$$u(x,t) = \frac{4}{\pi}\sum_{k=0}^{\infty} \mathrm{e}^{-(2k+1)^2\pi^2 t}\,\frac{\sin(2k+1)\pi x}{2k+1}. \tag{15.11}$$

It would be interesting and possibly revealing to animate temperature regimes within the rod as they evolve over time.

Load in the animation package with

```
In[1]:= <<Graphics/Animation.m
```

Approximate temperature u (15.11) using 4 term truncation with the script

```
In[2]:= n[k_]:= (2 k+1) Pi
In[3]:= e[t_,k_]:= Exp[-n[k]^2 t]
In[4]:= u[x_,t_]:= 4 Sum[e[t,k] Sin[n[k] x]/n[k],{k,0,4}]
```

Next we construct an animation sequence by stepping through time in steps of size $\Delta t = 0.05$ from $t = 0.1$ to $t = 0.4$:

```
In[5]:= frames = MoviePlot[u[x,t], {x,0,1},
          {t,0.1, 0.4, 0.05}, PlotRange->{0,1}]
```

which — after some considerable time spent assembling the frames — runs the movie again and again until told to desist. Several frames of the movie are shown superimposed in Figures 15.7. The switch `PlotRange->` holds axes scaling fixed during the animation.

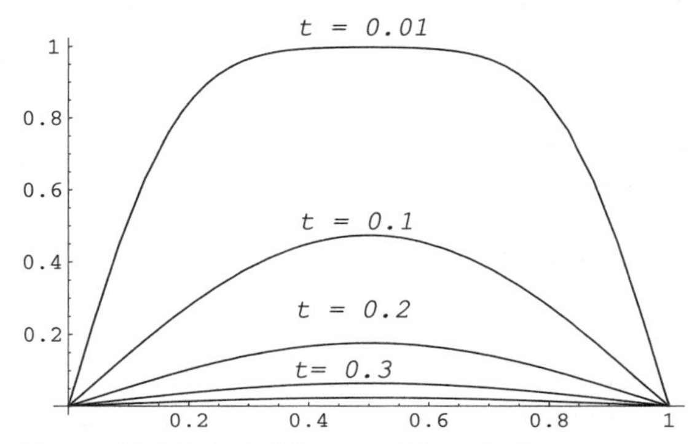

Figure 15.7 Selected frames of the relaxing temperatures.

To rerun the animation during the current session enter

```
In[6]:= ShowAnimation[frames].
```

To store the animated sequence as a file (say) *tempframes* type

```
In[7]:= Save["tempframes.m", frames].
```

To read in this graphic object for a rerun at a later session enter

```
In[]:= <<tempframes.m
```

then rerun with `ShowAnimation[frames]`.

To view one frame at a time, define a graphic-valued function $gr(t)$ labeled with the particular time t as follows:

```
In[8]:= gr[t_]:= Plot[u[x,t], {x,0,1},PlotLabel->
            StringForm["Time=``", N[t]], PlotRange->{0,1}]
```

Display snapshots, say at $t = 0.2$ with

```
In[9]:= gr[0.2]
```

then save for printing with

```
In[10]:= Display["snapshot", %]
```

The file `snapshot` may need additional massaging before printing, depending on the hardware.

Vibrations of a square drum (reprise)

Let us animate the (vertical) vibrations $u = u(x, y, t)$ of a square drum $0 < x, y < 1$ using the series solution

$$u = \sum_{m,n=1}^{\infty} c_{mn} \cos \pi t\sqrt{m^2 + n^2}\ \sin m\pi x\ \sin n\pi y \qquad (15.12)$$

obtained in Exercise 5.13.

Load the animation package with

```
In[1]:= <<Graphics/Animation.m
```

Let us assume the initial deformation is centered at (0.6,0.4) of radius 0.2 and is shaped like a rotated inverted quarter-cycle cosine:

$$u(x,y,0) = \begin{cases} -\cos \pi r/0.4 & \text{if } r = \sqrt{(x-0.6)^2+(y-0.4)^2} < 0.2 \\ 0 & \text{otherwise.} \end{cases} \tag{15.13}$$

This would be coded as

```
In[2]:= dist[x_,y_]:= Sqrt[(x-0.6)^2 + (y-0.4)^2]
In[3]:= ding[x_,y_]:= If[dist[x,y]< 0.2,
          -Cos[Pi dist[x,y]/0.4], 0]
```

Figure 15.8 shows this initial shape as generated by the command

```
In[4]:= Plot3D[ding[x,y],{x,0,1},{y,0,1}]
```

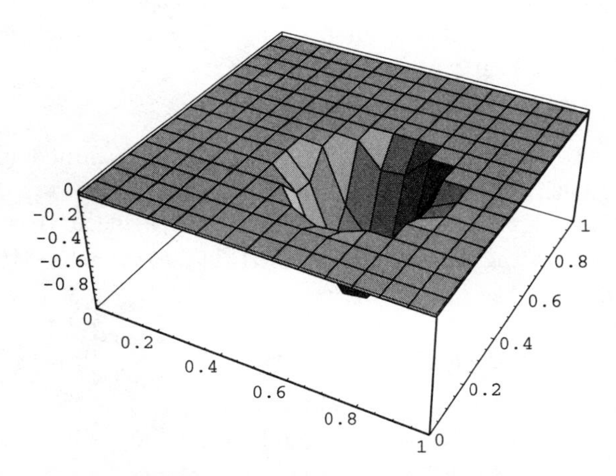

Figure 15.8 The initial deformation of the square drumhead.

Next we instruct the assistant to construct a table of the first 16 coefficients of the expansion (15.12) for the initial deformation (15.13) of Figure 15.8, namely

$$c_{mn} = 4\int_0^1\int_0^1 u(x,y,0)\ \sin m\pi x\ \sin n\pi y\ dxdy \tag{15.14}$$

with the command

```
In[5]:= c = Table[NIntegrate[ding[x,y]*Sin[m Pi x]
        *Sin[n Pi y],{x,0,1},{y,0,1}],{m,1,4},{n,1,4}]
```

Expect a long delay as the coefficients are estimated.

The partial sum for $1 \leq m, n \leq 4$ of (15.11) is generated with

```
In[6]:= vibs[x_,y_,t_]:= (4/Pi) Sum[c[[m,n]]
        *Cos[Pi t Sqrt[m^2 + n^2]]*Sin[m Pi x]*Sin[n Pi y],
        {m,1,4},{n,1,4}]
```

Finally we can animate the vibrations with

```
In[7]:= MoviePlot3D[vibs[x,y,t],{x,0,1},{y,0,1},
        {t,0.1, 2, 0.1}, PlotRange->{-0.1,0.1}]
```

Selected frames of this movie are shown in Figures 15.9a–h.

Figure 15.9 Snapshots of the vibrating square drum with initial shape shown in Figure 15.8.

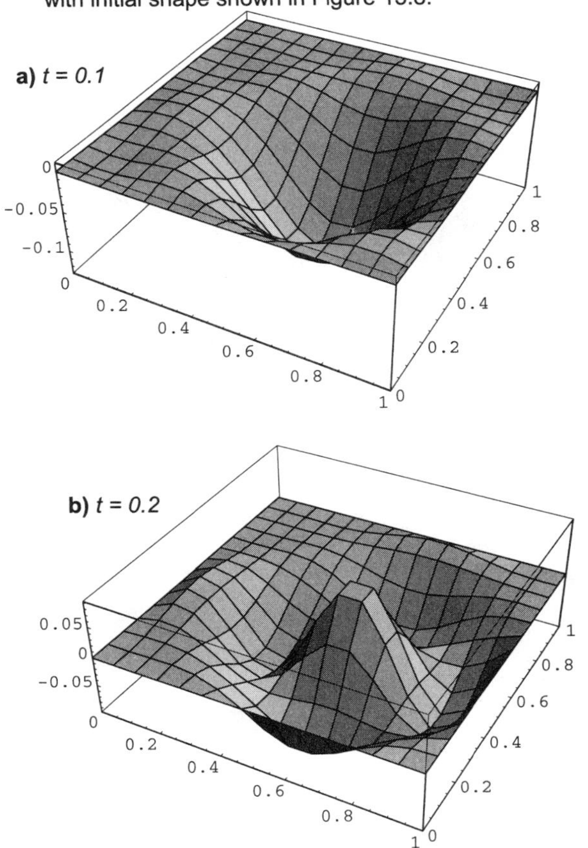

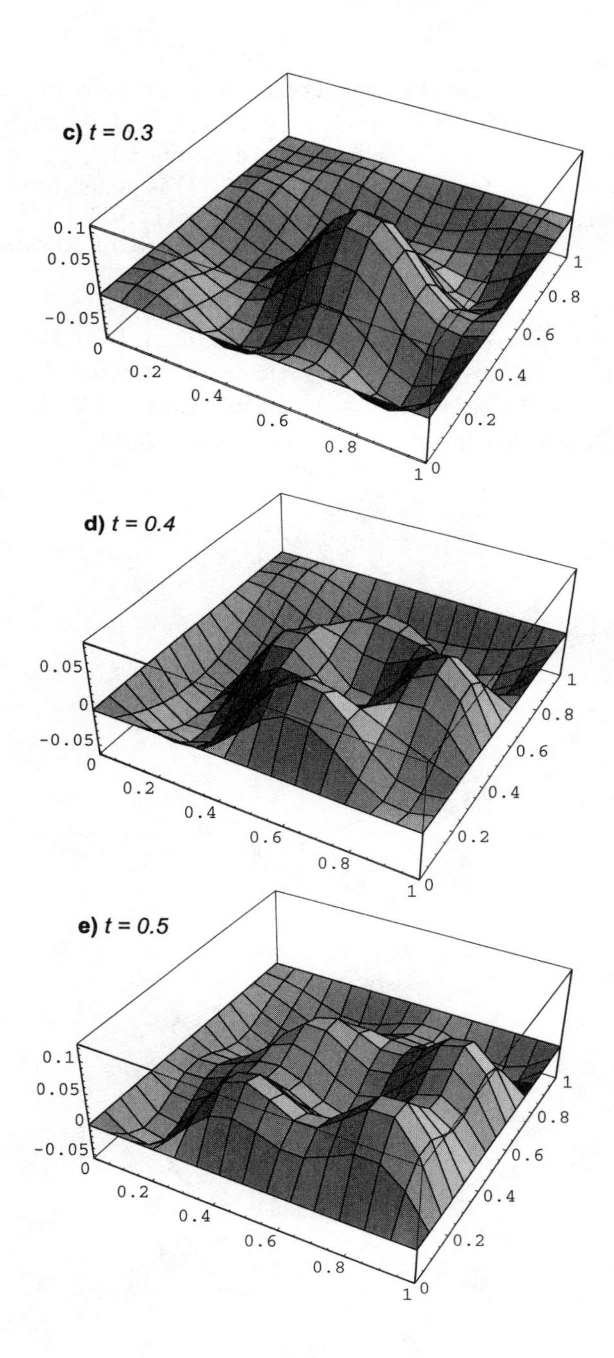
c) t = 0.3
0.1
0.05
0
-0.05
0
0.2
0.4
0.6
0.8
1
0
0.2
0.4
0.6
0.8
1
d) t = 0.4
0.05
0
-0.05
0
0.2
0.4
0.6
0.8
1
0
0.2
0.4
0.6
0.8
1
e) t = 0.5
0.1
0.05
0
-0.05
0
0.2
0.4
0.6
0.8
1
0
0.2
0.4
0.6
0.8
1

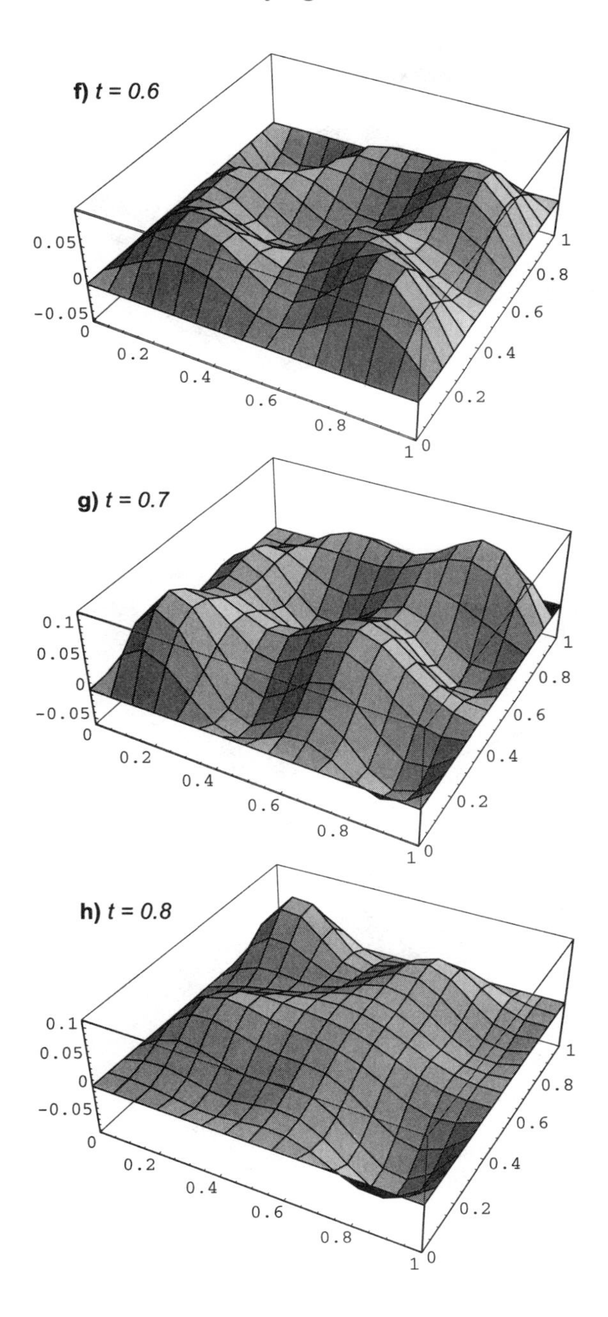
f) $t = 0.6$
0.05
0
-0.05
0
0.2
0.4
0.6
0.8
1
0
0.2
0.4
0.6
0.8
1
g) $t = 0.7$
0.1
0.05
0
-0.05
0
0.2
0.4
0.6
0.8
1
0
0.2
0.4
0.6
0.8
1
h) $t = 0.8$
0.1
0.05
0
-0.05
0
0.2
0.4
0.6
0.8
1
0
0.2
0.4
0.6
0.8
1

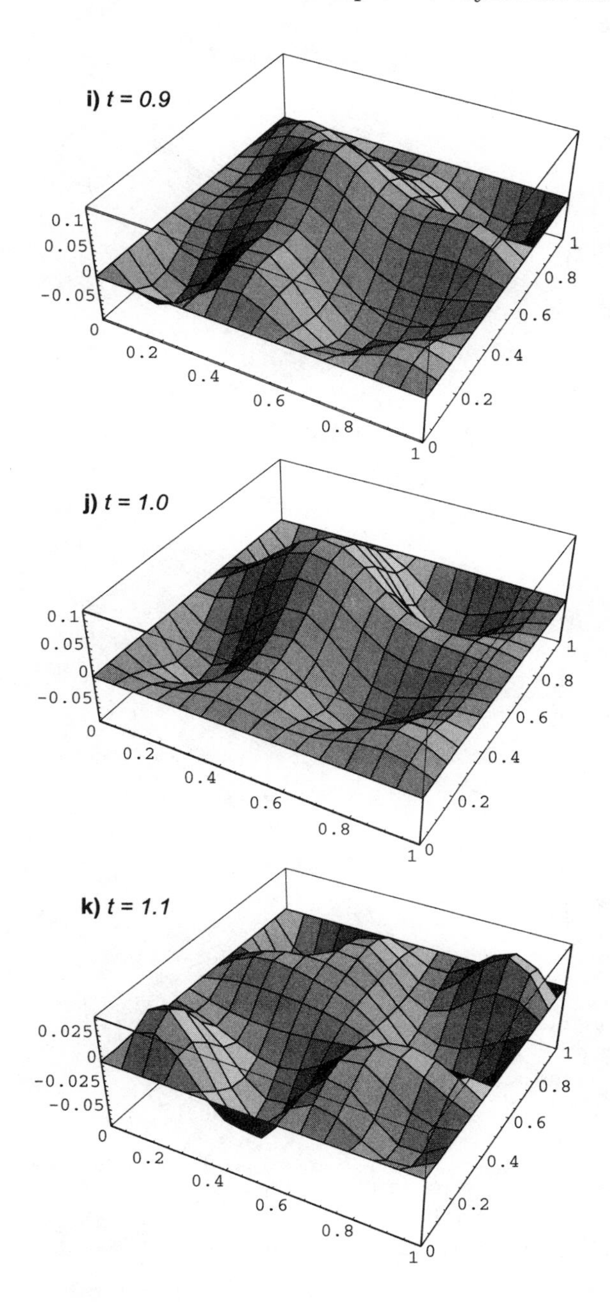
i) t = 0.9
0.1
0.05
0
-0.05
0
0.2
0.4
0.6
0.8
1
0
0.2
0.4
0.6
0.8
1
j) t = 1.0
0.1
0.05
0
-0.05
0
0.2
0.4
0.6
0.8
1
0
0.2
0.4
0.6
0.8
1
k) t = 1.1
0.025
0
-0.025
-0.05
0
0.2
0.4
0.6
0.8
1
0
0.2
0.4
0.6
0.8
1

Exercises

15.1 Find $H_{23}(x)$.

15.2 Find the 10 (real) roots of $T_{10}(x) = 0$.

15.3 Find the first 5 positive roots of $J_3(x) = 0$.

15.4 Graph $J_2(e^x)$.

15.5 Guess the coefficients of the Fourier expansion of $f(x) = x^4$ on the interval $[-\pi, \pi]$.

15.6 Obtain the first 15 terms of the expansion

$$\sin r^2 = \sum_{k=1}^{\infty} c_k J_0(\alpha_k r)$$

where the α_k are the positive roots of $J_0(x) = J_1(x)$.

15.7 Obtain the first 15 terms of the Legendre expansion

$$x^{1/3} = \sum_{n=0}^{\infty} c_n P_n(x)$$

on $-1 < x < 1$.

15.8 Check directly that $\nabla^2 u = 0$ for $x^2 + y^2 + z^2 > 0$ when

$$u = \operatorname{Re}(x + iy)^{-5}.$$

15.9 Is

$$u = z + \frac{\cos 5\theta}{r^5} + r \sin \theta$$

harmonic?

15.10 Find the first 7 eigenvalues of the Sturm-Liouville problem

$$y'' + 3y' = \lambda y$$

subject to

$$y(0) = 0 \text{ and } y(1) + y'(1) = 0.$$

15.11 Find the first 7 eigenvalues of the Sturm-Liouville problem

$$y'' + 3xy' = \lambda y$$

subject to

$$y(0) = 0 \text{ and } y(1) + y'(1) = 0.$$

15.12 Animate a superposition of temperature $u(x, t)$ and flux $v(x, t) = -u_x(x, t)$ within the relaxing rod (15.11) of BVP 1.

Hint: Use `In[]:= MoviePlot[{u[x,t],v[x,t]}, ... ]`

15.13 Animate the relaxing temperatures of Fourier's Ring Problem BVP 3 with initial temperature distribution

$$u(x, 0) = 1 \text{ when } 0 < x < \pi, \ 0 \text{ otherwise.}$$

15.14 Animate the vertical vibrations of a circular drum struck symmetrically at its center. See BVP 5 in §5.5.

15.15 Animate a quivering cube of Jello^{TM}. See Exercise 8.41.

15.16 Animate the vertical deflections of a circular drum when struck off-center. See BVP 5 in §10.2.

15.17 Animate the radial vibrations of a spherical bell (12.31).

Chapter 16
Operational Calculus

Let us sample Oliver Heaviside's powerful *operational methods*, called *frequency domain methods* by the engineering community. Heaviside never deigned to justify his stunning methods. As he so succinctly argued, "Shall I refuse my dinner because I do not fully understand the process of digestion? Does anybody fully understand anything?" See [Moore] and [Nahin]. Mathematicians of his time substantiated his approach via the Laplace transform, thereby initiating a central tool of engineering — *frequency domain design.* Subsequently arose *Nyquist analysis, root locus design, Bode analysis, Nichols Charts,* etc., the pantheon of 'Bell Lab Mathematics.'

I believe the Laplace transform interpretation of Heaviside's methods to be merely an early historical attempt, an attempt fraught with convergence, existence, and uniqueness questions. There is a modern interpretation, the *Mikusinski Operational Calculus,* free of these difficulties. Here is a brief account. Complete accounts can be found in the books by Mikusinski, Erdelyi, and Yosida. The traditional view is found in the books by Moore, Carslaw and Jaeger, Doetsch, and others listed in the references.

§16.1 Background

Signals with fast rise times reveal and carry much information. Think of radar, sonar, and digital transmission.

An idealized zero rise time signal is the *Heaviside unit step*

$$u_0(t) = \begin{cases} 1 & \text{if} \quad t \geq 0 \\ 0 & \text{if} \quad t < 0 \end{cases} . \tag{16.1}$$

Close the switch of Figure 16.1 to the *RC* circuit shown. Assume the capacitor is initially discharged and that the *time constant* $T = RC = 1$. The circuit is then often modeled by (Exercise 16.1)

$$\dot{y} + y = u_0 \tag{16.2}$$

subject to $y(0) = 0$. There is no need to look further — serious mathematical pathologies are present in this simple model (16.2).

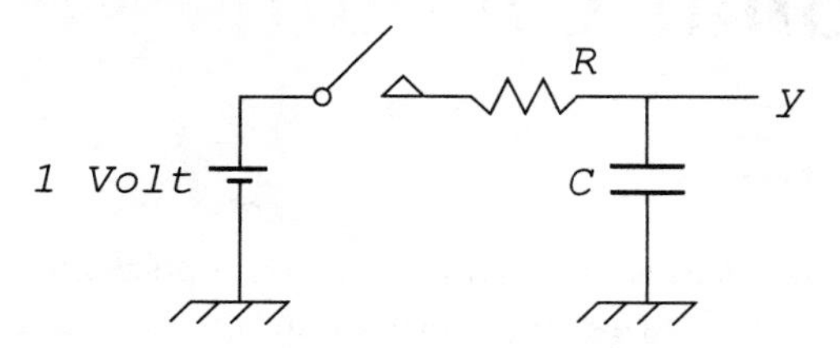

Figure 16.1 This simple *RC* charging circuit cannot be modeled by a classical ODE.

On the one hand, by multiplying (16.2) through by the integrating factor e^t, it appears that for $t \geq 0$,

$$y = e^{-t} * u_0 = \int_0^t e^{-\lambda} d\lambda = 1 - e^{-t}. \tag{16.3}$$

On the other hand a more careful look reveals that (16.3) is not a solution of (16.2) in the strict sense — a two sided derivative $\dot{y}$ cannot exist at $t = 0$ (Exercise 16.2). Alternatively, recall Darboux's result that a function z that is a derivative $z = \dot{y}$ enjoys the intermediate value property. Thus (16.2) cannot hold for an input u with a jump discontinuity. How do we escape this conundrum?

One escape might be to reformulate (16.2) as

$$\dot{y} + y =_{a.e.} u_0, \tag{16.4}$$

where equality holds almost everywhere. This is no escape at all — uniqueness is now lost. There are continuous non-constant *Cantor functions* $c(t)$ with

$$\dot{c}(t) =_{a.e.} 0 \tag{16.5}$$

(Exercise 16.3). So if $y(t)$ is one solution to (16.4), $y(t) + c(t)e^{-t}$ is another.

Rather than weakening equality we must instead *generalize differentiation* and be willing to accept *generalized functions* as solutions. Convolution is the key.

§16.2 Convolution Quotients

An *ordinary signal* is a real or complex valued function $f(t)$ such that

i) $f(t)$ is *causal*, i.e., $f(t) = 0$ if $t < 0$,

and

ii) $f(x)$ is *locally (absolutely) integrable*, i.e.,

$$\int_0^t |f(\lambda)| d\lambda < \infty$$

for all t.

We may convolve any two ordinary signals to obtain a third

$$h(t) = (f * g)(t) = \int_0^t f(t-\lambda) g(\lambda) d\lambda \tag{16.6}$$

(Exercise 16.4). There is the fundamental

Principle. (Titchmarsh)

$$f * g = 0 \text{ implies}(a.e.) \ f = 0 \text{ or } g = 0. \tag{16.7}$$

See [Mikusinski] or [Erdelyi].

This principle enables us to form the *quotient field* of the ordinary signals as follows. Consider as in ordinary arithmetic all formal quotients

$$\frac{f}{g}, \tag{16.8}$$

for $g \neq 0$, where equality between any two such symbols is defined by

$$\frac{f_1}{g_1} = \frac{f_2}{g_2} \text{ exactly when } f_1 * g_2 = g_1 * f_2. \tag{16.9}$$

The collection $\mathcal{F}$ of all equivalence classes of these symbols under equality is called the *field of all convolution quotients*, or the *Mikusinski generalized functions (signals)*.

Again as in ordinary arithmetic define

$$\frac{f_1}{g_1} + \frac{f_2}{g_2} = \frac{f_1 * g_2 + g_1 * f_2}{g_1 * g_2} \tag{16.10}$$

and

$$\frac{f_1}{g_1} \cdot \frac{f_2}{g_2} = \frac{f_1 * f_2}{g_1 * g_2}. \tag{16.11}$$

Just as in ordinary arithmetic, addition and multiplication is independent of the particular class representative (Exercise 16.5). For any $f \neq 0$, denote the generalized signal f/f using the symbol 1 or δ (the *Dirac delta*) interchangeably. Note that the generalized signal $1 = \delta$ is not an ordinary signal (Exercise 16.6). Set $0 = 0/g$, $-(f/g) = -f/g$, and for $f \neq 0$, $(f/g)^{-1} = g/f$.

The generalized signals $\mathcal{F}$ enjoy the *field properties*:

$$\begin{aligned}
\alpha + (\beta + \gamma) &= (\alpha + \beta) + \gamma & \alpha(\beta\gamma) &= (\alpha\beta)\gamma \\
\alpha + \beta &= \beta + \alpha & \alpha\beta &= \beta\alpha \\
\alpha + 0 &= \alpha & \alpha \cdot 1 &= \alpha \\
\alpha + (-\alpha) &= 0 & \alpha \cdot \alpha^{-1} &= 1 \ (\alpha \neq 0) \\
&\alpha(\beta + \gamma) = \alpha\beta + \alpha\gamma
\end{aligned} \tag{16.12}$$

as can be tediously verified (Exercise 16.7).

The ordinary signals are to be found naturally embedded within the field of generalized signals by the identification

$$f \mapsto \{f\} = \frac{f * g}{g} \tag{16.13}$$

for any g.

By this construction we have enlarged our signals to a much larger class $\mathcal{F}$. As in Arithmetic, where enlarging the integers to the rationals enables the solving of equations such as $3x = 5$, the enlargement of the signals to the generalized signals enables us to solve (16.2).

§16.3 Generalized Integration and Differentiation

The Heaviside unit step in $\mathcal{F}$ is denoted by

$$h = \{u_0\} \tag{16.14}$$

and multiplication by h is called *generalized integration*, since after all, for the ordinary signals,

$$\int_0^t f(\lambda)d\lambda = (u_0 * f)(t). \tag{16.15}$$

The multiplicative inverse of the Heaviside unit step is nowadays universally denoted by

$$s = h^{-1}. \tag{16.16}$$

Multiplication by s is called *generalized differentiation.*

Complicated functions can often be quite simply written in terms of the generalized function s.

Example 1. Since

$$h\{e^{-t}\} = \{\int_0^t e^{-\lambda}d\lambda\} = \{u_0 - e^{-t}\} = h - \{e^{-t}\},$$

$$\{e^{-t}\} = \frac{h}{h+1} = \frac{1/s}{1/s+1} = \frac{1}{s+1}.$$

More generally (Exercise 16.8),

$$\{e^{-at}\} = \frac{1}{s+a}. \tag{16.17}$$

Hence

$$\{\cosh at\} = \frac{\frac{1}{s-a} + \frac{1}{s+a}}{2} = \frac{s}{s^2 - a^2}. \tag{16.18}$$

Likewise

$$\{\sinh at\} = \frac{a}{s^2 - a^2}, \tag{16.19}$$

$$\{\cos at\} = \frac{s}{s^2 + a^2}, \tag{16.20}$$

and

$$\{\sin at\} = \frac{a}{s^2 + a^2}. \tag{16.21}$$

You may recognize these functions on the right as the *Laplace transform* of the respective time-domain signals — the *frequency domain* versions of the ordinary signals on the left. But from the Mikusinski viewpoint,

the time and frequency domains coincide.

Objects of $\mathcal{F}$ are both signals as well as linear operators on $\mathcal{F}$ (by multiplication).

Example 2. Suppose the ordinary signal f has a jump discontinuity at $t = 0$ but is classically continuously differentiable everywhere else. Then

$$s\{f(t)\} = \{\dot{f}(t)\} + f(0^+). \tag{16.22}$$

You may recognize this as the Laplace transform rule

$$L\dot{f}(t) = sF(s) - f(0^+).$$

To see (16.22), note that

$$\begin{aligned}\{f(t)\} &= \{\int_0^t \dot{f}(\lambda)dx + f(0^+)\} \\ &= h\{\dot{f}(t)\} + f(0^+)h,\end{aligned}$$

and now multiply through by s. So for example,

$$s\{e^{at}\} = a\{e^{at}\} + 1, \tag{16.23}$$

$$s\{\cos at\} = -a\{\sin at\} + 1, \tag{16.24}$$

$$s\{\sin at\} = a\{\cos at\}. \tag{16.25}$$

This generalized differentiation rule (16.22) yields the ordinary derivative with infinite spikes superimposed at jump discontinuities.

For example, the generalized derivative of the Heaviside unit step h is the spike

$$sh = \delta, \tag{16.26}$$

with value 0 everywhere but at $t = 0$.

§16.4 Generalized Solutions

By generalizing differentiation, our troublesome model of the charging capacitor (16.2) of Figure 16.1 becomes the correct model

$$s\{y\} + \{y\} = h$$

with generalized solution

$$\{y\} = \frac{1}{s(s+1)} = \frac{1}{s} - \frac{1}{s+1}$$

i.e.,

$$y = \begin{cases} 1 - e^{-t} & \text{if } t \geq 0 \\ 0 & \text{if } t < 0, \end{cases}$$

which now agrees with our convolution solution (16.3).

It would take a discussion of book length to carefully lay out the power of this approach. For further details I suggest first a look at [Erdelyi], then [Mikusinski] and [Yosida]. Allow me to summarize:

1) Any relation that holds for the Laplace transform holds in the field of Mikusinski convolution quotients.

2) Unlike the Laplace transform, every ordinary signal f has a 'transform,' namely the convolution quotient $\{f\}$. For example e^{t^2} has no Laplace transform yet in the Mikusinski scheme, it plays a role like any other ordinary signal.

3) ODEs with constant coefficients

$$\begin{aligned} a_n y^{(n)} + a_{n-1} y^{(n-1)} + \cdots + a_1 \dot{y} + a_0 y \\ = b_m u^{(m)} + b_{m-1} u^{(m-1)} + \cdots + b_0 u \end{aligned}$$

for a given *input* u will have a unique solution (*response*) Y found in the generalized signals

$$Y = \frac{s^m b_m + \cdots + s b_1 + b_0}{s^n a_n + \cdots + s a_1 + a_0} U.$$

If Y comes from an ordinary signal y, i.e. $Y = \{y\}$, then y is the solution of the ODE where $y^{(n-1)}$ is absolutely continuous. As an epigram, *such ODEs are uniquely solvable, but possibly by a generalized signal.* Moreover, in standard engineering jargon, the *response*

Y to the *input* U is given by multiplying (convolving) the input with the *transfer function (impulse response)*

$$P(s) = \frac{s^m b_m + \cdots + s b_1 + b_0}{s^n a_n + \cdots + s a_1 + a_0}.$$

4) Our ordinary signals $f(t)$ so far have been readouts from a single sensor. Consider distributed signals

$$f(x,t)$$

obtained as measurements taken at each point x of the spatial domain $\Omega \subset \mathbf{R}^n$. All the preceding goes through for these spatially distributed signals. Spatial partial differentiation is carried over into the field of all (time) convolution quotients.

Let us solve several problems operationally. Since this is only a survey, many details must be omitted.

§16.5 A Semi-Infinite Slab (BVP 18)

A homogeneous isotropic slab, initially at temperature 0, occupies the halfspace $x \geq 0$. The face at $x = 0$ is suddenly put and held at temperature 1. How will this sudden step change of temperature at the boundary propagate through the slab?

This is the undersea telegraph cable problem of Exercises 1.7–1.11. It is also related to BVP 8 on deep earth temperatures.

Analytically the problem is

$$\frac{\partial u}{\partial t} = \frac{\partial^2 u}{\partial x^2}, \quad 0 < x < \infty \tag{16.27}$$

subject to

i) $u(x, 0) = 0$,

ii) $u(0, t) = 1$,

iii) $u(\infty, t) = 0$.

As is traditional, use upper case letters for *frequency domain* objects:

$$\{u(x,t)\} = U(x,s).$$

Then condition i) and rule (16.22) brings the PDE (16.27) into the frequency domain as

$$sU = U'' \tag{16.28}$$

where the prime $'$ denotes differentiation with respect to x. Thus

$$U = c_1(s)e^{-x\sqrt{s}} + c_2(s)e^{x\sqrt{s}}. \tag{16.29}$$

By iii), $c_2 = 0$. By ii),

$$c_1(s) = \frac{1}{s}.$$

Thus our solution is

$$U = \frac{e^{-x\sqrt{s}}}{s}. \tag{16.30}$$

Consulting transform tables [Roberts and Kaufman] or [Doetsch], we find

$$u = \operatorname{erfc}(\frac{x}{2\sqrt{t}}), \tag{16.31}$$

which is Stokes's solution (Exercise 1.8) for the undersea telegraph cable step response.

More generally, if this slab is subjected to any causal boundary temperature variation $u(0,t) = f(t)$, the same computation reveals that temperatures u within the slab are given by

$$U(x,s) = e^{-\sqrt{s}x}F(s)$$

where $\{f\} = F$. But since

$$\frac{e^{-x\sqrt{s}}}{s} = \{\operatorname{erfc}(\frac{x}{2\sqrt{t}})\},$$

then by (16.22),

$$e^{-x\sqrt{s}} = s\{\operatorname{erfc}(\frac{x}{2\sqrt{t}})\} = \{\frac{x}{2\sqrt{\pi}}\frac{e^{-x^2/4t}}{t^{3/2}}\}$$

and hence for a general time-changing boundary condition $u(0,t) = f(t)$,

$$u = \frac{x}{2\sqrt{\pi}}\frac{e^{-x^2/4t}}{t^{3/2}} * f(t). \tag{16.32}$$

The impulse response, in this case

$$p(t) = \frac{x}{2\sqrt{\pi}}\frac{e^{-x^2/4t}}{t^{3/2}},$$

is often called the *Green's function* of the problem.

§16.6 Spilled Pollutants (BVP 19)

A pollutant has been spilled on the ground surface — see Figure 16.1. How will this pollutant disperse with time into the earth below?

We have seen in Exercise 1.2 that because mass flow rate is proportional to concentration gradient, the concentration u satisfies the equation of diffusion.

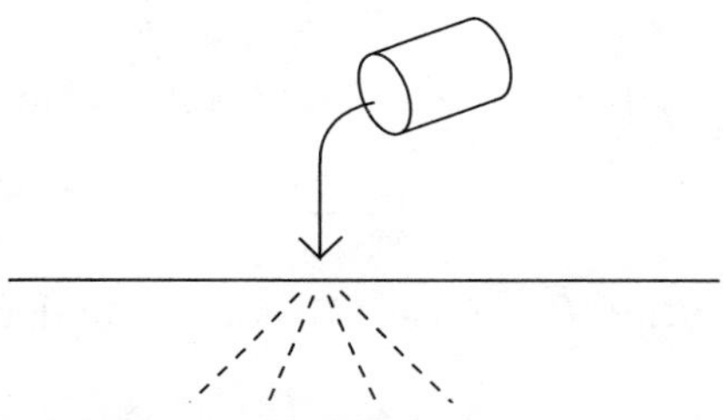

Assume the original ground spill was highly localized and spatially symmetric. Then some thought on symmetry (Exercise 16.10) will reveal that concentration u is independent of angular variables, i.e.,

$$u = u(\rho, t)$$

where ρ is the radial distance into the ground from the point of spill. Let $\dot{m}$ be the rate of pollutant mass entering the ground. Then the analytic model for concentration u is

$$\frac{\partial u}{\partial t} = a(\frac{\partial^2 u}{\partial \rho^2} + \frac{2}{\rho}\frac{\partial u}{\partial \rho}), \quad 0 \leq \rho < \infty, \tag{16.33}$$

subject to

i) $u(\rho, 0) = 0$,

ii) $2\pi\kappa \lim_{\rho\to 0} \rho^2 u_\rho(\rho, t) = -\dot{m}$, (Exercise 16.11)

iii) $u(\infty, t) = 0$,

where a is the diffusivity and κ the permeability of the earth to this compound. Going over the frequency domain, the model becomes

$$(s/a)U = U'' + \frac{2}{\rho}U' \tag{16.34}$$

where the prime $'$ denotes differentiation with respect to ρ. The general solution to (16.34) is (Exercise 16.12)

$$U = \frac{c_1(s)e^{-\rho\sqrt{s/a}} + c_2(s)e^{-\rho\sqrt{s/a}}}{\rho}. \tag{16.35}$$

Because of iii), $c_2 = 0$ and so

$$U = c_1(s)\frac{e^{-\rho\sqrt{s/a}}}{\rho}. \tag{16.36}$$

Imposing condition ii) yields (Exercise 16.13)

$$c_1(s) = \frac{\dot{m}}{2\pi\kappa s}. \tag{16.37}$$

Consequently our solution is

$$U = \frac{\dot{m}}{2\pi\kappa}\frac{e^{-\rho\sqrt{s/a}}}{\rho s} \tag{16.38}$$

giving as in the previous problem,

$$u = \frac{\dot{m}}{2\pi\kappa\rho}\text{erfc}(\frac{\rho}{2\sqrt{at}}). \tag{16.39}$$

Note as in the previous problem, the concentration response u to an arbitrary spill is given by the convolution of the spill with the impulse response, the generalized derivative of the step response (16.39) — see Exercise 16.14.

§16.7 The Closed-Loop Heat Pump Reprised (SSP 7)

Let us again solve for the transient ground temperatures during the heating season as the heat pump extracts heat from the earth via a vertical exchanger — see Figure 2.7.

As in §10.3, the temperature $u = u(r, t)$ about one unit length of the exchanger must obey

$$\frac{\partial u}{\partial t} = a(\frac{\partial^2 u}{\partial r^2} + \frac{1}{r}\frac{\partial u}{\partial r}), \quad 0 \leq r < \infty, \tag{16.40}$$

subject to

i) $u(r, 0) = B$,

ii) $2\pi\kappa \lim_{r\to 0} r u_r(r,t) = \dot{Q}$, (Exercise 16.15)

iii) $u(\infty, t) = B$,

where a and κ are the diffusivity and conductivity of the soil.

Let us for the moment assume unaffected ground temperature $B = 0$. Then in the frequency domain our problem becomes

$$(s/a)U = U'' + \frac{1}{r}U', \tag{16.41}$$

i.e.,

$$r^2U'' + rU - (s/a)r^2U = 0. \tag{16.42}$$

The general solution is of the form

$$U = a(s)J_0(ir\sqrt{s/a}) + b(s)Y_0(ir\sqrt{s/a}) \tag{16.43}$$

which we instead choose to write in the form

$$U = c_1(s)H_0^{(1)}(ir\sqrt{s/a}) + c_2(s)H_0^{(2)}(ir\sqrt{s/a}) \tag{16.44}$$

in terms of the *Hankel functions (Bessel functions of the third kind)*

$$H_0^{(1)}(x) = J_0(x) + iY_0(x) \tag{16.45a}$$

and

$$H_0^{(2)}(x) = J_0(x) - iY_0(x). \tag{16.45b}$$

Because of asymptotic growth properties [Stegun-Abramowitz, 9.2.4], condition iii) yields that $c_2(s) = 0$. Condition ii) implies (Exercise 16.16)

$$c_1(s) = \frac{\dot{Q}}{4\pi i\kappa s}. \tag{16.46}$$

Hence our solution is

$$U = \frac{\dot{Q}}{4\pi i\kappa}\frac{H_0^{(1)}(ir\sqrt{s/a})}{s}. \tag{16.47}$$

But the *modified Hankel function*

$$K_0(x) = \frac{\pi i}{2}H_0^{(1)}(ix) \tag{16.48}$$

occurs as a standard transform [Doetsch, 9.30]:

$$\{\frac{e^{x^2/4t}}{2t}\} = K_0(x\sqrt{s}). \tag{16.49}$$

Therefore (Exercise 16.17), after reinstating unaffected ground temperature B,

$$\begin{aligned} u &= B + \frac{\dot{Q}}{4\pi\kappa}\int_0^t \frac{e^{r^2/4a\lambda}}{t}d\lambda \\ &= B + \frac{\dot{Q}}{2\pi\kappa}\int_{\frac{r}{2\sqrt{at}}}^{\infty} \frac{e^{\beta^2}}{\beta}d\beta, \end{aligned} \tag{16.50}$$

known as the *Kelvin line source* solution (Exercise 16.18). Consult the *Design/Data Manual for Closed-loop Ground-coupled Heat Pump Systems* by Bose, Parker, and McQuiston to see how this solution (16.50) is leveraged into an entire design approach.

Exercises

16.1 Argue that when the switch is thrown in the circuit of Figure 16.1, the voltage y across the capacitor is determined by $RC\dot{y} + y = u_0$. (The correct model is actually $RCsY + Y = h$.)

16.2 Show that the causal signal of (16.3) is not classically differentiable at $t = 0$.

16.3 Construct a nonconstant nondecreasing continuous function $c(t)$ whose derivative exists and is 0 for almost every t.

Outline: Let $c(t) = 1/2$ on the interval $(1/3,\ 2/3)$, $c(t) = 1/4$ on $(1/9,\ 2/9)$, $c(t) = 3/4$ on $(7/9,\ 8/9)$, etc.

16.4 Show that the convolution of two ordinary signals is again an ordinary signal.

Hint: Employ Fubini's theorem of Appendix A.

16.5 Show that the operations of addition (16.10) and multiplication (16.11) are well defined.

Hint: $1/2 + 1/3 = 2/4 + 5/15$.

16.6 Show that the Dirac delta δ is not an ordinary signal — there is no ordinary signal e with $e * f = f$ for all ordinary signals f.

16.7 Establish the field properties (16.12) for the convolution quotients $\mathcal{F}$.

16.8 Verify (16.17).

16.9 Write a routine to graph the voltage response (16.32) of the undersea cable problem (BVP 18) to an input of one telegraph 'dit'

$$f(t) = \begin{cases} 1 & \text{if } 0 \leq t < 1 \\ 0 & \text{otherwise} \end{cases}$$

at a distance of $x = 1000$ down the cable. Note how the pulse has been smeared spatially and how rise time has increased.

16.10 Argue that a point source of spilled pollutant diffuses into the surrounding earth in a radially symmetric fashion $u = u(\rho, t)$.

16.11 Argue that boundary condition ii) of BVP 19 (16.33) correctly models a point spill of pollutant on the surface.

16.12 Establish (16.35).

16.13 Verify (16.37).

16.14 Find the concentration $u(\rho, t)$ of BVP 19 in response to a one-time spill whose input tapers off exponentially over time, i.e., where

$$\dot{m} = e^{-bt}.$$

Answer:

$$U = \frac{1}{2\pi\kappa\rho}\frac{e^{-\rho\sqrt{s/a}}}{s+b} = \{\frac{1}{4\pi^{3/2}\sqrt{a}}\frac{e^{-\rho^2/4at}}{t^{3/2}} * e^{-bt}\}.$$

16.15 Argue that boundary condition ii) of the problem (16.40) correctly models the extraction of heat from the plane from a single point.

16.16 Verify (16.46).

Hint: You will need the fact that $\lim_{z\to 0} zY_0(z) = 2/\pi, \quad -\pi < \arg z < \pi$. See [Abramowitz and Stegun, 9.1.13].

16.17 Verify (16.50).

16.18 Show by direct calculation that the Kelvin line source solution (16.50) solves the problem (16.40).

16.19 Operationally find the step response of an ideal transmission line of propagation velocity v. That is, solve the problem

$$\frac{\partial^2 u}{\partial t^2} = v^2 \frac{\partial^2 u}{\partial x^2}, \quad 0 < x < \infty$$

subject to $u(x,0) = 0,\ u_t(x,0) = 0,\ u(0,t) = 1,\ u(\infty,t) = 0.$

Answer: $U = s^{-1}e^{-sx/v} = \{u_0(t - x/v)\}$.

16.20 Find the frequency domain solution of

$$\frac{\partial u}{\partial t} = \frac{\partial^2 u}{\partial x^2}, \quad 0 < x < 1$$

subject to $u(x,0) = 0,\ u(0,t) = 0,\ u(1,t) = f(t).$

Answer:

$$U = \frac{\sinh x\sqrt{s}}{\sinh \sqrt{s}} F(s).$$

16.21 Find the impulse response (Green's function) $p(t)$ of Exercise 16.20, i.e., the function

$$\{p(t)\} = \frac{\sinh x\sqrt{s}}{\sinh \sqrt{s}},$$

via Separation of Variables as follows: solve for the step response (i.e., $f(t) \equiv 1$), then operate by s.

Answer:

$$p(t) = 2\pi \sum_{n=1}^{\infty} (-1)^{n+1} e^{-n^2\pi^2 t}\, n \sin n\pi x.$$

16.22 Solve Exercise 11.3 operationally.

16.23 Solve BVP 1 operationally.

16.24 Solve BVP 6 operationally.

16.25 Find the transfer function $P(s)$ of the Trombe wall (Figure 16.3) modeled by $\partial u/\partial t = \partial^2 u/\partial x^2$, $0 < x < 1$ subject to $u(x, 0) = 0$, $\kappa u_x(0, t) = -f(t)$, $u_x(1, t) = k(u(1, t) - u_0)$.

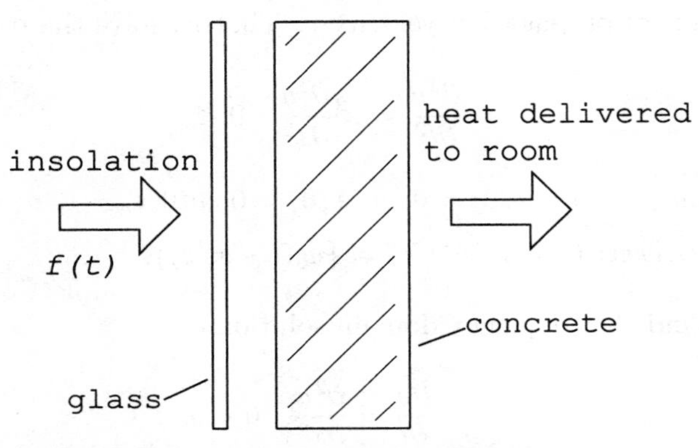

Figure 16.3 A Trombe wall. Sunlight is trapped by glass, converted to heat that diffuses through a masonary wall to be delivered by radiation and convection to a living space.

16.26 The pueblo-dwelling native Americans of the southwest built homes with adobe walls of a critical thickness — the walls delayed the sun's flux until it was needed to heat the interior during the cold nights. Model this wall. Using actual material properties, choose the optimal thickness for a 24-hour repetitive cycle. Compare your results against actual traditional construction.

Chapter 17
Fourier Integrals

Separation of variables can resolve problems even for unbounded spatial domains Ω. However the superposition of separated solutions to meet an awkward condition is now achieved via integration rather than summation. The method is fraught with delicate convergence questions.

Let us cavalierly work through several examples, later to reexamine their underpinnings.

§17.1 The Sector Revisited (SSP 2)

Reconsider the steady problem of a quarter-plane as shown in Figure 17.1:

$$\nabla^2 u = 0, \qquad 0 < x,\ y < \infty, \tag{17.1}$$

subject to

i) $u(0, y) = 1$,

ii) $u(x, 0) = 0$,

iii) u is bounded.

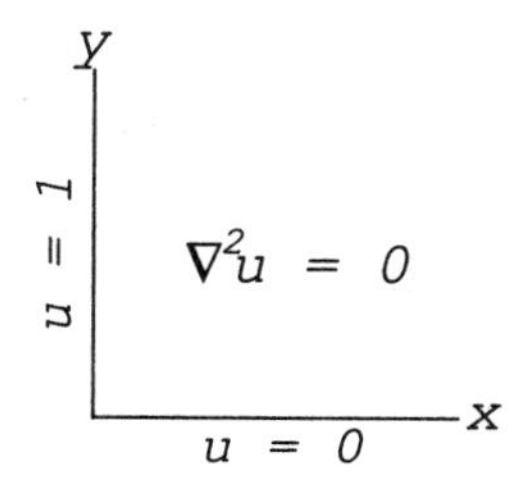

Figure 17.1 The steady Dirichlet problem on the infinite quarter plane. The solution is $u = 2\theta/\pi$.

We have seen in §2.2 that the solution is $u = 2\theta/\pi$.

Let us again solve this problem via separation of variables. Assume a separated solution

$$u = X(x)Y(y)$$

and find that

$$\frac{X''}{X} + \frac{Y''}{Y} = 0, \tag{17.2}$$

where the prime $'$ denotes differentiation with respect to the single appropriate variable. But then both terms of the left hand side of (17.2) are constant, i.e.,

$$-\frac{Y''}{Y} = \frac{X''}{X} = \lambda. \tag{17.3}$$

Suppose λ is positive

$$\lambda = \alpha^2, \tag{17.4}$$

(see Exercise 17.1). Then

$$X = c_1 e^{-\alpha x} + c_2 e^{\alpha x} \tag{17.5}$$

and

$$Y = a \cos \alpha y + b \sin \alpha y. \tag{17.6}$$

Insisting on bounded solutions (condition iii) means $c_2 = 0$. Imposing condition ii) yields $a = 0$. Thus a separated solution is

$$u = e^{-\alpha x} \sin \alpha y. \tag{17.7}$$

In order to meet the troublesome boundary condition i), let us continuously superimpose all the separated solutions (17.7) weighted by some $c(\alpha)$

$$u = \int_0^\infty c(\alpha) e^{-\alpha x} \sin \alpha y \, d\alpha. \tag{17.8}$$

Imposing the boundary condition i) upon (17.8) yields the requirement

$$1 = \int_0^\infty c(\alpha) \sin \alpha y \, d\alpha. \tag{17.9}$$

But there is the famous and important

Sinc formula. In the *Cauchy principal value* sense of the integral (explained below),

$$\int_0^\infty \frac{\sin \beta}{\beta} d\beta = \frac{\pi}{2}. \tag{17.10}$$

Proof. Exercise 17.16.

Corollary.

$$\frac{2}{\pi} \int_0^\infty \frac{\sin b\alpha}{\alpha} d\alpha = \operatorname{sgn} b. \tag{17.11}$$

Proof. Exercise 17.3.

So by taking

$$c(\alpha) = \frac{2}{\alpha\pi}$$

in (17.9), our steady problem (17.1) on the quarter-plane is 'solved' by

$$u = \frac{2}{\pi} \int_0^\infty e^{-\alpha x} \frac{\sin \alpha y}{\alpha} dy. \tag{17.12}$$

§17.2 A Doubly-Infinite Solid Rod (SSP 13)

Consider the homogeneous isotropic circular rod of radius 1 with the z-axis as axis, shown in Figure 17.2. Suppose the surface temperatures experience a step change from -1 to 1 at $z = 0$. Find the steady temperature $u = u(r, z)$ within the rod.

Analytically,

$$\nabla^2 u = 0, \quad r < 1, \qquad -\infty < z < \infty, \tag{17.13}$$

subject to

i) $u(1, z) = -1$ for $z < 0$,

ii) $u(1, z) = 1$ for $z > 0$,

iii) u is bounded.

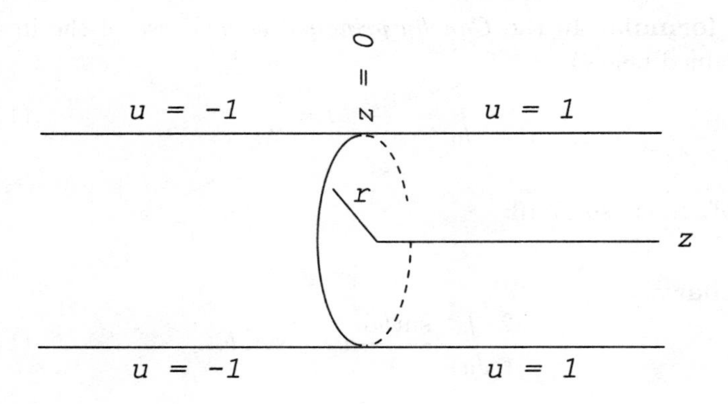

Figure 17.2 A doubly infinite solid circular rod with a step change in surface temperature at $z = 0$.

Assume variables separate

$$u = R(r)Z(z)$$

to obtain

$$\frac{R'' + (1/r)R'}{R} + \frac{Z''}{Z} = 0. \tag{17.14}$$

As in the previous problem, since u is bounded,

$$\frac{Z''}{Z} = -\alpha^2.$$

Since the solution is symmetric about $z = 0$, $Z(-z) = -Z(z)$, and hence

$$Z = \sin \alpha z. \tag{17.15}$$

Therefore

$$r^2R'' + rR' - \alpha^2 r^2 R = 0 \tag{17.16}$$

and hence because u is bounded,

$$R = I_0(\alpha r). \tag{17.17}$$

Superimposing these separated solutions weighted by $c(\alpha)$ yields

$$u = \int_0^\infty c(\alpha) I_0(\alpha r) \sin \alpha z \, d\alpha. \tag{17.18}$$

Again exploiting the sinc formula (17.11), we have our solution

$$u = \frac{2}{\pi} \int_0^\infty \frac{I_0(\alpha r)}{I_0(\alpha)} \frac{\sin \alpha z}{\alpha} d\alpha. \tag{17.19}$$

§17.3 A Semi-Infinite Slab Reprised (BVP 18)

Recall from §16.5 the problem

$$\frac{\partial u}{\partial t} = \frac{\partial^2 u}{\partial x^2}, \qquad 0 < x < \infty, \tag{17.20}$$

subject to

i) $u(x, 0) = 0$,

ii) $u(0, t) = 1$,

iii) $u(\infty, t) = 0$.

Note that this problem has no steady state (Exercise 17.5). An attempt to solve by Separation of Variables will not succeed (Exercise 17.6). Solve instead the transformed problem with

$$v = 1 - u.$$

That is, suppose

$$v = T(t)X(x)$$

to see that

$$\frac{\dot{T}}{T} = \frac{X''}{X} = \lambda = -\alpha^2,$$

and hence

$$v = e^{-\alpha^2 t} \sin \alpha x. \tag{17.21}$$

Superimposing these separated solutions gives

$$v = \int_0^\infty c(\alpha) e^{-\alpha^2 t} \sin \alpha x \, d\alpha. \tag{17.22}$$

To meet condition i) is to satisfy

$$1 = \int_0^\infty c(\alpha) \sin \alpha x \, d\alpha,$$

and hence, again by the sinc formula (17.11),

$$u = 1 - v = 1 - \frac{2}{\pi}\int_0^\infty e^{-\alpha^2 t}\frac{\sin \alpha x}{\alpha}\,d\alpha. \tag{17.23}$$

Contrast this solution with Stokes's solution obtained operationally in §16.5:

$$u = \operatorname{erfc}(\frac{x}{2\sqrt{t}}) = \frac{2}{\sqrt{\pi}}\int_{\frac{x}{2\sqrt{t}}}^\infty e^{-\beta^2}\,d\beta, \tag{17.24}$$

also verified in Exercise 1.8. Although appearing quite distinct, these two solutions are in fact identical (Exercise 17.18–17.21).

§17.4 Justification of the Method

Once a Fourier integral candidate solution u is obtained by separation of variables, you must verify that

i) u solves the PDE, and

ii) u satisfies the initial and boundary conditions.

Nearly always it is possible to verify i) by repeated applications of

Leibniz's Rule. Suppose

$$|g(x,y)| \le h(y)$$

and that

$$|\frac{\partial g(x,y)}{\partial x}| \le k(y)$$

for $a < x < b$, y in Y, where $h(y)$ and $k(y)$ are integrable over Y. Then for $a < x < b$,

$$\frac{d}{dx}\int_Y g(x,y)\,dy = \int_Y \frac{\partial g(x,y)}{\partial x}\,dy. \tag{17.25}$$

So for example, in the semi-infinite slab BVP 18 above, the solution (17.23)

$$u = 1 - \frac{2}{\pi}\int_0^\infty e^{-\alpha^2 t}\frac{\sin \alpha x}{\alpha}\,d\alpha$$

does indeed satisfy the PDE

$$\frac{\partial u}{\partial t} = \frac{\partial^2 u}{\partial x^2}, \qquad 0 < x < \infty,$$

for $t > \epsilon > 0$ since for all positive α,

$$g(x,t,\alpha) = e^{-\alpha^2 t}\frac{\sin \alpha x}{\alpha} << x\mathrm{e}^{-\alpha^2 \epsilon}, \ \frac{\partial g}{\partial t} << \alpha^2 x\mathrm{e}^{-\alpha^2 \epsilon},$$

$$\frac{\partial g}{\partial x} << e^{-\alpha^2 \epsilon}, \ \text{and} \ \frac{\partial^2 u}{\partial x^2} << \alpha \mathrm{e}^{-\alpha^2 \epsilon}.$$

(The symbol $f(\alpha) << g(\alpha)$ means that for some constant c, $|f(\alpha)| \leq c|g(\alpha)|$ for all large α). Therefore three applications of Leibniz's Rule will verify that (17.23) satisfies the heat equation (17.20) (Exercise 17.12). The initial and/or boundary conditions are usually the sticking point.

For example, again in the above BVP 18, the initial condition i) of (17.20) relied on the sinc formula (17.10):

$$1 = \frac{2}{\pi}\int_0^\infty \frac{\sin \alpha}{\alpha} d\alpha.$$

This is not a Lebesgue integral. The integrand is not absolutely integrable, i.e., (Exercise 17.14)

$$\int_0^\infty |\frac{\sin \alpha}{\alpha}| d\alpha = \infty. \tag{17.26}$$

We often must be content to satisfy initial and boundary conditions in the following, more forgiving sense of the integral:

§17.5 Cauchy Principal Value

We will say that an integral

$$\int_{-\infty}^{\infty} f(\beta)\, d\beta = I \tag{17.27}$$

converges in the Cauchy principal value sense or *has Cauchy principal value I* when f is absolutely integrable on every bounded interval $(-b,b)$ and when

$$\lim_{b \to \infty} \int_{-b}^{b} f(\beta)\, d\beta = I. \tag{17.28}$$

This weak notion of integrability can yield absurd formulae such as

$$\int_{-\infty}^{\infty} \beta^2 \sin\beta \, d\beta = 0.$$

Nevertheless Cauchy principal value is of central importance in pure and applied mathematics. See for instance Ayoub's *Introduction to the Analytic Theory of Numbers* or Papoulis's *Signal Analysis.*

There are two helpful lemmas.

Lemma A. Suppose $f(\beta)$ decreases monotonically to 0 as $\beta \to \infty$. Then

$$\int_0^{\infty} f(\beta) \sin\beta \, d\beta$$

converges in the Cauchy sense (assuming f vanishes for negative β), and in fact for non-negative integers n, we have the bounds

$$\int_0^{2n\pi} f(\beta) \sin\beta \, d\beta \le \int_0^{\infty} f(\beta) \sin\beta \, d\beta \le \int_0^{(2n+1)\pi} f(\beta) \sin\beta \, d\beta. \tag{17.29}$$

Proof. When $N\pi \le b < (N+1)\pi$,

$$\int_0^{b} f(\beta) \sin\beta \, d\beta$$

$$= \sum_{n=0}^{N-1} (-1)^n \int_{n\pi}^{(n+1)\pi} f(\beta) |\sin\beta| \, d\beta + \int_{N\pi}^{b} f(\beta) \sin\beta \, d\beta.$$

Since $f(\beta)$ is decreassing to 0, the summation is an alternating series of terms decreasing to zero, hence convergent by the alternating series test. The integral second term clearly vanishes at infinity. The inequality (17.29) is left as Exercise 17.15.

Lemma B. Suppose $f(\beta, t)$ decreases monotonically to 0 as $\beta \to \infty$ for each $t \ge 0$ and that $f(\beta, t)$ is jointly continuous. Then

$$\lim_{t \to 0+} \int_b^{\infty} f(\beta, t) \frac{\sin\beta}{\beta} d\beta = \int_b^{\infty} f(\beta, 0) \frac{\sin\beta}{\beta} d\beta. \tag{17.30}$$

Proof. Exercise 17.17.

§17.6 Application to BVP 18

Let us verify that the solution (17.23) of BVP 18 does indeed satisfy the initial and boundary conditions. These technical concerns are typical of the Method of Fourier Integrals.

We must establish the initial condition i) of (17.20):

$$\lim_{t\to 0}\int_0^\infty e^{-\alpha^2 t}\frac{\sin \alpha x}{\alpha}d\alpha = \int_0^\infty \frac{\sin \alpha x}{\alpha}d\alpha = \frac{\pi}{2}, \qquad (17.31)$$

or more explicitly, we must validate the exchange of limits

$$\lim_{t\to 0}\lim_{a\to\infty}\int_0^a e^{-\alpha^2 t}\frac{\sin \alpha x}{\alpha}d\alpha = \lim_{a\to\infty}\lim_{t\to 0}\int_0^a e^{-\alpha^2 t}\frac{\sin \alpha x}{\alpha}d\alpha.$$

We must also verify condition ii):

$$\lim_{x\to 0+}\int_0^\infty e^{-\alpha^2 t}\frac{\sin \alpha x}{\alpha}d\alpha = 0$$

and condition iii):

$$\lim_{x\to\infty}\int_0^\infty e^{-\alpha^2 t}\frac{\sin \alpha x}{\alpha}d\alpha = \frac{\pi}{2}.$$

To validate each of these interchanges of limits first make the substitution $\beta = \alpha x$, giving for $x,\ t > 0$,

$$\int_0^\infty e^{-\alpha^2 t}\sin\alpha\ d\alpha/\alpha = \int_0^\infty e^{-\beta^2 t/x^2}\sin\beta\ d\beta/\beta$$

$$= \int_0^b e^{-\beta^2 t/x^2}\sin\beta\ d\beta/\beta + \int_b^\infty e^{-\beta^2 t/x^2}\sin\beta d\ \beta/\beta. \qquad (17.32)$$

Apply Lemma B with $f(\beta, t/x^2) = e^{-\beta^2 t/x^2}$ to the second integral of (17.32). The interchange of limits in the first integral is justified by the Lebesgue Dominated Convergence Theorem (Appendix A) since we are dealing there with a Lebesgue, not Cauchy integral.

§17.7 Another Origin of the Method

An alternate path also leading to Fourier integral solutions of boundary value problems is suggested by a fundamental observation of Fourier. Certain absolutely integrable functions $f(x)$ on $(-\infty, \infty)$ possesses a *Fourier integral representation*

$$f(x) = \int_0^\infty a(\alpha) \cos \alpha x + b(\alpha) \sin \alpha x \;\; d\alpha, \tag{17.33}$$

where

$$a(\alpha) = \frac{1}{\pi} \int_{-\infty}^{\infty} f(x) \cos \alpha x \; dx \tag{17.34}$$

and

$$b(\alpha) = \frac{1}{\pi} \int_{-\infty}^{\infty} f(x) \sin \alpha x \; dx. \tag{17.35}$$

You may think of this representation as the limiting case of the Fourier series representation (7.24)–(7.27) of a function $f(x)$ on $(-N\pi, N\pi)$:

$$f(x) = \sum_{n=0}^{\infty} a_n \cos nx/N + b_n \sin nx/N.$$

Thus knowing of the existence of Fourier integral expansions, one might attempt a solution of say a steady problem $\nabla^2 u = 0$ such as (17.1) by assuming a solution form

$$u = \int_0^\infty a(x, \alpha) \cos \alpha y + b(x, \alpha) \sin \alpha y \; d\alpha, \tag{17.36}$$

imposing the PDE and the boundary conditions, then solving for $a(x, \alpha)$ and $b(x, \alpha)$.

Applying this approach to SSP 2 of (17.1), condition ii) would suggest $a(x, \alpha) = 0$. For $\nabla^2 u = 0$ we must have in (17.36) that

$$b_{xx}(x, \alpha) = -\alpha^2 b(x, \alpha),$$

and hence

$$b(x, \alpha) = c_1(\alpha) e^{-\alpha x} + c_2(\alpha) e^{\alpha x}.$$

The boundedness condition iii) suggests $c_2 = 0$. Finally imposing boundary condition i) yields by this alternate path the solution

$$u = \frac{2}{\pi} \int_0^\infty e^{-\alpha x} \frac{\sin \alpha y}{\alpha} d\alpha,$$

in full agreement with the solution (17.12) obtained by separation of variables. Both approaches should be in your repertoire. Note the similarity of this second approach to operational methods. Why? Because (17.33) is in fact a famous transform closely related to the Laplace transform.

§17.8 The Fourier Transform

The Fourier integral expansion (17.33) is actually a famous transform in disguise (Exercise 17.17): for absolutely integrable f, the *Fourier transform* of f is

$$\hat{f}(\omega) = \int_{-\infty}^{\infty} f(t) e^{-i\omega t} dt, \tag{17.37}$$

with its (delicately convergent, frequently divergent) inverse transform

$$f(t) = \frac{1}{2\pi} \int_{-\infty}^{\infty} \hat{f}(\omega) e^{i\omega t} d\omega. \tag{17.38}$$

The functions $f(t)$ and $\hat{f}(\omega)$ are called a *Fourier transform pair.*

The convergence and inversion questions are far more delicate for the Fourier transform than for the analogous Fourier series. Even L^2 results are more delicate, since on the infinite interval $(-\infty, \infty)$, L^2 functions are not necessarily L^1 (absolutely integrable). However the celebrated *Plancherel theorem* guarantees that the Fourier transform when restricted to $L^1 \cap L^2$ has a unique extension to a surjective isometry (norm preserving map) of all of $L^2(-\infty, \infty)$. This is still no guarantee of the convergence of the inversion integral (17.38) — see [Rudin]. However there is one reasonable result:

Dini's Criterion. Suppose both $f(t)$ and its Fourier transform $\hat{f}(\omega)$ are absolutely integrable. Then at each point $t = t_0$ where the difference quotient

$$Q(t) = \frac{f(t) - f(t_0)}{t - t_0}$$

is locally integrable, the inverse transform

$$\frac{1}{2\pi} \int_{-\infty}^{\infty} \hat{f}(\omega) e^{i\omega t_0} d\omega$$

converges even in the Lebesgue sense to $f(t_0)$.

Proof. Repeat the proof of Dini's Criterion of §7.2 with summation replaced by integration (Exercise 17.23). Again the result ultimately depends on the Riemann-Lebesgue Lemma.

This Criterion is occasionally of help when solving PDEs via Fourier integrals since it may guarantee that boundary conditions are realized — see Exercise 17.24.

For the more practically inclined I recommend the intuitive and applicable *The Fourier Integral and its Applications* by A. Papoulis.

Exercises

17.1 Show that a choice of $\lambda = -\alpha^2$ rather than (17.4) contradicts the bounded condition iii) of problem (17.1).

17.2 Obtain the bounded solution

$$u = \frac{2}{\pi} \sum_{n=1}^{\infty} \frac{1 - (-1)^n}{n} e^{-nx} \sin ny$$

of the steady problem $\nabla^2 u = 0$ on the semi-infinite strip $0 < x < \infty$, $0 < y < \pi$, subject to $x(0, y) = 1$ and $u(x, 0) = 0 = u(x, \pi)$. Mull over the qualitative differences between this solution and (17.12). Why does one require an integral, the other a series?

17.3 Argue that if the integral (17.11) exists in some sense, then it most likely takes on the values 1, 0, or -1 as displayed.

17.4 Show SSP 13 is also solved by

$$u = 1 - 2\sum_{n=1}^{\infty} \frac{J_0(\alpha_n r)}{\alpha_n J_1(\alpha_n)} e^{-\alpha_n z}$$

once extended oddly to $-\infty < z < 0$, where α_n are the positive zeros of J_0.

Hint: Treat z like t and solve for $v = 1 - u$.

17.5 Show the semi-infinite slab problem (17.20) has no steady state.

17.6 Attempt to solve the semi-infinite slab problem (17.20) by separation of variables $u = T(t)X(x)$. The attempt will fail. Why?

Answer: The boundary conditions are not homogeneous.

17.7 At first glance, the doubly infinite rod problem (17.12) may also have solutions of the form

$$u = \int_0^\infty c(\alpha) J_0(\alpha r) e^{-\alpha z}\, d\alpha$$

or

$$u = \int_0^\infty c(\alpha) J_0(\alpha r) \sinh \alpha z\, d\alpha.$$

Can there be solutions of these forms?

Answer: No. Why?

17.8 Solve

$$\frac{\partial u}{\partial t} = \frac{\partial^2 u}{\partial \rho^2} + \frac{2}{\rho}\frac{\partial u}{\partial \rho} + \frac{g}{\rho}, \qquad 0 < \rho < \infty,$$

subject to $u(\rho, 0) = 0$ and $u(\infty, t) = 0$. Give a physical interpretation of this problem.

Answer:

$$u = \frac{2g}{\pi} \int_0^\infty (1 - e^{-\alpha^2 t}) \frac{\sin \alpha\rho}{\alpha\rho}\, d\alpha.$$

17.9 Using the Fourier integral representation (17.33)–(17.35) solve the steady problem $\nabla^2 u = 0$ on the semi-infinite vertical strip $0 < x < 1,\ 0 < y$, subject to $u(0, y) = 0 = u(x, 0)$ and $u(1, y) = e^{-y}$.

Answer:

$$u = \frac{2}{\pi} \int_0^\infty \frac{\alpha}{1 + \alpha^2} \frac{\sinh \alpha x}{\sinh \alpha} \sin \alpha y\, d\alpha.$$

17.10 Solve the steady problem $\nabla^2 u = 0$ on the semi-infinite vertical strip $0 < x < 1,\ 0 < y$ subject to $u_x(0, y) = 0 = u(x, 0)$ and $u(1, y) = e^{-y}$.

Answer:

$$u = \frac{2}{\pi} \int_0^\infty \frac{\alpha}{1 + \alpha^2} \frac{\cosh \alpha x}{\cosh \alpha} \sin \alpha y \, d\alpha.$$

17.11 Solve the steady problem $\nabla^2 u = 0$ on the semi-infinite vertical strip $0 < x < 1,\ 0 < y$, subject to $u_x(0, y) = 0 = u_y(x, 0)$ and $u(1, y) = e^{-y}$.

Answer:

$$u = \frac{2}{\pi} \int_0^\infty \frac{1}{1 + \alpha^2} \frac{\cosh \alpha x}{\cosh \alpha} \cos \alpha y \, d\alpha.$$

17.12 Prove that Leibniz's Rule guarantees that (17.23) solves the PDE (17.20).

17.13 Carefully establish that (17.12) is indeed a solution to the problem (17.1). Establish both the PDE and the boundary conditions.

17.14 Show

$$\int_0^\infty \frac{|\sin \beta|}{\beta} d\beta = \infty.$$

Hint:

$$\int_0^\infty \frac{|\sin \beta|}{\beta} d\beta = \sum_{n=0}^{\infty} \int_{n\pi}^{(n+1)\pi} \frac{|\sin \beta|}{\beta} d\beta.$$

17.15 Prove the inequality (17.31).

17.16 Prove the sinc formula (17.10), i.e., in the Cauchy sense,

$$\int_0^\infty \frac{\sin \beta}{\beta} d\beta = \frac{\pi}{2}.$$

Outline: Apply Dini's Criterion of §7.2 to $f(x) = (\sin x/2)/(x/2)$ to obtain that $S_N(0) \to 1$ where

$$S_N(0) = \int_{-\pi}^{\pi} f(x) D_N(x) \, dx$$

$$= \frac{2}{\pi} \int_0^\pi \frac{\sin(N + 1/2)x}{x} dx = \int_0^{(N+1/2)\pi} \frac{\sin \beta}{\beta} d\beta.$$

A more elegant proof via contour integration can be found in any book on Complex Variables, e.g., [Churchill et al.].

17.17 Prove Lemma B.

17.18 Verify the formula

$$\int_0^\infty e^{-a\alpha^2}\cos b\alpha\,d\alpha = \frac{1}{2}\sqrt{\frac{\pi}{a}}e^{-b^2/4a}.$$

Put in other language and notation, establish the Fourier transform pair

$$e^{-at^2} \leftrightarrow \sqrt{\frac{\pi}{a}}e^{-\omega^2/4a}.$$

Outline: Set

$$y = \int_0^\infty e^{-a\alpha^2}\cos\alpha x\,d\alpha$$

and show by Leibniz's Rule and integration by parts that $y' = -xy/2a$. Hence $y = y(0)e^{-x^2/4a}$.

17.19 Establish the fundamental Fourier pair

$$\frac{\sin T\omega}{\pi\omega} \leftrightarrow p_T(t) = \begin{cases} 1 & \text{if } \; |t| < T \\ 0 & \text{otherwise} \end{cases}$$

As an epigram,
the rectangular pulse is paired with the sinc function.

17.20 Show for f real-valued and absolutely integrable,

$$\int_0^\infty f(\alpha)\frac{\sin 2T\alpha}{\alpha}d\alpha = \int_0^T \hat{f}(\beta)d\beta.$$

Hint: $2\int_0^\infty g(\alpha)\cos T\alpha\,d\alpha = \hat{g}(T)$. Also recall that product is paired with convolution.

17.21 Show the two solutions (17.23) and (17.24) of the semi-infinite slab coincide, i.e.,

$$1 - \frac{2}{\pi}\int_0^\infty e^{-\alpha^2 t}\frac{\sin\alpha x}{\alpha}d\alpha = frac2\sqrt{\pi}\int_{\frac{x}{2\sqrt{t}}}^\infty e^{-\beta^2}d\beta.$$

17.22 Deduce the Fourier integral formulae (17.33)–(17.35) from the Fourier transform formulae (17.37)–(17.38).

17.23 Prove Dini's Criterion for the Fourier transform.

17.24 Deduce from Dini's Criterion that $f(x) = e^{-|x|}$ everywhere equals its Fourier integral representation (17.33), i.e.,

$$e^{-|x|} = \frac{2}{\pi} \int_0^\infty \frac{\cos \alpha x}{1 + \alpha^2}\, d\alpha.$$

Chapter 18

Galerkin's Method

Separation of variables suggests an effective numerical method for solving boundary value problems. The method — developed by the Russian engineer B. G. Galerkin in 1915 — is an intimate companion of the modern Sobolev approach introduced in the next and final Chapter 19. Like many numerical methods, the Galerkin method transforms a problem in differential equations to a problem in numerical linear algebra.

§18.1 Truncation of Series Solutions

One obvious method for implementing a boundary value problem solution is *truncation*, where an orthogonal series solution obtained by separation of variables, say

$$u = \sum_{n=1}^{\infty} c_n q_n(t)\phi_n(x), \tag{18.1}$$

is merely lopped off to an N term series

$$\tilde{u} = \sum_{n=1}^{N} c_n q_n(t)\phi_n(x). \tag{18.2}$$

The advantage of this approach is complete control over error (Exercise 18.1). Note *the error is orthogonal to the approximate solution.* On the other hand it may not be possible to obtain such an orthogonal series solution for any one of many reasons. Moreover, even if such a series solution (18.1) is available, the functions $q_n(t)$ and/or $\phi_n(x)$ may be special functions and costly to implement (Exercise 18.2). Nevertheless truncation contains the germ of a very productive approach.

§18.2 Outline of the Galerkin Method

Write the PDE of a problem in the operator form

$$Lu = 0. \tag{18.3}$$

Examples: $Lu = \partial u/\partial t - \nabla^2 u, \quad Lu = \nabla^2 u, \quad Lu = \nabla^2 u - f.$

For a steady problem attempt a *trial (approximate)* solution to (18.3) of the form

$$\tilde{u} = u_0 + \sum_{j=1}^{N} c_j \phi_j(x), \tag{18.4}$$

where the *basis (trial)* functions ϕ_j satisfy homogeneous (usually 0) boundary conditions, and where u_0 is chosen so that the trial solution $\tilde{u}$ exactly satisfies the given boundary conditions of the problem. Impose the PDE on this trial solution to obtain the *residual (error)*

$$R = L\tilde{u}. \tag{18.5}$$

The Galerkin method requires the error to be orthogonal to the basis functions used in the trial solution, i.e.,

$$\langle \phi_i, R \rangle = 0, \quad \text{for } i = 1, 2, \ldots, N. \tag{18.6}$$

See Exercise 7.19. But (18.6) simply translates to N linear equations

$$\sum_{j=1}^{N} c_j \langle \phi_i, L\phi_j \rangle = -\langle \phi_i, Lu_0 \rangle, \qquad i = 1, 2, \ldots, N, \tag{18.7}$$

in the N unknowns c_j. Solve for these c_j.

For transient problems attempt a trial solution

$$\tilde{u} = u_0 + \sum_{j=1}^{N} c_j(t) \phi_j(x), \tag{18.8}$$

where again $\phi_j(x)$ satisfy homogeneous (usually 0) boundary conditions and where u_0 is chosen so that the trial solution satisfies the boundary conditions of the problem. Imposing the PDE yields a residual $R = L\tilde{u}$. Insisting that the residual be orthogonal to the first N spatial basis functions ϕ_i yields a system of N ordinary differential equations in t.

For example, for a diffusion problem

$$Lu = \frac{\partial u}{\partial t} - \nabla^2 u,$$

the resulting system is

$$\sum_{j=1}^{N} \dot{c}_j(t)\langle\phi_i, \phi_j\rangle = \sum_{j=1}^{N} c_j(t)\langle\phi_i, \nabla^2\phi_j\rangle + \langle\phi_i, \nabla^2 u_0\rangle. \tag{18.9a}$$

The initial conditions for this system are determined by a second application of the Galerkin approach: insist that the residual $I = u(x,0) - \tilde{u}(x,0)$ of the *initial* conditions also be orthogonal to the first N basis functions ϕ_i to obtain a simultaneous system of N linear equations

$$\sum_{j=1}^{N} c_j(0)\langle\phi_i, \phi_j\rangle = \langle\phi_i, u(x,0) - u_0(x)\rangle. \tag{18.9b}$$

Solve (18.9b) for these initial coefficients $c_j(0)$, then numerically integrate (18.9a) to obtain the proper $c_j(t)$ that initiate at $c_j(0)$.

In general, employing orthogonal trial functions ϕ_j simplifies calculations and often increases accuracy but may not be worth the trouble — usually simple polynomials suffice [Fletcher; Botha and Pinder]. The orthogonal polynomials are especially effective basis functions because of quadrature (Exercise 11.31) and other helpful properties [Ames; Twizell].

If the spatial domain Ω of the problem is awkwardly shaped, or if more solution accuracy is needed in certain portions of Ω, then Ω is subdivided into many subdomains Ω_k and basis functions are chosen on each subdomain that vanish on all but nearby subdomains — this is the method of *Galerkin finite elements.* In all cases the basis functions must be linearly independent for the resulting set of algebraic equations to be solvable. Moreover the span of all basis functions ϕ_j should be dense in $L^2(\Omega)$ or at least in a subspace containing the exact solution (Exercise 18.3).

§18.3 Steady Flow within a Square Tube (SSP 14)

Steady flow of a viscous incompressible fluid is governed by the Navier-Stokes equation displayed in Exercise 18.4. Far from any exit or entrance of fluid into a straight square tube, the nondimensional fluid velocity u down the tube is well modeled (Exercise 18.4) by Poisson's equation

$$\nabla^2 u = -1, \quad 0 < x, y < 1, \tag{18.10}$$

subject to 0 boundary conditions. By separation of variables the solution is (Exercise 18.5)

$$u = \frac{16}{\pi^4} \sum_{\substack{m,n=1 \\ m,n \text{ odd}}}^{\infty} \frac{\sin m\pi x \, \sin n\pi y}{m^3 n^2 + m^2 n^3}. \tag{18.11}$$

If we employ the eigenfunctions $\phi_{mn} = \sin m\pi x \, \sin n\pi y$ of the Laplacian as basis functions of the Galerkin trial solution

$$\tilde{u} = \sum_{m,n=1}^{N} c_{mn} \phi_{mn},$$

we merely obtain the first N^2 terms of the exact series solution (18.11) (Exercise 18.9). And at some computational cost since transcendental function support will be called by the CPU.

Instead try the simple polynomial basis functions

$$\phi_{mn}(x, y) = \alpha_m(x)\beta_n(y) = (x^m - x^{m+1})(y^n - y^{n+1}). \tag{18.12}$$

The residual is then

$$R = 1 + \sum_{m,n=1}^{N} c_{mn} \nabla^2 \phi_{mn} = 1 + \sum_{m,n=1}^{N} c_{mn} [\alpha''_m \beta_n + \alpha_m \beta''_n]. \tag{18.13}$$

Requiring the residual to be orthogonal to the basis,

$$\langle \phi_{ij}, R \rangle = \langle \alpha_i \beta_j, R \rangle = 0, \quad 1 \le i, j \le N,$$

leads to the requirement on the undetermined constants c_{mn} (Exercise 18.10):

$$\sum_{m,n=1}^{N} c_{mn}[f(i,m)g(j,n) + g(i,m)f(j,n)] = h(i,j), \tag{18.14}$$

where

$$f(i,j) = \frac{ij}{i+j-1} - \frac{i+j+2ij}{i+j} + \frac{i+j+ij+1}{i+j+1}, \tag{18.15}$$

$$g(i,j) = \frac{1}{i+j+1} - \frac{2}{i+j+2} + \frac{1}{i+j+3}, \tag{18.16}$$

and

$$h(i,j) = \frac{1}{(i+1)(i+2)(j+1)(j+2)}. \tag{18.17}$$

This daunting array of N^2 equations in N^2 unknowns is nevertheless a simple arithmetic matrix inversion problem (Exercise 18.11). You will find that even the one term ($N = 1$) Garlerkin approximate solution (Exercise 18.12)

$$\tilde{u} = 1.25(x - x^2)(y - y^2)$$

is good to 0.1% in kinetic energy. For $N = 3$ (nine terms), the kinetic energy error

$$\|u - \tilde{u}\|^2/\|u\|^2 \approx 0.04\%. \tag{18.18}$$

See Exercises 18.11–18.14. Interestingly the Galerkin trial solution is more accurate than truncation of the exact series solution (18.11) for the same small N.

§18.4 The Closed-Loop Heat Pump Reprised (SSP 7)

Recall in §15 we developed the rather nasty transient solution (15.27)

$$u = B - \frac{\dot{Q}}{2\pi\kappa}\log b/r \ +$$

$$\sum_{n=1}^{\infty} c_n e^{-\kappa\alpha_n^2 t/c\rho}(Y_0(\alpha_n a)J_0(\alpha_n r) - J_0(\alpha_n a)Y_0(\alpha_n r))$$

to the problem (10.22). Truncation would be *very* costly to implement.

Instead let us apply the Galerkin numerical method. Specifically let us solve

$$\frac{\partial v}{\partial t} = \frac{\partial^2 v}{\partial r^2} + \frac{1}{r}\frac{\partial v}{\partial r}, \quad a < r < 1 \tag{18.19}$$

subject to

i) $v(r,0) = \log r$,

ii) $v(1,t) = 0$,

iii) $v_r(a,t) = 0$,

where (scaled as in typical applications) $a = 0.01$.

Choose the simple basis functions

$$\phi_j(r) = 1 - r^{j+1} + (j+1)a^j r - (j+1)a^j = 1 - b_j + b_j r - r^{j+1}, \tag{18.20}$$

the trial solution

$$\tilde{v} = \sum_{j=1}^{N} c_j(t)\phi_j(x), \tag{18.21}$$

and the usual inner product

$$\langle f, g \rangle = \int_0^1 f(r)g(r)\, r dr. \tag{18.22}$$

I choose this basis because the functions ϕ_j are independent and complete (Exercise 18.15), they meet boundary conditions ii) and iii) exactly, and because they are easy to manipulate. We will meet the initial condition i) approximately by a second Galerkin computation.

Impose the Galerkin requirement $\langle \phi_i, R \rangle = 0$ on the residual R of (18.19) to obtain the relations (Exercise 18.16)

$$\sum_{j=1}^{N} \dot{c}_j(t)g(i,j) + \sum_{j=1}^{N} c_j(t)h(i,j) = 0, \quad i = 1,\ 2,\ 3,\ \ldots,\ N, \tag{18.23}$$

where

$$g(i,j) = \frac{(1-b_i)(1-b_j)}{2} + \frac{1 - 2b_ib_j + b_i + b_j}{3} - \frac{1-b_i}{j+3} - \frac{1-b_j}{i+3}$$
$$-\frac{b_i}{j+4} - \frac{b_j}{i+4} + \frac{b_ib_j}{4} + \frac{1}{i+j+4}, \tag{18.24}$$

where $h(i, j)$

$$= (j+1)^2\left(\frac{1-b_i}{j+1} + \frac{b_i}{j+2} - \frac{1}{i+j+2}\right) - b_j + \frac{b_i b_j}{2} + \frac{b_j}{i+2}, \quad (18.25)$$

and where $b_j = (j+1)a^j$.

We approximately meet the initial condition i) by insisting its residual,

$$I = \sum_{j=1}^{N} c_j(0)\phi_j(x) - \log r$$

also be orthogonal to the basis functions, i.e., $\langle \phi_i, I \rangle = 0$. Explicitly (Exercise 18.17),

$$\sum_{j=1}^{N} c_j(0) g(i,j) = -\frac{1}{4} + \frac{5b_i}{36} + \frac{1}{(i+3)^2}, \quad i = 1,\ 2,\ 3,\ \ldots,\ N. \quad (18.26)$$

So to finish, solve the system (18.26) for the initial conditions $c_j(0)$, then numerically solve the system (18.18). Suggestions for solving (18.18) are laid out in the exercises.

Alert. There is a price to pay for the use of simple polynomial basis functions — typically the matrix to be inverted is ill-conditioned [Golub and Ortega] for large N. Taking more terms (larger N) may not increase accuracy unless more care is taken during the numerical inversion [Chow, Dunninger, and Miklavcic].

§18.5 Vibrations of a Triangular Brace (BVP 20)

Think of a shaped metal plate used as a motor mount. The mount will be excited by external shocks into vibrating at its natural frequencies. A common design practice is to find the first mode shape ϕ_1 belonging to the lowest (dominant) frequency f_1, then simulate the strain on the motor mount as it vibrates in this dominant mode. If stresses exceed some fraction of material strength, the mount is redesigned. One strategy is to stiffen the mount so that its lowest natural frequency exceeds all conceivable excitation frequencies (since excitation tends to 'spill down' in frequency rather than 'spill up'). On the other hand, the motor and mount form a coupled system with coupled frequencies. Moreover a principal function of a mount is to decrease the transmission of shock to the frame, a criterion in conflict with higher stiffness. This conflict in design goals remains unresolved.

Let us compute the lowest frequency of vibration in the normal direction of a thin flat triangular plate. In the *first* approximation (see Exercise 15.11) the problem is the eigenequation of the vibrating stretched membrane

$$\nabla^2 u = \lambda u, \qquad 0 < x, y < x + y < 1, \tag{18.27}$$

subject to the zero boundary conditions

i) $u(x, 0) = 0$,

ii) $u(0, y) = 0$,

iii) $u(x, y) = 0$ on $x + y = 1$.

Rellich of course guarantees a complete orthogonal basis of such eigenmodes u. But what are they? The problem does not directly yield to separation of variables (Exercise 18.18). Let us instead proceed numerically via *Galerkin Finite Elements.*

The idea is simple. Subdivide the triangle Ω into congruent subtriangles (elements) Ω_k and number the interior vertices $v_1, v_2, \ldots, v_N$. Choose a *tepee* (pyramid) shaped continuous basis function ϕ_j that is 1 at the j-th vertex v_j, that vanishes on subtriangles that do not have v_j as a vertex, and is linear on each individual subtriangle. See Figure 18.1. Form the elementwise linear trial solution

$$\tilde{u} = \sum_{j=1}^{N} c_j \phi_j(x, y). \tag{18.28}$$

Proceed with the Galerkin method to obtain a system of equations for the unknowns c_j. Put geometrically, we estimate the shape of the eigenmode u with a 'geodesic' dome $\tilde{u}$ with heights c_j above the vertices (*nodes*) v_j on the floor below.

As we now discover in the course of this simple minded numerical attempt via linear extrapolation, *we are led naturally to the modern Sobolev viewpoint,* the nowadays standard view of boundary value problems.

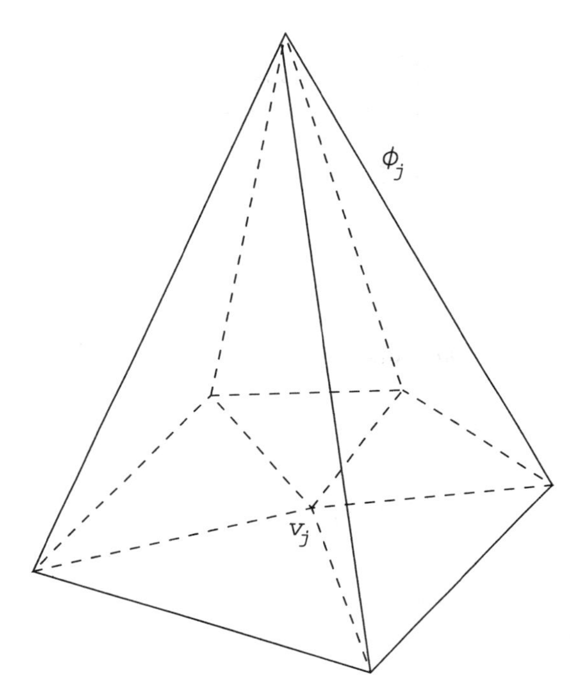

Figure 18.1 The basis function ϕ_j has value 1 above the vertex v_j on the plane below, but value 0 at all other vertices.

The residual of the trial solution (18.28) is $R = L\tilde{u} = \nabla^2\tilde{u} - \lambda\tilde{u}$ and so to impose the Galerkin requirement that the residual R be orthogonal to the first N basis functions means

$$\langle \phi_i, \nabla^2\tilde{u}\rangle = \lambda\langle \phi_i, \tilde{u}\rangle, \quad i = 1,\ 2,\ \ldots,\ N. \tag{18.29}$$

But on the interior of each subtriangle, $\nabla^2\tilde{u} \equiv 0$ since classical second derivatives of linear functions are zero. Therefore, unlike the previous two problems, the Galerkin requirement of orthogonality (18.29) at first glance leads to a system of equations with only the trivial trial solution $\tilde{u} = 0$.

But look again. Our trial solution $\tilde{u}$ has a 'geodesic' dome as graph — again view Figure 18.1. The gradient $\nabla\tilde{u}$ is constant within each subtriangle Ω_k. From a nonclassical viewpoint, the Laplacian $\nabla^2\tilde{u}$ should reflect the jumps in the coordinate values of the gradient $\nabla\tilde{u}$ at the boundary of each subtriangle. From this view, $\nabla^2\tilde{u}$ is the sum of Dirac delta functions supported on the edges of the subtriangles Ω_k whose amplitudes are determined by the difference in cross-sectional slopes in the x and y directions across the edges. Carefully integrating the basis functions ϕ_i against these delta functions will yield by a generalized integration by parts (see Exercise 18.33) that

$$\langle \phi_i, \nabla^2 \tilde{u} \rangle \doteq -[\phi_i, \tilde{u}] \tag{18.30}$$

where the inner product

$$[\phi, \psi] = \int_\Omega \nabla\phi \cdot \nabla\psi \, d\Omega \tag{18.31}$$

is known as the *Dirichlet* inner product. (Much more on this in Chapter 24).

So the Galerkin requirement (18.29) on accuracy becomes the *Sobolev requirement*

$$[\phi_i, \tilde{u}] = -\lambda \langle \phi_i, \tilde{u} \rangle, \quad i = 1,\ 2,\ \ldots,\ N. \tag{18.32}$$

These equations, now involving only classical derivatives, can be solved for the approximate eigenmodes $\tilde{u}$ and their eigenvalues.

More explicitly, (18.32) translates to the N equations

$$\sum_{j=1}^{N} c_j [\phi_i, \phi_j] = -\lambda \sum_{j=1}^{N} c_j \langle \phi_i, \phi_j \rangle, \quad i = 1,\ 2,\ 3,\ \ldots,\ N, \tag{18.33}$$

which in matrix form is the finite dimensional doubly symmetric eigenvalue problem

$$K\mathbf{c} = -\lambda M\mathbf{c}. \tag{18.34}$$

Finish by applying one of the many numerical eigenvalue algorithms (e.g., `eig[K,-M]` in MatLab) to extract the (approximate) lowest (dominant) frequency $f_1 = \sqrt{-\lambda_1}/2\pi$. See Strang [1986], Golub and van Loan [1989], and Exercise 19.15.

The values of the inner products $[\phi_i, \phi_j]$ and $\langle \phi_i, \phi_j \rangle$ of (18.33) are surprisingly easy to code. Each pyramidal basis function ϕ_j is the sum (*a.e.*) of its n_j triangular faces ϕ_{jk} (Figure 18.1), i.e.,

$$\phi_j = \sum_{k=1}^{n_j} \phi_{jk},$$

where each face ϕ_{jk} lives (is supported) on one of the n_j subtriangles Ω_k with v_j as a vertex, has value 1 at v_j, is 0 at the other two vertices, and vanishes off the subtriangle Ω_k. The inner products of (18.33) are then sums of $[\phi_{ik}, \phi_{jk}]$ or $\langle \phi_{ik}, \phi_{jk} \rangle$, each of which is one of 4 values that can be precomputed using one subtriangle in standard position. Work Exercise 18.24, then 18.25 to acquire computational insight. The bulk of any routine is bookkeeping — keeping track of which vertex belongs to what subtriangle.

An overview of finite elements for the beginner is found in the wonderful *An Introduction to Applied Mathematics* by Gilbert Strang. The definitive book on this subject is *An Analysis of the Finite Element Method* by Strang and Fix.

Exercises

18.1 Estimate the error when the series solution (5.11) to BVP 1 is truncated to 4 terms.

Answer:

$$\|\frac{4}{\pi}\sum_{k=4}^{\infty} e^{-(2k+1)^2\pi^2 t} \frac{\sin(2k+1)\pi x}{2k+1}\|^2 < \frac{2e^{-162\pi^2 t}}{\pi} \sum_{k=4}^{\infty} \frac{1}{(2k+1)^2}.$$

18.2 Outline a routine for evaluating the solution of Exercise 15.18

$$u = \sum_{n=1}^{\infty} e^{-\alpha_n^2 t} \frac{J_0(\alpha_n r)}{\alpha_n J_1(\alpha_n)}$$

by truncation to N terms at arbitrary $r < b = 1$ and t.

18.3 Why is it important that Galerkin basis functions be dense?

18.4 Steady flow of a viscous incompressible fluid is modeled by the **Navier-Stokes equation**

$$\nabla p = -\rho(V \cdot \nabla)V + \mu \nabla^2 V$$

where p is pressure, V is fluid velocity, ρ is density, and μ is viscosity. For steady flow within a straight square tube argue that the component u of velocity down the tube is, in nondimensional variables, well modeled by Poisson's

$$\nabla^2 u + 1 = 0$$

with zero boundary conditions. See Exercise 15.22.

18.5 Verify that indeed that the series (18.11) solves Poisson's equation (18.10). Carefully check all details.

18.6 More generally than Exercise 18.5, prove that Rellich guarantees that any Poisson problem

$$\nabla^2 u = f$$

subject to zero boundary conditions on a bounded domain Ω has a unique solution in $L^2(\Omega)$ whenever f belongs to $L^2(\Omega)$.

18.7 More explicitly, show that the solution u in Exercise 18.6 is

$$u = -R(0)f = \sum_{n=1}^{\infty} \frac{\langle f, \phi_n \rangle}{\lambda_n} \phi_n$$

where ϕ_n are the normalized eigenfunctions of the Laplacian belonging to the eigenvalues λ_n.

18.8 In particular, show that the solution u to

$$\nabla^2 u = f, \quad 0 < x, y < 1,$$

subject to zero boundary conditions can be written as a Hilbert-Schmidt integral. The kernel is called the *Green's function* for the problem.

Answer:

$$u(x, y) = \int_0^1 \int_0^1 f(\alpha, \beta) \kappa(\alpha, \beta, x, y) \, d\alpha \, d\beta$$

where the kernel

$$\kappa(\alpha, \beta, x, y) = -\frac{4}{\pi^2} \sum_{m,n=1}^{\infty} \frac{\sin m\pi\alpha \ \sin m\pi x \ \sin n\pi\beta \ \sin n\pi y}{m^2 + n^2}.$$

18.9 Show that when the eigenfunctions of an orthogonal series solution are used as the basis functions of the Galerkin method, the resulting solution is merely the truncated exact series solution.

18.10 Verify (18.14)–(18.17).

Hint: $\langle \alpha_i, \alpha_m'' \rangle = -\langle \alpha_i', \alpha_m' \rangle$.

18.11 Write a routine for solving the linear equations (18.14) and obtaining the Galerkin solution to SSP 14.

Outline: Obtain the doubly indexed array, order lexicographically, do forward elimination (best with scaled partial pivoting), back substitute, then doubly reindex.

18.12 Hand calculate the one term Galerkin approximation of the solution to BVP 14 (18.10) using the basis functions (18.12).

Answer: $\tilde{u} = 1.25(x - x^2)(y - y^2)$.

18.13 Write a routine that truncates the exact series solution (18.11).

18.14 Compare the accuracy of the Galerkin solution obtained in Exercise 18.11 with the truncated solution of the previous exercise. Estimate the percent variation in kinetic energy $\|u - \tilde{u}\|^2/\|u\|^2$ by

$$\frac{\sum_{i,j}(u_{ij} - \tilde{u}_{ij})^2}{\sum_{i,j} u_{ij}^2}$$

at the vertices of a (say) 10 by 10 grid on the unit square.

18.15 Show that the basis functions (18.20) are independent and complete under the inner product (18.22).

18.16 Verify (18.18).

18.17 Verify (18.24) and (18.25).

18.18 Write a routine for inverting N by N matrices. Apply the routine to the *Gramian* $G = \langle \phi_i, \phi_j \rangle$ with entries given in (18.24). Note how G becomes ill-conditioned for large N, how the entries in the inverse G^{-1} become huge. See [Golub and Ortega].

18.19 Verify then solve the system (18.26) for the initial conditions.

18.20 **(The Improved Euler (Heun) Algorithm)** Solve a system $\dot{x} = F(x)$ in time steps of size $h = \Delta t$ by the rule

$$x_{n+1} = x_n + h\frac{m_1 + m_2}{2}$$

where

$$m_1 = F(x_n) \text{ and } m_2 = F(x_n + h\, m_1).$$

Apply this method to solve the system (18.19). Compare your results with (16.50).

18.21 Repeat the previous exercise using — in place of the Improved Euler method — the more accurate four step **Runge-Kutta** method. (The pseudocode for the Runge-Kutta algorithm can be found in any modern beginning ODE book).

18.22 Outline an alternate more careful approach for solving the system (18.23) that does not entail the inversion of the ill-conditioned Grammian G. Consult [Golub and van Loan] for suggestions.

18.23 Show there are no separated solutions $u(x, y) = X(x)Y(y)$ to the eigenproblem (18.27). However argue that $\phi = \sin 2\pi(1 - x)\ \sin \pi y - \sin \pi(1 - x)\ \sin 2\pi y$ is the dominant mode with eigenvalue $\lambda = -5\pi^2$.

18.24 Hand calculate an approximation of the dominant eigenvalue λ of of the triangular plate (18.27) determined by the Sobolev requirement (18.34) using the 9 subtriangles shown in Figure 18.2.

Answer: $\tilde{\lambda} = -\frac{[\phi_1, \phi_1]}{\langle \phi_1, \phi_1 \rangle} = -72.$

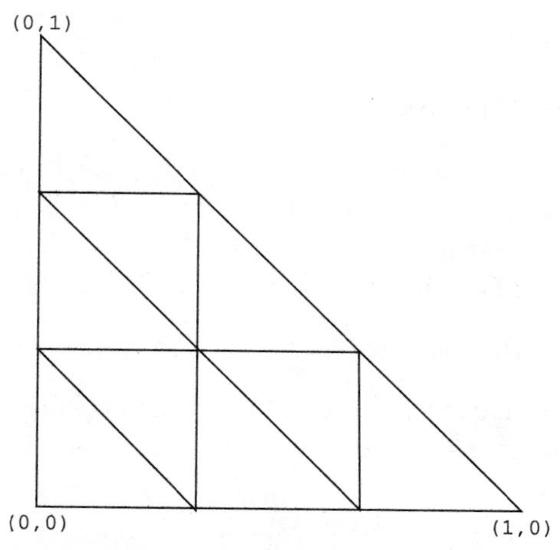

Figure 18.2 The elements of Exercise 18.24.

18.25 Redo the previous exercise using the 16 subtriangles of Figure 18.3.

Answer:

$$\begin{pmatrix} 4 & -1 & -1 \\ -1 & 4 & 0 \\ -1 & 0 & 4 \end{pmatrix} \begin{vmatrix} c_1 \\ c_2 \\ c_3 \end{vmatrix} = -\frac{\lambda}{192} \begin{pmatrix} 6 & 1 & 1 \\ 1 & 6 & 1 \\ 1 & 1 & 6 \end{pmatrix} \begin{vmatrix} c_1 \\ c_2 \\ c_3 \end{vmatrix},$$

i.e., $\tilde{\lambda} = 192(28 - \sqrt{224})/40 \approx 62.5602$.

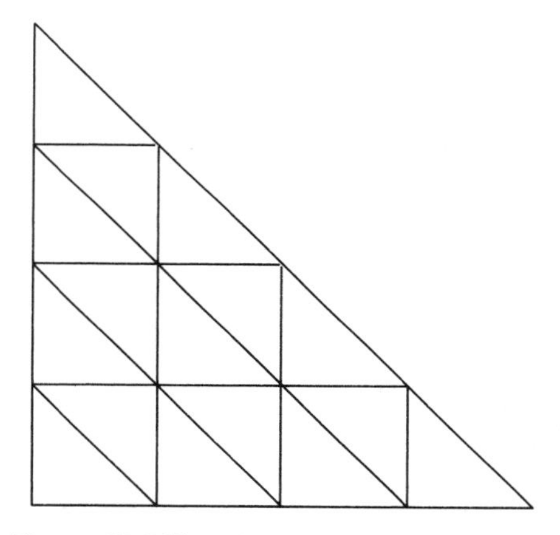

Figure 18.3 The elements of Exercise 18.25.

18.26 Code and solve the finite element relation (18.33). Find the approximate value of the dominant frequency. Graph the trial eigenmode $\tilde{u}$. The exact answer is $\lambda_1 = -5\pi^2$.

18.27 Solve $y''+y = 1,\ 0 < x < 1$, subject to $y(0) = 0 = y(1)$ by the Galerkin method using the basis functions $\phi_j(x) = x^j - x^{j+1}$. Compare your result with the exact solution. Experiment with taking a large number of terms N. Note how accuracy fails to improve due to ill conditioning.

18.28 Solve $y'' + y = 1,\ \ 0 < x < 1$, subject to $y(0) = 0 = y(1)$ by **shooting** for the right end condition: replace the second derivative y'' by the second divided difference

$$\frac{\Delta^2 y_i}{\Delta x^2} = \frac{y_{i+1} - 2y_i + y_{i-1}}{\Delta x^2}$$

where $\Delta x = 1/n$ and $y_i \approx y(i\Delta x)$. Replace the ODE by the finite divided differences to obtain the explicit

$$y_{i+1} = \Delta x^2 + (2 - \Delta x^2)y_i - y_{i-1}.$$

Start at the left end with $y_0 \approx 0$ and $y_1 = 0$, recursively solve for the next y_{i+1}, and compare the error y_n with 0 at the right end. Adjust the size of y_0 and Δx until the required accuracy is achieved at the right end. Compare the effort required with the preceding Galerkin approach.

18.29 Solve BVP 1 using the Galerkin method with the basis functions $\phi_j = x^j - x^{j+1}$.

18.30 Solve BVP 1 using the **Crank-Nicolson** algorithm: replace the partial $\partial u/\partial t$ by the finite divided difference $(u^{j+1} - u^j)/\Delta t$ and the second spatial partial $\partial^2 u/\partial x^2$ by the average of the second divided difference of the present and the future temperature

$$\frac{\Delta^2 u_i^{j+1} + \Delta^2 u_i^j}{2\Delta x^2}$$

where Δ^2 is as in Exercise 18.28. The PDE becomes the implicit

$$u_i^{j+1} - u_i^j = h[u_{i+1}^{j+1} - 2u_i^{j+1} + u_{i-1}^{j+1} + u_{i+1}^j - 2u_i^j + u_{i-1}^j]$$

where $h = \Delta t/2\Delta x^2$. This is a system of the form

$$(I - H)u^{j+1} = (I + H)u^j$$

where the matrix H is banded tridiagonal. A mere dozen lines of code will solve this problem [Golub and Ortega]. Turn first to this method when a diffusion problem must be solved in a hurry — it is stable, accurate, and easy to code.

18.31 Solve the one dimensional eigenvalue problem $y'' = \lambda y$ subject to the zero boundary conditions $y(0) = 0 = y(1)$ by Galerkin's method using the basis elements $\phi_j = x^j - x^{j+1}$, $j = 0, 1, 2, \ldots$.

Outline: Solve the generalized eigenvalue problem $Su = -\lambda Gu$ where $S = ([\phi_i, \phi_j])$ and $G = (\langle \phi_i, \phi_j \rangle)$.

18.32 Solve the one dimensional eigenvalue problem $y'' = \lambda y$ subject to the zero boundary conditions $y(0) = 0 = y(1)$ by Galerkin finite elements.

Outline: Partition the interval [0,1] into N subintervals each of length $\Delta x = 1/N$ with endpoints $x_0 = 0 < x_1 < \ldots < x_N = 1$.

At each node x_j form the sawtooth basis function

$$\phi_j(x) = \begin{cases} (x - x_{j-1})/\Delta x & \text{if } x_{j-1} \le x \le x_j \\ 1 - (x - x_j)/\Delta x & \text{if } x_j \le x \le x_{j+1} \\ 0 & \text{otherwise} \end{cases} .$$

See Figure 18.4. Attempt the trial solution

$$\tilde{u} = \sum_{j=1}^{N-1} c_j \phi_j .$$

The Sobolev requirement

$$\sum_{j=1}^{N-1} c_j [\phi_i, \phi_j] = -\lambda \sum_{k=1}^{N-1} c_j \langle \phi_i, \phi_j \rangle$$

translates to the system

$$c_{i+1} - 2c_i + c_{i-1} = \lambda \frac{\Delta^2 x}{6} (c_{i+1} + 4c_i + c_{i-1}).$$

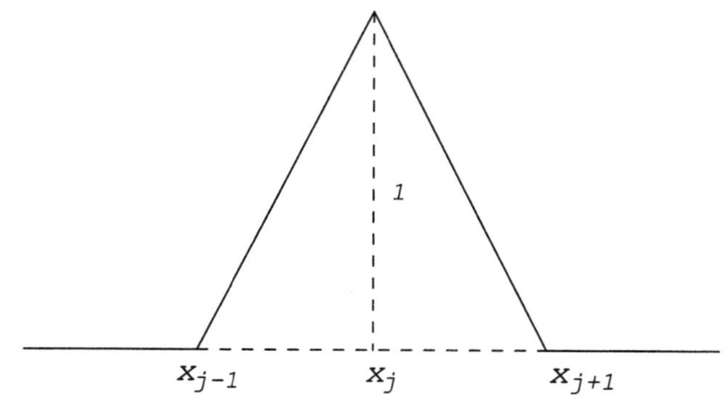

Figure 18.4 The basis element ϕ_j of Exercise 18.32 is 1 above x_j but 0 at every other node.

18.33 Show that in the sense of generalized derivatives, if $\tilde{u}$ is the piecewise linear trial solution of the previous exercise, then

$$\langle \psi, \nabla^2 \tilde{u} \rangle = -[\psi, \tilde{u}]$$

for any continuously differentiable ψ with zero boundary values.

18.34 (S.W. Shaw) Approximate the solution of the nonlinear problem $u_t(x,t) = u_{xx}(x,t)) + \epsilon u^2(x,t)$ on $0 < x < 1$ subject to $u(0,t) = 0 = u(1,t)$ where initially $u(x,0) = 1$.

Approach: Choose the eigenmodes of the linear problem as basis functions, i.e., use $\phi_j(x) = \sin j\pi x$.

18.35 Solve for the lateral deflection $y = y(x)$ of a retaining wall given by $y'''' = kxy$ subject to $y(0) = y'(0) = 0$ (at the surface) and $y(1) = y'(1) = 0$ (at the footing).

Suggestion: Use cubic splines as basis elements with a weak reformulation.

Chapter 19
Sobolev Methods

This will be a brief introduction to the modern and nowadays standard approach to Boundary Value Problems. Like the Heaviside-Mikusinski methods, the Sobolev methods employ generalized nonvalued derivatives. Because the Sobolev approach requires nonelementary Functional Analysis, much of this section can only be sketched. We will only examine problems whose spatial operator is the Laplacian with zero boundary conditions.

In brief, the Sobolev approach weakens the requirements on the notion of 'solution' using weak convergence.

§19.1 Dirichlet Inner Product

We saw for Galerkin finite elements the usefulness of the bilinear form

$$[\phi, \psi] = \int_\Omega \nabla\phi \cdot \nabla\psi \, d\Omega. \tag{19.1}$$

For those functions ϕ, ψ for which the symbols make sense (Exercise 19.1),

$$[\phi_1 + \phi_2, \psi] = [\phi_1, \psi] + [\phi_2, \psi], \tag{19.2a}$$

$$[c\phi, \psi] = c[\phi, \psi], \tag{19.2b}$$

$$[\phi, \psi] = [\psi, \phi], \tag{19.2c}$$

$$[\phi, \phi] \geq 0. \tag{19.2d}$$

For instance, these properties hold for all functions ϕ, ψ defined on a bounded domain Ω whose first partial derivatives are bounded.

This form (19.1) lacks only the positive definiteness of an inner product. After all, $[\phi, \phi] = 0$ for constant functions ϕ. So hereafter allow the form (19.1) to be applied only to functions ϕ, ψ of

$$C_0^\infty(\Omega),$$

the space of all infinitely continuously differentiable functions on Ω that vanish off some compact (closed and bounded) subset K of Ω.

The definiteness property

$$[\phi, \phi] = 0 \text{ implies } \phi = 0 \tag{19.3}$$

now obtains (Exercise 19.2). Therefore

the form (19.1) *becomes an inner product on* $C_0^\infty(\Omega)$.

This inner product is called the *Dirichlet* or *energy inner product.* The word 'energy' is often used since in distributed vibrational problems $\ddot{u} = \nabla^2 u$, the form $[u, u] = -\langle \nabla^2 u, u \rangle$ is potential energy — see Exercises 6.46–6.48.

§19.2 Dominance of the Dirichlet Inner Product

The Dirichlet inner product $[.,.]$ dominates the ordinary Hilbert inner product

$$\langle \phi, \psi \rangle = \int_\Omega \phi \, \psi \, d\Omega. \tag{19.4}$$

Lemma A. For functions ϕ in $C_0^\infty(\Omega)$ where Ω is a bounded domain,

$$\langle \phi, \phi \rangle \leq c^2 [\phi, \phi], \tag{19.5}$$

where the constant c is the least width of Ω. Consequently any sequence convergent with respect to the Dirichlet norm is necessarily convergent with respect to the ordinary Hilbert norm.

Proof. Suppose the domain Ω can be enclosed between two planes $x_1 = a_1$ and $x_1 = b_1$. Extend ϕ to be zero off Ω. By the Fundamental Theorem of Calculus

$$\phi^2 = (\int_{a_1}^{x_1} \frac{\partial \phi}{\partial x_1} \, dx_1)^2.$$

Hence by Cauchy's inequality (7.1),

$$\phi^2 \leq \int_{a_1}^{x_1} (\frac{\partial \phi}{\partial x_1})^2 dx_1 \int_{a_1}^{x_1} 1^2 \, dx_1 \leq (b_1 - a_1) \int_{a_1}^{b_1} (\frac{\partial \phi}{\partial x_1})^2 dx_1.$$

Thus

$$\begin{aligned}\langle \phi, \phi \rangle = \int_\Omega \phi^2 \, d\Omega &\le (b_1 - a_1) \int_\Omega \int_{a_1}^{b_1} (\frac{\partial \phi}{\partial x_1})^2 dx_1 \, d\Omega \\ &= (b_1 - a_1) \int_{a_1}^{b_1} \int_\Omega (\frac{\partial \phi}{\partial x_1})^2 d\Omega dx_1 \\ &= (b_1 - a_1)^2 \int_\Omega (\frac{\partial \phi}{\partial x_1})^2 d\Omega \le (b_1 - a_1)^2 [\phi, \phi].\end{aligned}$$

Note that this proof requires only that Ω be bounded in one coordinate and so applies to certain unbounded domains.

§19.3 Sobolev space $W_0^{1,2}(\Omega)$

Definition. The inner product

$$\langle \phi, \psi \rangle_0 = \langle \phi, \psi \rangle + [\phi, \psi] \tag{19.6}$$

on $C_0^\infty(\Omega)$ is called the *Sobolev inner product.*

Definition. The completion of $C_0^\infty(\Omega)$ under the Sobolev norm $\|\phi\|_0 = \langle \phi, \phi \rangle_0^{1/2}$ is the *Sobolev space* $W_0^{1,2}(\Omega)$.

Notation. Abbreviate $W_0^{1,2}(\Omega)$ to $W_0(\Omega)$.

The Galerkin finite element approximations $\tilde{u}$ of BVP 20 reside in this Hilbert space $W_0(\Omega)$.

Lemma B. The Sobolev space $W_0(\Omega)$ is naturally embedded in the Hilbert space $L^2(\Omega)$. The embedding is continuous.

Proof. Because the Sobolev norm dominates the Hilbert norm, i.e., $\|\phi\| \le \|\phi\|_0$, any Sobolev norm Cauchy convergent sequence of functions from $C_0^\infty(\Omega)$ is Hilbert norm Cauchy convergent. But $L^2(\Omega)$ is complete. Thus there is the natural injection (*embedding*)

$$I : W_0(\Omega) \to L^2(\Omega),$$

where I is the identity map $I\phi = \phi$. But since

$$\|I\phi\| = \|\phi\| \le \|\phi\|_0,$$

this embedding is continuous (bounded) from the Sobolev to the Hilbert norm.

A milestone of applied mathematics is Rellich's 1930 discovery that this embedding is compact for bounded domains Ω.

§19.4 Rellich's Compact Embedding Theorem

We have seen throughout this book the power and scope of Rellich's Principle of §6.7 on the Laplacian with zero boundary conditions. It is in turn a direct consequence of the following fundamental result, only one of a class of embedding results of Rellich and Sobolev.

Rellich's Compact Embedding Theorem. For a bounded domain Ω, the injection (embedding)

$$I : W_0(\Omega) \to L^2(\Omega) \tag{19.7}$$

is compact, i.e., it is sequentially continuous, weak to strong. Put more prosaically, given a sequence of functions ϕ_k in $W_0(\Omega)$ for which

$$\langle \phi_k, \psi \rangle_0 \to 0$$

for every ψ in $W_0(\Omega)$, then

$$\|\phi_k\| \to 0.$$

Loosely speaking, since the Sobolev norm controls the size of the derivative as well, even *weak* Sobolev convergence is stronger than ordinary Hilbert strong convergence.

Note that I have shifted to an equivalent definition of *compact* operator — an operator that is sequentially continuous, weak to strong. Compare with the definition of §6.6 (Exercise 19.4).

Rellich's Theorem is proven from the following refinement of (19.5).

Lemma C. (Poincaré's Inequality) For the special case Ω a cube in $\mathbf{R}^n$ of side length d,

$$\|\phi\|^2 \leq \langle \phi, 1 \rangle^2 / d^n + nd^2[\phi, \phi]/2 \tag{19.8}$$

for all ϕ with continuous partial derivatives on $\bar{\Omega}$.

Proof. Exercise 19.5.

Proof Sketch of Rellich's Compact Embedding Theorem. Because $C_0^\infty(\Omega)$ is dense in $W_0(\Omega)$, we may assume the sequence ϕ_k is drawn from $C_0^\infty(\Omega)$. Inscribe the domain Ω in a cube C. Extend the functions ϕ_k in $C_0^\infty(\Omega)$ to be zero off Ω. By rescaling and translation we may assume the cube C is the unit cube $0 < x_i < 1$, $i = 1, 2, \ldots, n$. Partition the unit cube C in the usual way into N^n congruent cubes of side length $\Delta x = 1/N$. Apply Poincaré's inequality on each subcube then sum over all subcubes to obtain the global inequality

$$\|\phi_k\|^2 \leq (1/\Delta x^n) \sum_{j=1}^{N^n} \langle \phi_k, 1_j \rangle^2 + n\Delta x^2 [\phi_k, \phi_k]/2, \tag{19.9}$$

where 1_j denotes the characteristic function of the jth subcube C_j. Because a weakly convergent sequence is bounded, we may make the second term on the right of (19.9) arbitrarily small by choosing small Δx. Because the embedding of $W_0(\Omega)$ into $L^2(\Omega)$ is bounded, it is weakly continuous. Hence ϕ_k converges weakly to 0 in $L^2(\Omega)$. Thus each of the N^n terms of the summation on the right of (24.9) can be made arbitrarily small for all sufficiently large k. Hence $\|\phi_k\| \to 0$.

Corollary A. The Laplacian $A = \nabla^2$ with zero boundary conditions on a bounded domain has a compact resolvent $R(\lambda)$.

Proof Sketch. It is enough (Exercise 19.6) to show $R(0) = R = (0 - A)^{-1}$ exists and is compact. Suppose ϕ and ψ are in $C_0^\infty(\Omega)$. As we have done over and over again in special cases throughout this book, integration by parts yields the fundamental identity

$$\langle \nabla^2 \phi, \psi \rangle = \int_\Omega \nabla^2 \phi \psi \, d\Omega = -\int_\Omega \nabla \phi \cdot \nabla \psi \, d\Omega = -[\phi, \psi]. \tag{19.10}$$

Or put another way, if $f = A\phi = \nabla^2\phi$, then

$$\langle f, \psi \rangle = -[\phi, \psi].$$

Conversely, the linear functional

$$\zeta_f(\psi) = \langle f, \psi \rangle$$

is bounded on $W_0(\Omega)$. Therefore by the Riesz Representation Theorem, for some ϕ in $W_0(\Omega)$,

$$\langle f, \psi \rangle = -[\phi, \psi],$$

since the Dirichlet and Sobolev norms are equivalent. Define a map R from all of $L^2(\Omega)$ to $W_0(\Omega)$ by the rule

$$\langle f, \psi \rangle = [Rf, \psi]. \tag{24.11}$$

Because $C_0^\infty(\Omega)$ is dense in $L^2(\Omega)$, (a significant detail), this definition yields an injective bounded operator R that inverts $-A$ for at least all ϕ in $C_0^\infty(\Omega)$. Declare this bounded operator R to be the resolvent of A at $\lambda = 0$, thereby well defining and conditioning A.

Finally, R is compact. For if f_k converges weakly to 0 in $L^2(\Omega)$ and $Rf_k = \phi_k$, then

$$\langle f_k, \psi \rangle = [Rf_k, \psi] = [\phi_k, \psi] \to 0$$

and hence by Rellich's Compact Embedding Theorem,

$$\|\phi_k\| = \|Rf_k\| \to 0.$$

Or from another view, the map

$$R : L^2(\Omega) \to L^2(\Omega)$$

can be factored through $W_0(\Omega)$ where the second map is compact. See Figure 19.1.

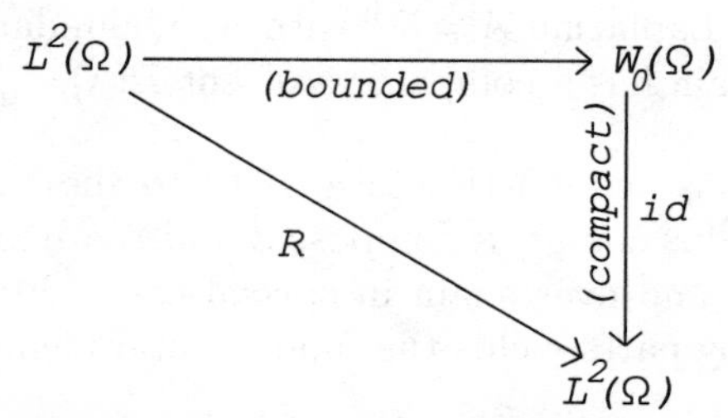

Figure 19.1 The resolvent R is compact since it can be factored through $W_0(\Omega)$.

Corollary B. The Laplacian $A = \nabla^2$ with zero boundary values on a bounded domain Ω possesses a complete set of orthogonal eigenfunctions.

Proof. Since $R = R(0)$ is Hermitian and compact, the spectral theorem (§6.6) applies.

Corollary C. The eigenfunctions ϕ of the Laplacian $A = \nabla^2$ with zero boundary values on a bounded domain Ω are infinitely differentiable on Ω. The eigenfunctions are continuous on $\bar{\Omega}$ and zero on the boundary when the boundary $\partial\Omega$ of Ω is an analytic arc.

Proof. See the Sobolev Embedding Theorem [Adams; Sobolev].

§19.5 Weak Restatement of Problems

Suppose we are presented with a problem whose spatial operator is the Laplacian ∇^2 with zero boundary conditions on the bounded domain Ω.

Examples:

$$\frac{\partial u}{\partial t} = \nabla^2 u, \quad \frac{\partial^2 u}{\partial t^2} = \nabla^2 u, \quad \nabla^2 u = f. \tag{19.12}$$

Certainly if we have a solution u to say the first of these examples, then of course

$$\langle \frac{\partial u}{\partial t}, \psi \rangle = \langle \nabla^2 u, \psi \rangle \tag{19.13}$$

for any ψ and conversely. But this is too much to ask.

The *weak restatement* of each of the above example problems (19.12) is respectively

$$\frac{d}{dt}\langle u, \psi \rangle = -[u, \psi], \quad \frac{d^2}{dt^2}\langle u, \psi \rangle = -[u, \psi], \quad [u, \psi] = -\langle f, \psi \rangle \tag{19.14}$$

for all ψ in $W_0(\Omega)$. We then search for a solution u in $W_0(\Omega)$. This was exactly our strategy in §18.5 during the solution of BVP 20 with Galerkin finite elements. This weakening of the requirements on solutions u is often described as "throwing off (spatial) derivatives onto the test function ψ."

Let us now look at one of our previously 'solved' problems that had neither a classical nor strong solution. Our 'solutions' were in fact solutions to the weak restatement.

§19.6 The Plucked String Revisited (BVP 4)

We saw in Chapter 4 that the plucked string has neither classical nor strong solution. The corner where the string is initially plucked will split into two corners that propagate, thus ruling out the existence of both the classical second spatial partial $\partial^2 u/\partial x^2$ as well as the strong second time derivative $\ddot{u}$. See Figure 4.4. Yet we seem to have in hand a series solution

$$u = \sum_{n=1}^{\infty} c_n \cos n\pi t \, \sin n\pi x, \tag{19.15}$$

where (Exercise 5.18)

$$c_n = \frac{8\sin(n\pi/2)}{n^2\pi^2}, \tag{19.16}$$

obtained by separation of variables that agrees both with D'Alembert's solution (Exercise 4.16) as well as with the numerical solution obtained by finite divided differences (Exercise 5.18). How can this be a solution to a problem without a solution?

The series (19.15) is in fact the unique solution to the weak restatement of BVP 4, i.e., the solution of the problem

$$\frac{d^2}{dt^2}\langle u, \psi\rangle = -[u, \psi], \quad \psi \text{ in } W_0(0,1), \tag{19.17}$$

subject to the initial and boundary conditions given in (5.26). Let us carefully check that this is indeed the case.

First, note that the eigenfunctions $\phi_n = \sin n\pi x$ of $A = \nabla^2$ also form a complete orthogonal basis of $W_0(0,1)$ under the Sobolev inner product — see Exercise 19.14. Second, note that (Exercise 19.10)

$$\langle \phi_n, \phi_n \rangle_0 = \frac{1}{2} + n^2\pi^2, \tag{19.18}$$

and hence (Exercise 19.11) the elements of $W_0(0,1)$ are exactly the series

$$\psi = \sum_{n=1}^{\infty} d_n \sin n\pi x,$$

where

$$\sum_{n=1}^{\infty} (1/2 + n^2\pi^2) d_n^2 < \infty. \tag{19.19}$$

Consequently the purported solution $u(x,t)$ of (19.15)–(19.16) is an element of $W_0(0,1)$ for every t and hence

$$[u, \psi] = \frac{\pi^2}{2} \sum_{n=1}^{\infty} c_n d_n n^2 \cos n\pi t, \tag{19.20}$$

where the series (19.20) is absolutely and uniformly convergent in t. But then we may twice differentiate term-by-term the series

$$\langle u, \psi \rangle = \frac{1}{2} \sum_{n=1}^{\infty} c_n d_n \cos n\pi t$$

to obtain that indeed

$$\frac{d^2}{dt^2} \langle u, \psi \rangle = -[u, \psi].$$

Thus u is a (weak) solution of the weak reformulation (19.17).

By taking $\psi = \sin n\pi x$ we see u is the only weak solution (Exercise 19.12).

Similar arguments show the solution (5.31) of the circular drum, the solution (12.43) of the spherical bell, etc are each the unique solution to the corresponding weakly restated problem.

As a rule, because of the inherent truncations involved,

convergent numerical methods necessarily yield the weak solution.

An introduction to the modern Sobolev theory can be found in Sobolev's *Applications of Functional Analysis in Mathematical Physics*, in the more modern but difficult book by Gilbarg and Trudinger, or in the very readable [Evans].

Exercises

19.1 Show that for functions whose first partial derivatives are bounded on the bounded domain Ω, the form (19.1) satisfies (19.2a)–(19.2d), i.e., the form (19.1) is a *symmetric positive semidefinite bilinear form.*

19.2 Prove that the form (19.1) is an inner product on $C_0^\infty(\Omega)$ by proving the definiteness property (19.3).

19.3 Display a function of $L^2(0,1)$ not in $W_0^1(0,1)$.

Answer: $\phi = x^{-1/4}$. Alternatively see Exercise 19.11.

19.4 Prove the following two notions agree for an everywhere defined linear map $T : X_1 \to X_2$ from one Hilbert space X_1 to a second X_2:

I) An operator T is *compact* if it is sequentially continuous weak to strong.

II) An operator T is *compact* if for any bounded sequence f_k, the sequence $g_k = Tf_k$ contains a (strongly) convergent subsequence.

Deduce that in either case T is bounded.

Hint: Assume the following standard facts from functional analysis: i) *the closed unit ball of a separable Hilbert space is weakly compact*, and ii) *weakly convergent sequences are bounded.*

19.5 Prove Poincaré's inequality (19.8).

Outline: For two vector variables x, x' roaming within the cube $\Omega \subset \mathbf{R}^n$: $0 \le x_i, x_i' \le d$, compute the integral

$$\int_{\Omega\times\Omega} (\phi(x) - \phi(x'))^2 dx dx'$$

in two ways: First expand the integrand $(\phi(x) - \phi(x'))^2 = \phi(x)^2 - 2\phi(x)\phi(x') + \phi(x')^2$. Second, expand using a telescoping series

$$\phi(x) - \phi(x') = \phi(x) - \phi(x_1', x_2, \ldots, x_n) + \phi(x_1', x_2, \ldots, x_n)$$

$$-\phi(x_1', x_2', x_3, \ldots, x_n) + \phi(x_1', x_2', x_3, \ldots, x_n) - \ldots - \phi(x'),$$

use the fundamental theorem of calculus to write each difference as the integral of the partial, apply Cauchy's inequality twice, then integrate. See [Ladyzhenskaya, p. 26].

19.6 Prove the **First Resolvent Formula**: for every two values λ_1 and λ_2 where the resolvent $R(\lambda)$ of A exists,

$$R(\lambda_1) - R(\lambda_2) = -(\lambda_1 - \lambda_2)R(\lambda_1)R(\lambda_2).$$

Deduce via Exercise 13.12 that if the resolvent is compact at one value, it is compact for all values where it exists.

19.7 Show using the fundamental theorem of calculus as in the proof of Lemma A that all functions of $W_0(a, b)$ are continuous and zero at the boundary (the interval endpoints).

Hint: Convergence in Dirichlet norm implies uniform convergence when the spatial dimension $n = 1$.

19.8 In contrast to the previous exercise, show that some of the eigenfunctions of the Laplacian with zero boundary conditions on the punctured disk $\Omega : 0 < r < 1$ are not zero at the origin.

Thus the eigenfunctions guaranteed by Rellich's Theorem are not necessarily classical eigenfunctions.

Outline: Let $\Omega = \{r < 1\}$ and $\Omega_0 = \{0 < r < 1\}$. Show that $W_0(\Omega_0)$ is weakly dense in $W_0(\Omega)$ by showing $\langle \phi, \beta\psi \rangle_0 \to \langle \phi, \psi \rangle_0$ for ϕ, ψ in $C_0^\infty(\Omega)$ and $\beta = \exp(-\epsilon/(r - \epsilon))$ for $r > \epsilon$, 0 otherwise. So if $[Rf, \psi] = \langle f, \psi \rangle$ holds for ψ in $C_0^\infty(\Omega_0)$, it must hold for ψ in $C_0^\infty(\Omega)$. Therefore the resolvent for the punctured disk Ω_0 is the resolvent for the unpunctured disk Ω. Thus $J_0(\alpha r)$ is an eigenfunction that does not vanish at $r = 0$ for α a zero of J_0.

19.9 Prove that the eigenfunctions of the Laplacian with zero boundary conditions on the bounded domain Ω form a complete orthogonal basis of $W_0(\Omega)$ (under the Sobolev norm).

Outline: A careful examination of the proof of Corollary A shows the resolvent R is a compact Hermitian operator when cut back to $W_0(\Omega)$ since

$$[\phi_k, \phi_k] = [Rf_k, \phi_k] = \langle f_k, \phi_k \rangle \leq \|f_k\| \|\phi_k\|.$$

19.10 Verify (19.18).

19.11 Show $W_0(0,1)$ contains

$$\psi = \sum_{n=1}^{\infty} d_n \sin n\pi x \text{ exactly when } \sum_{n=1}^{\infty} n^2 d_n^2 < \infty,$$

while dom∇^2 contains ψ exactly when

$$\sum_{n=1}^{\infty} n^4 d_n^2 < \infty.$$

Generalize this result for the eigenfunctions of the Laplacian with zero boundary conditions on any bounded domain Ω.

19.12 Prove the uniqueness of the weak solution (19.15) to the weak restatement (19.17).

Hint: Any solution u must be of the form

$$u = \sum_{n=1}^{\infty} q_n(t) \sin n\pi x.$$

Now apply the restatement (19.17) with $\psi = \sin n\pi x$.

19.13 Argue in three brief lines that the Poisson problem

$$\nabla^2 u = f$$

subject to zero boundary conditions on a bounded domain Ω has a unique weak solution u.

19.14 A steady Dirichlet problem $\nabla^2 u = 0$ on a bounded domain Ω is often solved as follows: find a function g with the correct boundary values and with bounded second partial derivatives on Ω. Let $f = -\nabla^2 g$ (in the classical sense). Then solve the Poisson problem $\nabla^2 u = f$ where the Laplacian has zero boundary conditions. (See Exercises 18.6–18.8). Thus $u + f$ will solve the original Dirichlet problem. Prove this solution $u + f$ is unique.

19.15 Show $\phi = \log r$ is in $W_0(\mathbf{D})$ where $\mathbf{D} = \{r < 1\}$. Thus the functions of $W_0(\Omega)$ need not be continuous or even bounded. Contrast this with Exercise 19.7.

Appendix A
Measure and Integration

For the details behind this sketch see P. R. Halmos's *Measure Theory* or Rudin's *Real and Complex Analysis.*

§A.1 Lebesgue measure

The *Borel subsets* of $\mathbf{R}^n$ form the smallest family $\mathcal{B}$ of subsets E that includes all open sets and at the same time is closed under countable union, countable intersection, and complementation. All familiar sets are Borel sets.

Fact. There is exactly one *measure* μ on the Borel sets E of $\mathbf{R}^n$ that enjoys the three properties

I. $\mu(E) \geq 0$,

II. if the sequence of sets $E_1, E_2, E_3, \ldots$ is pairwise disjoint, i.e., if $E_i \cap E_j = \emptyset$ for $i \neq j$, then

$$\mu\left(\bigcup_{n=1}^{\infty} E_n\right) = \sum_{n=1}^{\infty} \mu(E_n),$$

III. μ agrees with the usual measure of volume on rectangular parallelopipeds.

A *(Lebesgue) measurable set* is any subset E of the form

$$E = B \cup N$$

where B is a Borel set and where N is any subset of a Borel set of measure 0. Extend the measure μ to all measurable sets by setting

$$\mu(E) = \mu(B).$$

Throughout this book, *set* means *measurable set.*

A property is said to hold *almost everywhere (a.e.)* if it holds at all points except possibly on a set of measure 0.

A function $f : \mathbf{R}^n \to \mathbf{R}$ is *measurable* if the inverse image under f of every interval is measurable. Throughout this book, *function* means measurable function.

§A.2 The Lebesgue Integral

Let f be any real-valued (measurable) function on $\mathbf{R}^n$. For $\epsilon > 0$ let I_k be the half-open interval $I_k = [k\epsilon, (k+1)\epsilon)$ and let E_k be the inverse image of I_k under f:

$$E_k = f^{-1}(I_k).$$

Then the limit

$$\lim_{\epsilon \to 0} \sum_{k=-\infty}^{\infty} k\epsilon\, \mu(E_k)$$

(when extant) is the *Lebesgue integral* of f and is denoted by the symbol

$$\int_{\mathbf{R}^n} f(x)\, dx.$$

(I will employ the handy differential notation dx rather than the cumbersome $dx_1\, dx_2 \cdots\, dx_n$ whenever there is no possibility for confusion.) For functions defined only on subsets X, extend f to be zero off X and define

$$\int_X f(x)\, dx = \int_{\mathbf{R}^n} f(x)\, dx.$$

The Lebesgue integral enjoys all the standard properties of an integral: if f and g are integrable, then

$$\int_X (f(x) + g(x))\, dx = \int_X f(x)\, dx + \int_X g(x)\, dx,$$

$$\int cf(x)\, dx = c \int_X f(x)\, dx,$$

and

$$\int_X |f(x)|\, dx \geq 0$$

with equality when and only when $f(x) = 0$ almost everywhere.

The Lebesgue integral suffers from none of the defects of the elementary Riemann integral of Calculus. Redefining a function at one, finitely many, countably many, or even at any set of points with measure 0 will not affect integrability nor the value of the integral. When a (proper) Riemann integral exists, so will the Lebesgue integral and their values will agree. A function f is integrable exactly when it is absolutely integrable, i.e., when $|f|$ is integrable. More generally, if $|f| \leq g$ where g is integrable, then f is integrable.

Lebesgue is once to have said, "My integral is computed like bank tellers total up currency — one divides the bills into stacks of the same denomination, then sums."

§A.3 Three Norms

Suppose we are presented with two shapes, two signals, two profiles f and g, snapshots of a natural phenomenon. How large is the error between these two profiles? How far apart are they, one from the other? There are three reasonable and physically natural interpretations:

I. Measure their maximum separation:

$$\|f - g\|_\infty = \sup_{a \leq x \leq b} |f(x) - g(x)|,$$

II. measure the absolute area between them:

$$\|f - g\|_1 = \int_a^b |f(x) - g(x)| dx,$$

III. measure their square deviation:

$$\|f - g\|_2 = (\int_a^b |f(x) - g(x)|^2 dx)^{1/2}.$$

All three are attractive and useful notions. For example, if $u(x) \geq 0$ is the temperature of a homogeneous rod $0 \leq x \leq 1$, then

$$c\rho \|u\|_1 = \int_0^1 c\rho u(x) dx$$

measures the thermal energy contained within the rod relative to the reference temperature 0. Or say $v(t)$ is the voltage across a resistor R during the time interval $0 \leq t \leq T$, then

$$\|v\|_2^2 / R = \int_0^T v^2(t)\, dt / R$$

is the energy (in Joules) consumed by the resistor during this time period.

These *norms* have natural generalizations to measurable functions f on any domain $\Omega \subset \mathbf{R}^n$:

$$\|f\|_\infty = \sup_{x\in\Omega} |f(x)|,$$

$$\|f\|_1 = \int_\Omega |f(x)|dx,$$

and

$$\|f\|_2 = (\int_\Omega |f(x)|^2 dx)^{1/2}.$$

These norms give rise to associated *function spaces* $L^\infty(\Omega)$, $L^1(\Omega)$, and $L^2(\Omega)$ consisting of all real valued functions f on Ω where the respective norm is finite.

Alert. There are delicate technical issues hidden within the above development. As in scientific data collection we ignore obvious measurement noise. Two functions f_1 and f_2 that differ in value at one, finitely many, or even at infinitely many points x with measure 0 are considered to be the same function. A more precise version of the L^∞ norm would then employ the *essential supremum* [Halmos, 1961]. Moreover I remind the reader that *function* means *measurable function* and that integration is in the absolutely integrable Lebesgue sense. I suggest consultation with a well trained mathematician as doubtful details arise. Measure and Integration is itself a profession.

Let us find properties common to these three normed spaces. Let X be one of the function spaces $L^\infty(\Omega)$, $L^1(\Omega)$, or $L^2(\Omega)$ with the associated norm $\|\cdot\|$.

Result A. The function space X is closed under addition and scalar multiplication.

The result is straightforward for $L^\infty(\Omega)$ and $L^1(\Omega)$. The result for $L^2(\Omega)$ follows from the powerful Hölder inequality.

Result B. The function space X is a vector space.

Definition. A sequence $\{f_n\}_{n=1}^{\infty}$, of functions f_n drawn from X is said to *converge to the limit* f_0 in X if

$$\lim_{n\to\infty} \|f_n - f_0\| = 0.$$

This is convergence in the *norm sense* and should be thought of as convergence in generalized energy.

Warning. Convergence in one norm does not imply convergence in another.

Definition. An infinite series of functions f_n drawn from X

$$\sum_{n=1}^{\infty} f_n$$

converges (sums) to F when the sequence of partial sums

$$F_N = \sum_{n=1}^{N} f_n$$

converges (in norm) to F. We then write

$$F = \sum_{1}^{\infty} f_n.$$

In reality the above definition extends the notion of equality.

Completeness Principle. X is *complete*, i.e., Cauchy convergent sequences converge.

Definition. A sequence $\{f_n\}_{n=1}^{\infty}$ in X is *Cauchy convergent* when for every $\epsilon > 0$, there is some index N so that

$$m, n \geq N \text{ implies } \|f_m - f_n\| < \epsilon.$$

(Eventually all terms of the sequence are arbitrarily close).

The Completeness Principle follows from two famous advanced results: the *Stone-Weierstrass Theorem* and *Lusin's Theorem* — results on approximations of and by continuous functions [Rudin].

Of the three spaces above, the *Hilbert space* $X = L^2(\Omega)$ is by far the most mathematically tractable space. It possesses many additional useful properties arising from its *inner product*

$$\langle f, g \rangle = \int_\Omega f(x)g(x)dx.$$

§A.4 Lebesgue Dominated Convergence Theorem

Theorem. Suppose the sequence of functions f_n converges pointwise almost everywhere on X to a function f. Also suppose each $|f_n| \leq g$ on X where g is integrable over X. Then the limit function f is integrable over X and

$$\lim_{n\to\infty} \int_X f_n(x)\, dx = \int_X f(x)\, dx.$$

§A.5 Fubini's Theorem

Theorem.(Tonneli) For $f(x,y) \geq 0$,

$$\int_{X\times Y} f(x,y)\, dx\, dy = \int_X (\int_Y f(x,y)\, dy)\, dx = \int_Y (\int_X f(x,y)\, dx)\, dy.$$

Corollary. (Fubini) Suppose $f(x,y)$ is integrable over $X \times Y$. Then both

$$g(x) = \int_Y f(x,y)\, dy \text{ and } h(y) = \int_X f(x,y)\, dx$$

are integrable (a.e.) and

$$\int_{X\times Y} f(x,y)\, dx\, dy = \int_X (\int_Y f(x,y)\, dy)\, dx = \int_Y (\int_X f(x,y)\, dx)\, dy.$$

Appendix B
Quantum Mechanics

Here is a brief conceptual introduction to quantum mechanics — notions that continue to motivate the study of boundary value problems.

§B.1 Classical Mechanics of a Single Particle

Consider a single particle of mass m roaming in space with displacement (vector) q at time t. Let $p = m\dot{q}$ denote its momentum. Let this particle be influenced by a single external force $F = F(q)$ which depends on displacement q but not on time t nor on velocity $v = \dot{q}$. Assume the system is Hamiltonian, i.e., lossless. Then by Newton,

Observed Fact. $\dot{p} = F(q)$.

Assume moreover that F is linear, that

$$F(q) = Aq,$$

where A is a (constant) 3×3 matrix. Thus motion is governed by

$$\dot{p} = Aq. \tag{B.1}$$

By a mere change of notation over to *phase space* of all pairs

$$\psi = \begin{vmatrix} q \\ p \end{vmatrix},$$

the equation of motion becomes Hamilton's

$$J\dot{\psi} = H\psi, \tag{B.2}$$

where

$$J = \begin{pmatrix} 0 & -I \\ I & 0 \end{pmatrix}$$

and

$$H = \begin{pmatrix} -A & 0 \\ 0 & I/m \end{pmatrix}.$$

Then because $J^{-1} = -J$, all solutions of the equation of motion (B.2) are of the form

$$\psi = e^{-JHt}\psi_0 \tag{B.3}$$

where ψ_0 is the initial state.

Assume further as in application that A is Hermitian negative definite with an orthonormal basis of eigenvectors ϕ_k belonging to the eigenvalues $-\omega_k$. Then phase space X can be factored into mutually orthogonal 2 dimensional subspaces X_k spanned by

$$\begin{vmatrix} \phi_k \\ 0 \end{vmatrix} \text{ and } \begin{vmatrix} 0 \\ \phi_k \end{vmatrix}.$$

Note that each 2-d subspace X_k is an invariant subspace of JH upon which JH has two distinct eigenvalues $\pm i\sqrt{\omega/m}$. Thus the transition operator (semigroup) e^{-JHt} becomes diagonal and trajectories decouple:

$$\psi = \sum_k c_k e^{it\sqrt{\omega_k/m}}\psi_{k1} + d_k e^{-it\sqrt{\omega_k/m}}\psi_{k2}. \tag{B.4}$$

Each trajectory is then a superposition of many periodic trajectories, each of constant norm and energy — a vindication of Ptolemy.

§B.2 Measurement

Additively factor $H = T + V$ where

$$T = \begin{pmatrix} 0 & 0 \\ 0 & I/m \end{pmatrix}$$

and

$$V = \begin{pmatrix} -A & 0 \\ 0 & 0 \end{pmatrix}.$$

Then the operator T *observes (measures) kinetic energy*:

$$\langle T\psi, \psi\rangle/2 = p^2/2m = mv^2/2$$

and V *observes (measures) potential energy*:

$$\langle V\psi, \psi\rangle/2 = -\langle Aq, q\rangle/2 = \text{force} \times \text{displacement}.$$

The Hamiltonian H observes total energy. Note that total energy is conserved along each trajectory since $J^* = -J$ and hence from (B.2),

$$\langle H\psi, \psi\rangle^{\bullet} = \langle H\dot{\psi}, \psi\rangle + \langle H\psi, \dot{\psi}\rangle$$

$$= \langle H(-JH)\psi, \psi\rangle + \langle H\psi, -JH\psi\rangle = -\langle JH\psi, H\psi\rangle - \langle H\psi, JH\psi\rangle$$

$$= -\langle JH\psi, H\psi\rangle + \langle JH\psi, H\psi\rangle = 0.$$

Exercise. For any (Hermitian) observer A, along each trajectory

$$\langle A\psi, \psi\rangle^{\bullet} = -\langle [A, H]\psi, \psi\rangle$$

where

$$[A, B] = AJB - BJA.$$

Exercise. Let $\mathcal{H} = \langle H\psi, \psi\rangle/2$. Show for $k = 1,\ 2,\ 3$ that along each trajectory,

$$\frac{\partial \mathcal{H}}{\partial p_k} = \dot{q_k} \ \text{ and } \ \frac{\partial \mathcal{H}}{\partial q_k} = -\dot{p_k}. \tag{B.5}$$

§B.3 Classical Distributed Vibrations

The identical analysis above obtains for lossless vibrations of a one degree of freedom linear *distributed* system $m\ddot{q} = Aq$ where displacement from equilibrium $q = q(x, t)$ lies in the (spatial) Hilbert space $L^2(\Omega)$, where stiffness A is negative definite Hermitian, and where mass is uniformly distributed with density m. For example think of vertical displacements q of the vibrating string or drum, or longitudinal displacements q of a beam, or dominant transverse electric intensity within a waveguide, where in each case $A = \nabla^2$. (See BVPs 4 and 5, Exercises 4.1–4.3). Or instead think of transverse vibrations of the Euler-Bernoulli beam or of a plate where $A = -\nabla^2\nabla^2 = -\Delta^2$. In all these cases $\psi = \langle q, p\rangle^T$ is a spatially distributed element of $L^2(\Omega) \times L^2(\Omega)$.

But there is an alternate analysis that leads to a more seamless transition to Quantum Mechanics. Redimensionalize so that the domain Ω has measure 1. Consider the cumulative distribution function

$$R(y,t) = \mu\{x; q(x,t) \le y\} \tag{B.6}$$

where μ is Lebesgue measure. Then by the very definition of the Lebesgue integral,

$$\int_\Omega q(x,t)\,dx = \int_{-\infty}^{\infty} y\,dR = \int_{-\infty}^{\infty} y\rho(y,t)\,dy$$

where $\rho(y,t) = \partial R(y,t)/\partial y$. Since $\rho \ge 0$, set $\psi(y,t) = \sqrt{\rho(y,t)}$ to obtain

$$\int_\Omega q(x,t)\,dx = \int_{-\infty}^{\infty} y\psi(y,t)^2\,dy = \langle y\psi, \psi\rangle.$$

This can be read as *the mean displacement is observed by multiplication by y.* It is then easy to see with the substitution $y \to y^{1/n}$ that

$$\int_\Omega q(x,t)^n\,dx = \int_{-\infty}^{\infty} y^n\psi(y,t)^2\,dy$$

and hence for any polynomial $f(y)$ that the expected value of $f(q(x))$ is

$$\int_\Omega f(q)\,dx = \langle f(y)\psi, \psi\rangle. \tag{B.7}$$

§B.4 Quantum Mechanics of a Single Particle

The following ideas are the result of several painful shifts of paradigm that occurred during the early 20th century [van der Waerden].

Suppose a single particle of mass m is confined to lie somewhere within the domain Ω. For conceptual simplicity assume Ω is a bounded subset of the real line. Although the particle is a discrete entity, its actual location, its momentum, energy, etc are uncertain — think of individual crimes making up a crime wave [Davies]. The most that can be known of the particle is its *wave function* $\psi = \psi(x,t)$, a function of norm 1 in the spatial complex Hilbert space $L^2(\Omega)$, a wave undulating with time. The non-negative quantity $\rho = \psi\bar{\psi}$ is to be thought of as a time varying probability density that characterizes what is knowable about the particle.

von Neumann's Axiom: Any measurement (observation) of a property of this particle is performed by an instrument whose mathematical analog is a Hermitian operator, say A, yielding on average the measurement $\langle A\psi, \psi\rangle$, i.e., the expected value

$$\langle A\psi, \psi\rangle = \int_\Omega A\psi\bar{\psi}\ dx.$$

For example, using standard notation,

$$q = \langle Q\psi, \psi\rangle = \langle x\psi, \psi\rangle \text{ is the (expected) location,}$$
$$p = \langle P\psi, \psi\rangle \text{ is momentum,}$$
$$E = \langle H\psi, \psi\rangle \text{ is total energy,}$$
$$k.e. = \langle T\psi, \psi\rangle \text{ is kinetic energy,}$$
$$p.e. = \langle V\psi, \psi\rangle \text{ is potential energy,}$$

and so forth.

Repacing the skew symmetric J by the complex number i, the analog of Hamilton's Law (B.2) is *Schroedinger's Equation:*

$$i\hbar\dot{\psi} = H\psi = T\psi + V\psi, \tag{B.8}$$

where, as we shall see, $T = -(\hbar^2/2m)\nabla^2$, and where $h = 2\pi\hbar$ is *Planck's constant.*

The trajectories must then be

$$\psi = e^{-itH/\hbar}\psi_0 \tag{B.9}$$

where ψ_0 is the initial state. Eigenfunctions ψ of H are called *stationary states* since if $H\psi = \lambda\psi$,

$$e^{-itH/\hbar}\psi = e^{-it\lambda/\hbar}\psi,$$

hence the probability density $\rho = \psi\overline{\psi}$ is constant over time.

§B.5 Poisson Bracket

Definition. Given two Hermitian operators A and B, form the third Hermitian operator

$$[A,B] = i(AB - BA). \tag{B.10}$$

This operator, the *Poisson bracket* of A and B, is a measure of the degree to which A and B fail to commute.

Theorem A. For any Hermitian instrument A, the trajectory of its expected measurement is determined by

$$\langle A\psi, \psi\rangle^{\cdot} = -\langle [A,H]\psi, \psi\rangle / \hbar. \tag{B.11}$$

Proof. From (B.6),

$$\langle A\psi, \psi\rangle^{\bullet} = \langle A\dot{\psi}, \psi\rangle + \langle A\psi, \dot{\psi}\rangle = \langle AH\psi/i\hbar, \psi\rangle + \langle A\psi, H\psi/i\hbar\rangle$$

$$= -i\langle AH\psi, \psi\rangle/\hbar + i\langle HA\psi, \psi\rangle/\hbar = -\langle [A,H]\psi, \psi\rangle/\hbar.$$

Note that we have assumed $(A\psi)^{\cdot} = A\dot{\psi}$. Theorem A suggests the following operation on operators.

Definition. To each Hermitian operator A is associated the Hermitian operator

$$\dot{A} = -[A,H]/\hbar. \tag{B.12}$$

Corollary. $\langle \dot{A}\psi, \psi\rangle = \langle A\psi, \psi\rangle^{\cdot}$.

Operators A with $\dot{A} = 0$ are *stationary*, they commute with H and hence yield *integrals* — the expected values $\langle A\psi, \psi\rangle$ are constant along trajectories.

Metarule. *Classical Mechanics consists of relationships between measured values, while Quantum Mechanics consists of the analogous relationships between the instruments taking the measurements.*

With this rule in mind, define momentum as follows.

Definition. The momentum operator is the Hermitian

$$P = m\dot{Q}, \tag{B.13}$$

where displacement Q is multiplication by x.

Corollary.

$$P = -m[x, H]/\hbar. \tag{B.14}$$

Everything comes down to one assumption, the *commutation relation*:

$$[P, Q] = \hbar. \tag{B.15}$$

§B.6 Consequences of the Commutation Relation

Theorem B. The commutation relation $[P, Q] = \hbar$ implies

$$P = -i\hbar\frac{\partial}{\partial x} + \kappa(x) \tag{B.16}$$

and conversely.

Note that by (B.14), $[P, Q] = \hbar$ translates to

$$[[x, H], x] = -\frac{\hbar^2}{m}. \tag{B.17}$$

First we need two lemmas.

Lemma A. If $[K, x] = 0$, then K is multiplication by $\kappa(x)$.

Proof Sketch of Lemma A. $[K, x] = 0$ means $Kx\psi = xK\psi$ for all states ψ. Thus $Kx^n\psi = x^nK\psi$ and hence K commutes with all polynomials $f(x)$ giving $Kf(x) = Kf(x)\cdot 1 = f(x)K1 = f(x)\kappa(x) = \kappa(x)f(x)$. Thus K acts as multiplication by $\kappa(x)$ on at least polynomial states. Delicate matters now arise concerning domains of definition of admissible Hermitian operators. See [Reed and Simon]. Unbounded domains Ω require a weighted measure satisfying some asymptotic condition like (11.9).

Lemma B. If $[K, x] = k$, then $K = -ik\partial/\partial x + \kappa(x)$.

Proof. Note that $K_0 = -ik\partial/\partial x$ does indeed satisfy $[K_0, x] = k$. Thus by Lemma A, $K - K_0$ must be a multiplication operator $\kappa(x)$.

Theorem B now follows from (B.15) and Lemma B.

Occam's razor would hold that the multiplier $\kappa(x)$ in (B.16) is 0. There are arguments involving invariance under change of coordinates that will indeed yield that result [Weyl, Mackey]. Or one might argue physically that since momentum and displacement should be independent variables, $\kappa(x) = 0$.

Assumption. The multiplier $\kappa(x)$ in (B.16) is zero.

Corollary.

$$P = -i\hbar\frac{\partial}{\partial x}. \tag{B.18}$$

Again following the Metarule define

$$T = P^2/2m \quad \text{and} \quad H = T + V.$$

It is an easy exercise to deduce

$$T = -\frac{\hbar^2}{2m}\frac{\partial^2}{\partial x^2}. \tag{B.19}$$

But the central question remains — why should the commutation relation (B.15) hold?

§B.7 Arguments for the Commutation Relation

The strongest argument for the commutation relation is the many experimentally confirmed predictions that follow from it — see [Weyl] and [Pauling and Wilson]. However I find the following argument persuasive.

By the Metarule, $P = m\dot{Q}$, $H = T + V$, and $T = P^2/2m$. Thus follows a relation similar to, and deducible from the commutation relation.

Result.

$$[P^2, Q] = 2\hbar P. \tag{B.20}$$

Proof. Since $[Q, V] = 0$,

$$P = m\dot{Q} = -m[Q, H]/\hbar = -m[Q, T + V]/\hbar = -[Q, T]/\hbar$$

$$= -m[Q, P^2/2m]/\hbar = [P^2, Q]/2\hbar = i(P^2Q - QP^2)/2\hbar.$$

Our goal is to deduce

$$[P, Q] = \hbar$$

from (B.20). More technically, we are investigating the free algebra in the noncommuting indeterminates P and Q modulo prime ideals containing $[P^2, Q] - 2\hbar P$. See [Lam].

Set $Z = [P, Q] - \hbar$ and note (B.18) yields easily

$$ZP = -PZ.$$

Assume a relation dual to (B.20), viz.,

$$[P, Q^2] = 2\hbar Q.$$

Again it follows that

$$ZQ = -QZ.$$

But then Z^2 commutes with both P and Q and 'hence' Z^2 is a scalar multiple of the identity, i.e.,

$$Z^2 = c^2.$$

Hence

$$(Z - c)(Z + c) = 0.$$

But the *Weyl algebra* of all polynomial operators in P and Q has no zero divisors [Lam]. Hence $Z = 0$.

Another persuasive argument flows from the Uncertainty Principle.

§B.8 The Uncertainty Principle

Let us again think about measurement to motivate some statistical notions. Unlike (say) the multiplication operator $Q = x$ observing displacement, many instruments (Hermitian operators) have discrete point spectrum. Suppose A is such an operator. Let $\psi_{A1}, \psi_{A2}, \dots$ be the normalized eigenfunctions of A with eigenvalues $\lambda_{A1}, \lambda_{A2}, \dots$. Assume the eigenfunctions ψ_{Ak} of A form a complete orthonormal basis for the Hilbert space $X = L^2(\Omega)$.

Observed Fact. When a state ψ is observed by an instrument A with a complete discrete point spectrum, the state ψ 'collapses' to a multiple of one of the eigenfunctions ψ_{Ak} of A and yields the measured value λ_{Ak} with probability $|\langle \psi_{Ak}, \psi \rangle|^2$.

Thus after many repeated trials the expected value (mean) of the measurement of the state ψ by A is

$$\langle A\psi, \psi \rangle = \sum_{k=1}^{\infty} \lambda_{Ak} |\langle \psi_{Ak}, \psi \rangle|^2,$$

with second moment

$$\|A\psi\|^2 = \sum_{k=1}^{\infty} \lambda_{Ak}^2 |\langle \psi_{Ak}, \psi \rangle|^2,$$

and standard deviation

$$\sigma(A) = \sqrt{\|A\psi\|^2 - \langle A\psi, \psi \rangle^2}. \tag{B.21}$$

The General Heisenberg Uncertainty Principle. For any two (Hermitian) instruments A and B and state ψ,

$$\sigma(A)\sigma(B) \geq \langle [A, B]\psi, \psi \rangle / 2. \tag{B.22}$$

Proof. This requires nothing more than Cauchy's inequality. Set $\alpha = \langle A\psi, \psi \rangle$ and $\beta = \langle B\psi, \psi \rangle$. Let $A_1 = A - \alpha$ and $B_1 = B - \beta$. Note that

$$\langle [A, B]\psi, \psi \rangle = \langle [A_1, B_1]\psi, \psi \rangle.$$

Moreover

$$|\langle [A_1 B_1]\psi, \psi\rangle| = |\langle A_1\psi, B_1\psi\rangle - \overline{\langle A_1\psi, B_1\psi\rangle}| = |2i\mathrm{Im}\langle A_1\psi, B_1\psi\rangle|$$
$$\leq 2|\langle A_1\psi, B_1\psi\rangle| \leq 2\|A_1\psi\| \cdot \|B_1\psi\|.$$

Thus

$$|\langle [A, B]\psi, \psi\rangle| \leq 2\|A_1\psi\| \cdot \|B_1\psi\|.$$

But

$$\|A_1\psi\|^2 = \langle (A-\alpha)\psi, (A-\alpha)\psi\rangle = \ldots = \|A\psi\|^2 - \langle A\psi, \psi\rangle^2 = \sigma^2(A)$$

and likewise for B.

In particular,

$$\sigma(P)\sigma(Q) \geq \langle [P, Q]\psi, \psi\rangle / 2. \tag{B.23}$$

If one performs a Gedanken experiment where our moving particle is emitting a signal of known signature, then its position and momentum could be estimated from reception delay and doppler shift. But there is a famous analogous inequality of Fourier Analysis (the *uncertainty principle*, also a consequence of Cauchy's inequality) which guarantees that the product of the 'duration' of a signal with its frequency 'spread' must be at least a certain universal nonzero constant [Papoulis, 1962]. Moreover the inequality is sharp — Gaussian signals actually achieve the lower bound.

It is reasonable, since classical trajectories minimize action, that quantum trajectories minimize uncertainty. Thus for the fundamental variables P and Q, trajectories should realize some universal lower bound $\hbar/2$ of uncertainty, i.e.,

$$[P, Q] = \hbar$$

and

$$\sigma(P)\sigma(Q) \geq \frac{\hbar}{2}. \tag{B.24}$$

References

I. H. Abbot and A. E. von Doenhoff, *Theory of Wing Sections*, Dover, New York, 1949.

M. L. Abell and J. P. Braselton, *Differential Equations with Mathematica,* Academic Press, San Diego, 1993.

M. Abramowitz and I. A. Stegun, *Handbook of Mathematical Functions,* Dover, New York, 1965.

R. A. Adams, *Sobolev Spaces*, Academic Press, Boston, 1978.

W. S. Ames, *Numerical Methods for Partial Differential Equations*, 2nd ed., Academic Press, New York, 1977.

ASHRAE, *HVAC Applications*, the American Society of Heating, Refrigerating and Air-Conditioning Engineers Publications, Atlanta, 1999.

ASHRAE, *HVAC Systems and Equipment*, the American Society of Heating, Refrigerating and Air-Conditioning Engineers Publications, Atlanta, 2000.

R. Ayoub, *An Introduction to the Analytic Theory of Numbers,* Amer. Math. Soc., Providence, RI, 1963.

S. Axler, P. Bourdon, and W. Ramey, *Harmonic Function Theory,* Springer-Verlag, New York, 1992.

G. K. Batchelor, *An Introduction to Fluid Dynamics,* Cambridge Univ. Press, Cambridge, 1970.

H. Bateman, *The Bateman Manuscript Project: Higher Transcendental Functions,* A. Erdelyi, ed., vols. 1–3, McGraw-Hill, New York, 1953–1955.

F. Bauer and J. A. Nohel, *Quantitative Ordinary Differential Equations,* Dover, New York, 1987.

F. M. Bessel, "Untersuchungen des Theils der planetarischen Stoerungen, welcher aus der Bewegung der Sonne entsteht," *Berliner Abhandlungen*, 1824, pp. 1–52.

R. B. Bird, W.E. Stewart, and E.N. Lightfoot, *Transport Phenomena,* John Wiley and Sons, New York, 1960.

R. L. Bisplinghoff, H. Ashley, and R. L. Halfman, *Aeroelasticity,* Addison-Wesley, Cambridge, MA, 1955.

J. E. Bose, J. D. Parker, and F. C. McQuiston, *Design/Data Manual for Closed-Loop Ground-Coupled Heat Pump Systems,* ASHRAE Publications, Atlanta, 1985.

J. F. Botha and G. F. Pinder, *Fundamental Concepts in the Numerical Solution of Differential Equations,* Wiley-Interscience, New York, 1983.

F. Bowman, *Introduction to Bessel Functions*, Dover, New York, 1958.

F. Brauer and J.A. Nohel, *The Quantitative Theory of Ordinary Differential Equations,* Dover, NY.

W. C. Brown, *Matrices and Vector Spaces,* Marcel Dekker, New York, 1991.

P. L. Butzer and H. Berens, *Semi-Groups of Operators and Approximations,* Springer-Verlag, New York, 1967.

H. S. Carslaw and J.C. Jaeger, *Operational Methods in Applied Mathematics,* 2nd ed., Dover, New York, 1947.

H.S. Carslaw and J.C. Jaeger, *Conduction of Heat in Solids,* Oxford Univ. Press, London, 1959.

Y. Chait, M. Miklavcic, C.R. MacCluer, and C.J. Radcliffe, " A natural modal expansion for the flexible robot arm problem via a self-adjoint formulation," *IEEE Transactions on Robotics and Automation,* Vol. 6, No. 5, October 1990, pp. 601–603.

W. Cheney and D. Kincaid, *Numerical Mathematics and Computing*, 2nd ed., Brooks/Cole Publ., Pacific Grove, CA, 1985.

S. N. Chow, D. R. Dunninger, and M. Miklavcic, " Galerkin approximations for singular linear elliptic and semilinear parabolic problems," *Applicable Analysis,* Vol. 40, 1991, pp. 41–52.

R. Courant and D. Hilbert, *Methods of Mathematical Physics,* Wiley and sons, New York, 1989.

L. Carleson, "On convergence and growth of partial sums of Fourier series," *Acta Math.*, Vol. 116, 1966, pp. 135–157.

R. V. Churchill, J.W. Brown, and R.F. Verhey, *Complex Variables and Applications*, 3ed., McGraw-Hill, New York, 1976.

P. C. W. Davies, *Quantum Mechanics*, Chapman and Hall, New York, 1989.

G. Doetsch, *Tabellen zur Laplace-Transformation,* Springer-Verlag, Berlin, 1947.

J. A. Duffie and W. A. Beckman, *Solar Engineering of Solar Processes*, Wiley, New York, 1980.

A. Erdelyi, *Operational Calculus and Generalized Functions,* Holt-Rinehart-Winston, New York, 1962.

C. A. J. Fletcher, *Computational Galerkin Methods,* Springer-Verlag, New York, 1984.

J. B. J. Fourier, *Théorie Analytique de la Chaleur,* 1822.

L. Fox, *The Numerical Solution of Two-Point Boundary Problems in Ordinary Differential Equations,* Dover, New York, 1990.

B. G. Galerkin, *Vestnik Inzhenerov, Tech.,* 19, 1915, pp. 897–908.

D. Gilbarg and N. S. Trudinger, *Elliptic Partial Differential Equations of Second Order,* Springer-Verlag, New York, 1977.

D. T. Gillespie, *A Quantum Mechanics Primer,* Halstead Press, John Wiley and Sons, New York, 1974.

J.A. Goldstein, *Semigroups of Linear Operators,* Oxford Univ. Press, Oxford, 1985.

G.H. Golub and C. van Loan, *Matrix Computations,* 2nd. ed., John Hopkins Press, Baltimore, 1989.

G. H. Golub and J. M. Ortega, *Scientific Computing and Differential Equations,* Academic Press, Boston, 1992.

C. Gordon, D. L. Webb, and S. Wolpert, "One cannot hear the shape of a drum," *Bulletin of the Amer. Math. Soc.*, July, 1992, pp. 134–138.

M. Hajmirzaahmad and A. M. Krall, "Singular second-order operators: the maximal and minimal operators, and selfadjoint operators in between," *SIAM Review,* Vol. 34, No. 4, December 1992, pp. 614–634.

P. R. Halmos, *Introduction to Hilbert Space,* 2ed., Chelsea, New York, 1957.

P. R. Halmos, *Measure Theory*, Van Nostrand, New York, 1961.

P. R. Halmos, *A Hilbert Space Problem Book*, 2ed., Springer-Verlag, New York, 1982.

J. P. Den Hartog, *Strength of Materials*, Dover, New York, 1949.

E. Hille and R.S. Phillips, *Functional Analysis and Semi-Groups,* Amer. Math. Soc. Publ., Providence, RI, 1957.

J. Hladik, *La Transformation de Laplace à Plusieurs Variables,* Masson et Cie, Paris, 1969.

H. Hochstadt, *The Functions of Mathematical Physics,* Dover, New York, 1986.

S. S. Holland, *Applied Analysis by the Hilbert Space Method*, Marcel Decker, New York, 1990.

A. J. Hull and C. J. Radcliffe, "Experimental verification of the nonself-adjoint state space duct model," *ASME J. Vibrations and Acoustics,* Vol. 114, July 1992, pp. 404–408.

C. Johnson, *Numerical Solutions of Partial Differential Equations by the Finite Element Method*, Cambridge University Press, Cambridge, 1990.

M. Kac, "Can one hear the shape of a drum?," *Amer. Math. Monthly*, 73, 1966, pp. 1–23.

Y. Katznelson, *An Introduction to Harmonic Analysis*, Dover, New York, 1976, p. 59.

M. S. Klamkin, *Mathematical Modelling: Classroom Notes in Applied Mathematics,* SIAM Publications, Philadelphia, 1987, pp. 106–107.

H. Kolsky, *Stress Waves in Solids,* Dover, New York, 1963.

T. Kuhn, *The Structure of Scientific Revolution*, 2ed., Univ. of Chicago Press, Chicago, 1970.

T. Kusada, O. Piet, and J. W. Bean, "Annual variations of temperature field and heat transfer under heated ground surfaces," *NBS Building Science Series 156*, June, 1983.

A. H. Lachenbruch and B. V. Marshall, "Changing climate: geothermal evidence from permafrost in the Alaska arctic," *Science,* Vol. 234, 1986, pp. 689–696.

O. A. Ladyzhenskaya, *The Boundary Value Problems of Mathematical Physics,* Applied Mathematical Sciences 49, Springer-Verlag, New York, 1985.

T. Y. Lam, *A First Course in Noncommutative Rings,* Springer-Verlag, New York, 1991.

N. N. Lebedev, *Special Functions and their Applications,* Dover, New York, 1972.

T. Lewis, ed., *Global and Planetary Change,* Vol. 6, Nos. 2–4, December, 1992.

C. R. MacCluer and Y. Chait, "Choosing an inner product that separates variables," *SIAM Review*, Vol. 33, No. 4, Sept. 1991, pp. 467–471.

C. R. MacCluer, "Chaos in linear distributed systems," *ASME J. Dynamic Systems, Measurement, and Control,* Vol. 114, 1992, pp. 322–324.

C. R. MacCluer, *Industrial Mathematics,* Prentice Hall, Upper Saddle River, NJ, 2000.

G. W. Mackey, *The Mathematical Foundations of Quantum Mechanics*, W.A. Benjamin Inc., New York, 1963.

A. D. May, *Traffic Flow Fundamentals,* Prentice Hall, Upper Saddle River, NJ, 1990.

L. Meirovitch, *Elements of Vibration Analysis*, McGraw-Hill, New York, 1978.

J. Mikusinski, *Operational Calculus,* vol. I, 2ed., Pergamon Press, Oxford, 1983.

J. Mikusinski and T. K. Boehme, *Operational Calculus,* vol. II, 2ed., Pergamon Press, Oxford, 1987.

W. Miller, *Symmetry and Separation of Variables,* Addison-Wesley, Reading, MA, 1977.

W. Miller, E. G. Kalnins, and G.C. Williams, "Recent Advances in the use of separation of variables methods in general relativity," *Royal Soc. of London,* (to appear).

L. M. Milne-Thomson, *Theoretical Aerodynamics*, Macmillan, London, 1948.

D. H. Moore, *Heaviside Operational Calculus,* Amer. Elsevier, New York, 1971.

Béla Sz.-Nagy, *Introduction to Real Functions and Orthogonal Expansions,* Oxford Univ. Press, New York, 1965.

P. J. Nahin, *Oliver Heaviside: Sage in Solitude*, IEEE Press, New York, 1987.

A. W. Naylor and G. R. Sell, *Linear Operator Theory in Engineering and Science,* Applied Mat. Sciences **40**, Springer-Verlag, 2nd ed., New York, 1982.

J. von Neumann, *Mathematical Foundations of Quantum Mechanics*, Princeton Univ. Press, Princeton NJ, 1955.

P. V. O'Neil, *Advanced Engineering Mathematics*, 2ed., Wadsworth Publ., Belmont, CA, 1987.

J. M. Ortega, *Numerical Analysis, a second course,* SIAM Publications, Philadelphia, 1990.

W. J. Orvis, *1-2-3 for Scientists and Engineers*, Sybex, San Francisco, 1987.

M. N. Özişik, *Boundary Value Problems of Heat Conduction*, Dover, New York, 1968.

L. Pauling and E. B. Wilson, Jr., *Introduction to Quantum Mechanics with Applications to Chemistry,* Dover, New York, 1985.

S. M. Perlow, "Intermodulation distortion in resistive mixers," *RCA Review,* Vol. 35, March 1974, pp. 25–47.

K. E. Petersen, *Brownian Motion, Hardy spaces, and Bounded Mean*, Cambridge University Press, Cambridge, 1977.

M. Planck, *Treatise on Themodynamics*, 3ed., Dover, New York, 1945.

A. Papoulis, *The Fourier Integral and its Applications,* McGraw-Hill, New York, 1962.

A. Papoulis, *Signal Analysis,* McGraw-Hill, New York, 1977.

M. H. Protter, "Can one hear the shape of a drum? Revisited," *SIAM Review*, Vol. 29, No. 2, June 1987, pp. 185–197.

C. J. Radcliffe and C. D. Mote, Jr., "Identification and Control of Rotating Disk Vibration", *ASME Journal of Dynamics, Systems, Measurement and Control*, Vol. 105, March 1983, pp. 39-45.

J. W. S. Rayleigh, *Theory of Sound*, Dover, New York, 1945.

S. Ramo, *Fields and Waves in Communication Electronics*, 2ed., Wiley and sons, New York, 1984.

M. Reed and B. Simon, *Methods of Modern Physics*, vol. I, *Functional Analysis*, Academic Press, New York, 1972.

F. Rellich, “Ein Satz über mittlere Konvergenz,” *Nach. Gott.*, 1930, pp. 30–35.

R. D. Richmeyer and K.W. Morton, *Difference Methods for Initial-Value Problems*, Interscience Publishers, New York, 1967.

M. Riesz and B. Nagy, *Functional Analysis*, Dover, New York, 1955.

W. Rudin, *Principles of Mathematical Analysis*, McGraw Hill, New York, 1964.

W. Rudin, *Real and Complex Analysis*, McGraw Hill, New York, 1987.

D. L. Russell, “On the positive square root of the fourth derivative operator,” *Quarterly of Applied Mathematics,* Vol. XLVI, No. 4, December 1988, pp. 751–773.

H. Sagan, *Boundary and Eigenvalue Problems in Mathematical Physics*, Dover, New York, 1989.

H. M. Schey, *Div, Grad, Curl, and all that,* W.W. Norton, New York, 1973.

S. L. Sobolev, *Applications of Functional Analysis in Mathematical Physics,* Amer. Math. Soc., Providence, 1963.

S. L. Sobolev, *Partial Differential Equations of Mathematical Physics,* Pergamon Press, Oxford, 1964.

R. V. Southwell, ”On the Free Transverse Vibration of a Uniform Circular Disc Clamped at its Centre: and on the Effects of Rotation,” *Proceedings of the Royal Society of London*, Series A, Vol. 101, 1921, pp. 133-153.

G. Strang, *An Introduction to Applied Mathematics,* Wellesley-Cambridge Press, Wellesley, MA, 1986.

G. Strang and G. J. Fix, *An Analysis of the Finite Element Method,* Prentice-Hall, Englewood, NJ, 1973.

R. E. Svetic, et al., “A metabolic force for gene clustering,” *Bull. Math. Biol.*, Vol. 66, Issue 3 , May 2004, pp. 559-581.

G. Szego, *Orthogonal Polynomials,* Amer. Math. Soc. Publications, 4ed, Providence, RI, 1975.

S. Timoshenko, *Mechanics of Materials,* van Nostrand Reinhold, New York, 1972.

E. H. Twizell, *Computational Methods for Partial Differential Equations,* Wiley and sons, New York, 1984.

B. L. van der Waerden, *Sources of Quantum Mechanics*, Dover, New York, 1968.

G. N. Watson, *A Treatise on the Theory of Bessel Functions,* 2ed, Cambridge Univ. Press, London, 1952.

G. N. Watson, *Bessel Functions*, 2ed., Macmillian, New York, 1945.

H. F. Weinberger, *A First Course in Partial Differential Equations*, Blaisdell Publishers, Waltham, MA, 1965.

Wentzel-Kramers-Brillouin: G. Wentzel, *Z. Physik,* **38**, p.518 (1926); H.A. Kramers, *Z. Physik,* **39**, p.518 (1926); L. Brillouin, *Compt. Rend.,* **183**, p.24 (1926).

H. Weyl, *The Theory of Groups and Quantum Mechanics,* Dover, NY, 1950.

K. Yoshida, *Functional Analysis,* Springer-Verlag, 6ed., New York, 1980.

K. Yosida, *Operational Calculus,* Springer-Verlag, App. Math. Sci. vol. 55, New York, 1984.

N. Young, *An Introduction to Hilbert Space,* Cambridge Univ. Press, 1988.

R. M. Young, *An Introduction to Nonharmonic Fourier Series,* Academic Press, New York, 1980.

C. Zener, *Proc. Royal Soc.,* **A145**, 1934, p. 523.

Index

T

A CATALOG OF SELECTED

DOVER BOOKS

IN SCIENCE AND MATHEMATICS

Mathematics

FUNCTIONAL ANALYSIS (Second Corrected Edition), George Bachman and Lawrence Narici. Excellent treatment of subject geared toward students with background in linear algebra, advanced calculus, physics, and engineering. Text covers introduction to inner-product spaces, normed, metric spaces, and topological spaces; complete orthonormal sets, the Hahn-Banach Theorem and its consequences, and many other related subjects. 1966 ed. 544pp. 6⅛ x 9¼. 40251-7

ASYMPTOTIC EXPANSIONS OF INTEGRALS, Norman Bleistein & Richard A. Handelsman. Best introduction to important field with applications in a variety of scientific disciplines. New preface. Problems. Diagrams. Tables. Bibliography. Index. 448pp. 5⅜ x 8½. 65082-0

VECTOR AND TENSOR ANALYSIS WITH APPLICATIONS, A. I. Borisenko and I. E. Tarapov. Concise introduction. Worked-out problems, solutions, exercises. 257pp. 5⅝ x 8¼. 63833-2

THE ABSOLUTE DIFFERENTIAL CALCULUS (CALCULUS OF TENSORS), Tullio Levi-Civita. Great 20th-century mathematician's classic work on material necessary for mathematical grasp of theory of relativity. 452pp. 5⅝ x 8¼. 63401-9

AN INTRODUCTION TO ORDINARY DIFFERENTIAL EQUATIONS, Earl A. Coddington. A thorough and systematic first course in elementary differential equations for undergraduates in mathematics and science, with many exercises and problems (with answers). Index. 304pp. 5⅜ x 8½. 65942-9

FOURIER SERIES AND ORTHOGONAL FUNCTIONS, Harry F. Davis. An incisive text combining theory and practical example to introduce Fourier series, orthogonal functions and applications of the Fourier method to boundary-value problems. 570 exercises. Answers and notes. 416pp. 5⅜ x 8½. 65973-9

COMPUTABILITY AND UNSOLVABILITY, Martin Davis. Classic graduate-level introduction to theory of computability, usually referred to as theory of recurrent functions. New preface and appendix. 288pp. 5⅜ x 8½. 61471-9

ASYMPTOTIC METHODS IN ANALYSIS, N. G. de Bruijn. An inexpensive, comprehensive guide to asymptotic methods–the pioneering work that teaches by explaining worked examples in detail. Index. 224pp. 5⅜ x 8½ 64221-6

APPLIED COMPLEX VARIABLES, John W. Dettman. Step-by-step coverage of fundamentals of analytic function theory–plus lucid exposition of five important applications: Potential Theory; Ordinary Differential Equations; Fourier Transforms; Laplace Transforms; Asymptotic Expansions. 66 figures. Exercises at chapter ends. 512pp. 5⅜ x 8½. 64670-X

INTRODUCTION TO LINEAR ALGEBRA AND DIFFERENTIAL EQUATIONS, John W. Dettman. Excellent text covers complex numbers, determinants, orthonormal bases, Laplace transforms, much more. Exercises with solutions. Undergraduate level. 416pp. 5⅜ x 8½. 65191-6

CATALOG OF DOVER BOOKS

INTRODUCTORY REAL ANALYSIS, A.N. Kolmogorov, S. V. Fomin. Translated by Richard A. Silverman. Self-contained, evenly paced introduction to real and functional analysis. Some 350 problems. 403pp. 5⅜ x 8½. 61226-0

APPLIED ANALYSIS, Cornelius Lanczos. Classic work on analysis and design of finite processes for approximating solution of analytical problems. Algebraic equations, matrices, harmonic analysis, quadrature methods, more. 559pp. 5⅜ x 8½. 65656-X

AN INTRODUCTION TO ALGEBRAIC STRUCTURES, Joseph Landin. Superb self-contained text covers "abstract algebra": sets and numbers, theory of groups, theory of rings, much more. Numerous well-chosen examples, exercises. 247pp. 5⅜ x 8½. 65940-2

QUALITATIVE THEORY OF DIFFERENTIAL EQUATIONS, V. V. Nemytskii and V.V. Stepanov. Classic graduate-level text by two prominent Soviet mathematicians covers classical differential equations as well as topological dynamics and ergodic theory. Bibliographies. 523pp. 5⅜ x 8½. 65954-2

THEORY OF MATRICES, Sam Perlis. Outstanding text covering rank, nonsingularity and inverses in connection with the development of canonical matrices under the relation of equivalence, and without the intervention of determinants. Includes exercises. 237pp. 5⅜ x 8½. 66810-X

INTRODUCTION TO ANALYSIS, Maxwell Rosenlicht. Unusually clear, accessible coverage of set theory, real number system, metric spaces, continuous functions, Riemann integration, multiple integrals, more. Wide range of problems. Undergraduate level. Bibliography. 254pp. 5⅜ x 8½. 65038-3

MODERN NONLINEAR EQUATIONS, Thomas L. Saaty. Emphasizes practical solution of problems; covers seven types of equations. ". . . a welcome contribution to the existing literature. . . . "–*Math Reviews*. 490pp. 5⅜ x 8½. 64232-1

MATRICES AND LINEAR ALGEBRA, Hans Schneider and George Phillip Barker. Basic textbook covers theory of matrices and its applications to systems of linear equations and related topics such as determinants, eigenvalues, and differential equations. Numerous exercises. 432pp. 5⅜ x 8½. 66014-1

MATHEMATICS APPLIED TO CONTINUUM MECHANICS, Lee A. Segel. Analyzes models of fluid flow and solid deformation. For upper-level math, science, and engineering students. 608pp. 5⅜ x 8½. 65369-2

ELEMENTS OF REAL ANALYSIS, David A. Sprecher. Classic text covers fundamental concepts, real number system, point sets, functions of a real variable, Fourier series, much more. Over 500 exercises. 352pp. 5⅜ x 8½. 65385-4

SET THEORY AND LOGIC, Robert R. Stoll. Lucid introduction to unified theory of mathematical concepts. Set theory and logic seen as tools for conceptual understanding of real number system. 496pp. 5⅜ x 8¼. 63829-4